Mixed Methods Research for Improved Scientific Study

Mette L. Baran
Cardinal Stritch University, USA

Janice E. Jones
Cardinal Stritch University, USA

A volume in the Advances in Knowledge Acquisition, Transfer, and Management (AKATM) Book Series

Information Science REFERENCE
An Imprint of IGI Global

Published in the United States of America by
Information Science Reference (an imprint of IGI Global)
701 E. Chocolate Avenue
Hershey PA, USA 17033
Tel: 717-533-8845
Fax: 717-533-8661
E-mail: cust@igi-global.com
Web site: http://www.igi-global.com

Library of Congress Cataloging-in-Publication Data

Names: Baran, Mette, 1965- editor. | Jones, Janice, 1959- editor.
Title: Mixed methods research for improved scientific study / Mette Baran and Janice Jones, editors.
Description: Hershey, PA : Information Science Reference, 2016. | Includes bibliographical references and index.
Identifiers: LCCN 2015050773| ISBN 9781522500070 (hardcover) | ISBN 9781522500087 (ebook)
Subjects: LCSH: Mixed methods research. | Interdisciplinary research--Methodology. | Research--Methodology.
Classification: LCC H62 .M53423 2016 | DDC 001.4/2--dc23 LC record available at http://lccn.loc.gov/2015050773

This book is published in the IGI Global book series Advances in Knowledge Acquisition, Transfer, and Management (AKATM) (ISSN: 2326-7607; eISSN: 2326-7615)

British Cataloguing in Publication Data
A Cataloguing in Publication record for this book is available from the British Library.

For electronic access to this publication, please contact: eresources@igi-global.com.

Advances in Knowledge Acquisition, Transfer, and Management (AKATM) Book Series

Murray E. Jennex
San Diego State University, USA

ISSN: 2326-7607
EISSN: 2326-7615

Mission

Organizations and businesses continue to utilize knowledge management practices in order to streamline processes and procedures. The emergence of web technologies has provided new methods of information usage and knowledge sharing.

The **Advances in Knowledge Acquisition, Transfer, and Management (AKATM) Book Series** brings together research on emerging technologies and their effect on information systems as well as the knowledge society.**AKATM** will provide researchers, students, practitioners, and industry leaders with research highlights surrounding the knowledge management discipline, including technology support issues and knowledge representation.

Coverage

- Cognitive Theories
- Cultural Impacts
- Information and Communication Systems
- Knowledge acquisition and transfer processes
- Knowledge management strategy
- Knowledge Sharing
- Organizational learning
- Organizational Memory
- Small and Medium Enterprises
- Virtual Communities

IGI Global is currently accepting manuscripts for publication within this series. To submit a proposal for a volume in this series, please contact our Acquisition Editors at Acquisitions@igi-global.com or visit: http://www.igi-global.com/publish/.

The Advances in Knowledge Acquisition, Transfer, and Management (AKATM) Book Series (ISSN 2326-7607) is published by IGI Global, 701 E. Chocolate Avenue, Hershey, PA 17033-1240, USA, www.igi-global.com. This series is composed of titles available for purchase individually; each title is edited to be contextually exclusive from any other title within the series. For pricing and ordering information please visit http://www.igi-global.com/book-series/advances-knowledge-acquisition-transfer-management/37159. Postmaster: Send all address changes to above address.

Titles in this Series

For a list of additional titles in this series, please visit: www.igi-global.com

Handbook of Research on Innovations in Information Retrieval, Analysis, and Management
Jorge Tiago Martins (The University of Sheffield, UK) and Andreea Molnar (University of Portsmouth, UK)
Information Science Reference • copyright 2016 • 580pp • H/C (ISBN: 9781466688339) • US $325.00 (our price)

Societal Benefits of Freely Accessible Technologies and Knowledge Resources
Oswaldo Terán (Universidad de los Andes, Venezuela) and Jose Aguilar (Universidad de los Andes, Venezuela)
Information Science Reference • copyright 2015 • 361pp • H/C (ISBN: 9781466683365) • US $175.00 (our price)

Contemporary Issues Surrounding Ethical Research Methods and Practice
Chi B. Anyansi-Archibong (North Carolina A&T State University, USA)
Information Science Reference • copyright 2015 • 376pp • H/C (ISBN: 9781466685628) • US $205.00 (our price)

Organizational Knowledge Dynamics Managing Knowledge Creation, Acquisition, Sharing, and Transformation
Constantin Bratianu (Bucharest University of Economic Studies, Romania)
Information Science Reference • copyright 2015 • 349pp • H/C (ISBN: 9781466683181) • US $210.00 (our price)

Strategic Data-Based Wisdom in the Big Data Era
John Girard (Middle Georgia State College, USA) Deanna Klein (Minot State University, USA) and Kristi Berg (Minot State University, USA)
Information Science Reference • copyright 2015 • 311pp • H/C (ISBN: 9781466681224) • US $205.00 (our price)

Handbook of Research on Maximizing Cognitive Learning through Knowledge Visualization
Anna Ursyn (University of Northern Colorado, USA)
Information Science Reference • copyright 2015 • 572pp • H/C (ISBN: 9781466681422) • US $325.00 (our price)

Information Seeking Behavior and Technology Adoption Theories and Trends
Mohammed Nasser Al-Suqri (Sultan Qaboos University, Oman) and Ali Saif Al-Aufi (Sultan Qaboos University, Oman)
Information Science Reference • copyright 2015 • 321pp • H/C (ISBN: 9781466681569) • US $200.00 (our price)

Handbook of Research on Scholarly Publishing and Research Methods
Victor C.X. Wang (Florida Atlantic University, USA)
Information Science Reference • copyright 2015 • 582pp • H/C (ISBN: 9781466674097) • US $335.00 (our price)

IGI GLOBAL
DISSEMINATOR OF KNOWLEDGE
www.igi-global.com

701 E. Chocolate Ave., Hershey, PA 17033
Order online at www.igi-global.com or call 717-533-8845 x100
To place a standing order for titles released in this series, contact: cust@igi-global.com
Mon-Fri 8:00 am - 5:00 pm (est) or fax 24 hours a day 717-533-8661

Editorial Advisory Board

Table of Contents

Foreword xvi

Preface xviii

Acknowledgment xxiv

Section 1
Research Paradigms

Chapter 1
The Nature of Research Methodologies: Terms and Usage within Quantitative, Qualitative, and Mixed Methods 1
Ben Tran, Alliant International University, USA

Chapter 2
Moving from Tension to Texture: The Paradigmatic Roots of Mixed Methods Research 28
Preston B. Cosgrove, Cardinal Stritch University, USA
Peter M. Jonas, Cardinal Stritch University, USA

Chapter 3
Mixed Method Research: A Concept 39
Aroop Mukherjee, Universiti Putra Malaysia, Malaysia
Nitty Hirawaty Kamarulzaman, Universiti Putra Malaysia, Malaysia

Section 2
Designing a Mixed Methods Research Study

Chapter 4
It is All in the Design: Creating the Foundations of a Mixed Methods Research Study 66
Mette L. Baran, Cardinal Stritch University, USA

Chapter 5
Designs of Mixed Method Research 79
Amir Manzoor, Bahria University, Pakistan

Section 3
Sampling in Research

Chapter 6
Sampling in Research 107
Never Mujere, University of Zimbabwe, Zimbabwe

Section 4
Data Collection Using Innovative Tools

Chapter 7
Analyzing Qualitative Data: Visualizing Lived Experiences through Poems and Photography 123
Carolyn N. Stevenson, Kaplan University, USA

Section 5
Analyzing Data

Chapter 8
Analyzing Quantitative Data 149
Sema A. Kalaian, Eastern Michigan University, USA
Rafa Kasim, Indiana Tech University, USA

Chapter 9
Analyzing Quantitative Data in Mixed Methods Research for Improved Scientific Study 165
Christopher Boachie, Central University College, Ghana

Chapter 10
How Marketers Conduct Mixed Methods Research: Incorporating the Exploratory Sequential Design with the Hierarchy of Effects Model 187
Roger Baran, DePaul University, USA

Chapter 11
Making Sense of All the Words: Analyzing Qualitative Data 197
Janice E. Jones, Cardinal Stritch University, USA
A. J. Metz, University of Utah, USA

Section 6
Data Analysis: Examples of Research Studies using Mixed Methods

Chapter 12
Examining Online Communication: A Method for the Quantitative Analysis of Qualitative Data 213
Michael G. Hughes, HumRRO, USA
Jennifer A. Griffith, Alfred University, USA
Cristina Byrne, University of Oklahoma, USA
Darin S. Nei, Hogan Assessment Systems, USA
Lauren Harkrider Beechly, IBM, USA
Thomas A. Zeni, East Central University, USA
Amanda Shipman, Kenexa, USA
Shane Connelly, University of Oklahoma, USA
Michael D. Mumford, University of Oklahoma, USA

Chapter 13
Mentoring and Support for the edTPA: A Mixed Methods Program Review of edTPA Support Practices 236
Randa Suleiman, Cardinal Stritch University, USA
Clavon Byrd, Cardinal Stritch University, USA

Chapter 14
Morphological Ontology Design Engineering: A Methodology to Model Ill-Structured Problems 262
Joey Jansen van Vuuren, Council for Scientific and Industrial Research, South Africa
Louise Leenen, Council for Scientific and Industrial Research, South Africa
Marthie M. Grobler, Council for Scientific and Industrial Research, South Africa
Ka Fai Peter Chan, Council for Scientific and Industrial Research, South Africa
Zubeida C. Khan, Council for Scientific and Industrial Research, South Africa

Section 7
Conducting Research from Start to Finish

Chapter 15
Creating and Implementing a Mixed Method Research Study 293
Mette L. Baran, Cardinal Stritch University, USA
Janice E. Jones, Cardinal Stritch University, USA

Compilation of References 303

About the Contributors 326

Index 333

Detailed Table of Contents

Foreword xvi

Preface xviii

Acknowledgment xxiv

Section 1
Research Paradigms

Research paradigms address the philosophical dimensions of social sciences. A research paradigm is a set of fundamental assumptions and beliefs as to how the world is perceived which then serves as a thinking framework that guides the behavior of the researcher. This chapter explains the different points of view on paradigms. The historical background of mixed methods research, including the debate among the three methodology communities, is described. In addition, the nature of mixed methods and its rational is discussed.

Chapter 1
The Nature of Research Methodologies: Terms and Usage within Quantitative, Qualitative, and Mixed Methods 1
Ben Tran, Alliant International University, USA

Mixed methods research is, generally speaking, an approach to knowledge (theory to practice) that attempts to consider multiple viewpoints, perspectives, positions, and standpoints. As such, before the advent of mixed methods, many studies used multiple methods to achieve the benefits of triangulation without restricting themselves to any paradigmatic membership or methodological category. Today, the primary philosophy of mixed research is that of pragmatism. This chapter will cover the history and the foundation of research methodologies and explain the purpose of research within various methodologies. This chapter will also explain the various terminologies used within research and research design as well as the meaning of these terminologies. This chapter will not cover statistics, however, mixed methods, methodology, research, and paradigm, statistical research methodology will be touched upon.

Chapter 2
Moving from Tension to Texture: The Paradigmatic Roots of Mixed Methods Research 28
Preston B. Cosgrove, Cardinal Stritch University, USA
Peter M. Jonas, Cardinal Stritch University, USA

Much like a jigsaw puzzle box top guides one in how to connect the pieces, an individual's research paradigm operates as a conscious or subconscious influence in conducting a research project. This chapter starts by making the argument for the critical role of research paradigms before moving into a thorough investigation of the paradigmatic origins of the qualitative-quantitative "debate." While mixed-methods research is often seen as the mediator in the dispute, the authors then articulate four broad ways in which mixed methods research addresses the paradigm divide at the heart of qualitative and quantitative research. The result is paradigmatically complex, but offers researchers flexibility as they seek to address their research question.

Chapter 3
Mixed Method Research: A Concept 39
Aroop Mukherjee, Universiti Putra Malaysia, Malaysia
Nitty Hirawaty Kamarulzaman, Universiti Putra Malaysia, Malaysia

Mixed methods have emerged as the third research community in the social and behavioural sciences during the past decades, joining quantitative and qualitative methods of scholarly inquiry. Mixed methods research, research paradigm, methodology, and action research have encouraged the combined use of quantitative and qualitative research to answer complex questions in recent years. Mixed methods research integrates both methods, the quantitative and the qualitative, to present research findings within a single system process. The chapter aims to provide an insight between mixed method research and action research, which includes the basic foundation of mixed method research and research paradigm. The chapter will discuss the concept of action research and how mixed method is applied to action research and its processes. A brief idea about the future plan of action required for mixed methods research to attain better research designs and processes is also discussed in the chapter.

Section 2
Designing a Mixed Methods Research Study

Researchers need to make numerous methodical decisions including the research design which is the blueprint for the study. This chapter outlines the design decisions that need to be made. The design is chosen based on which strategy is best suited to answer the research question(s). The three basic mixed methods designs are discussed: Parallel Convergent, Sequential (Explanatory or Exploratory), and Embedded Design. Additional decisions need to be made about the implementation of the study including the sequencing of the data collection, weight of the two methods (quantitative and qualitative), and when and how to converge the data.

Chapter 4
It is All in the Design: Creating the Foundations of a Mixed Methods Research Study 66
Mette L. Baran, Cardinal Stritch University, USA

This chapter introduces the various design choices researchers need to decide on prior to conducting the study. In the first section of this chapter a detailed description of research design is followed by an explanation that the type of information that is collected is based on whether the research question is descriptive, explanatory, or exploratory. The major strategic implementation methods for quantitative, qualitative, and mixed methods are then discussed. The three strategies for mixed methods research Parallel Convergent, Sequential, and Embedded Design are presented in detail along with the rationale for their use. Finally, in the last section, the strands or sequencing of the data collection phase of the study is explained.

Chapter 5
Designs of Mixed Method Research 79
Amir Manzoor, Bahria University, Pakistan

Mixed methods research is becoming an increasingly popular approach in the discipline fields of sociology, psychology, education and health sciences. Calls for the integration of quantitative and qualitative research methods have been advanced in these fields. A key feature of mixed methods research is its methodological pluralism, which frequently results in research which provides broader perspectives than those offered by mono-method designs. The central premise of mixed methods is that the use of quantitative and qualitative approaches in combination provides a better understanding of research problems and complex phenomena than either approach alone. The purpose of this chapter is to review designs of mixed methods research. The study surveys the common designs of mixed methods research and examine the main characteristics of each in terms of purposes, strengths, and issues, and posits suggestions on the application of these designs.

Section 3
Sampling in Research

This chapter explains the role of sampling in research. Sampling plays an important role in any research study and careful consideration needs to be placed on who to include as participants as part of the design process. Researchers need to determine sample sizes for the quantitative and qualitative data and whether or not to include the same participants for both strands of the study. In addition, decisions around random sampling or purposeful sampling must be considered.

Chapter 6
Sampling in Research.. 107
Never Mujere, University of Zimbabwe, Zimbabwe

Research is aimed at the discovery and interpretation of facts, revision of accepted theories or laws in the light of new facts or practical application of such new or revised theories or laws. A sample provides needed information about the population quickly. However, there is no guarantee that any sample will be precisely representative of the population from which it comes. It is cheaper to observe a part rather than the whole. This chapter is a discussion on sampling in research and it is mainly designed to equip researchers with knowledge of the general issues to consider when sampling. The purpose of sampling in research, dangers of sampling and how to minimize them, types of sampling and guides for deciding the sample size are discussed. For a clear flow of ideas, a few definitions of the terms used are given. They highlight the types and methods of sampling, sampling errors and discusses techniques of sample size determination. Different types of sampling technique, how to carry them out, and their advantages and disadvantages are also introduced.

Section 4
Data Collection Using Innovative Tools

Collecting data is critical to the design of the study and the researcher needs to determine whether the data should be collected concurrently--namely quantitative and qualitative data at the same time, or sequentially--collecting and analyzing one type of data at a time. While the traditional forms of data collection have been stable over the years--collecting archival data, surveys/questionnaires, interviews, and observation--alternative forms of data collection tools are starting to emerge. This allows the researcher to explore an issue at a more holistic level.

Chapter 7
Analyzing Qualitative Data: Visualizing Lived Experiences through Poems and Photography 123
Carolyn N. Stevenson, Kaplan University, USA

Use of photography and poetry offer a way for participants to express lived experiences through a visual and written means of self-expression. These forms of data collection can provide a rich, thick description of those often overlying on the peripherals of society. Traditional means of qualitative research such as interview and observation can at times create a barrier between the researcher and the participants because of the face-to-face interaction. Participants may be uncomfortable expressing authentic feeling during a formal interview process. By offering participants the opportunity to personally select descriptive photographs and articulate expression in their own voice through poetry, an authentic expression occurs.

Section 5
Analyzing Data

Researchers need to give great consideration to the data analysis process, mastering both the deductive and the inductive analytical stages. There are two sources of interpretation and the researcher needs to move swiftly between a pragmatic positivist approach to a more interpretary stance relying on observations and words to better understand the context surrounding the study in order to build theory. Mixed methods researchers operate between statistical enumeration and analysis seeking to confirm hypotheses while coding qualitative data to detect patterns; hence they are required to demonstrate a repertoire for methodologies.

Chapter 8
Analyzing Quantitative Data... 149
Sema A. Kalaian, Eastern Michigan University, USA
Rafa Kasim, Indiana Tech University, USA

The main purpose of this chapter is to present a conceptual and practical overview of some of the basic and advanced statistical tools for analyzing quantitative data. Analyzing quantitative data involves two broad analytical methods that serve two main purposes, which are descriptive and inferential statistical methods. The chapter covers both descriptive and inferential quantitative methods. It covers some of the descriptive statistical methods such as mean, median, mode, variance, standard deviation, and graphical methods (e.g., histograms). It also covers inferential statistical methods such as correlation, simple regression, multiple regression, t-test for two independent samples, t-test for two dependent samples, and analysis of variance (ANOVA).

Chapter 9
Analyzing Quantitative Data in Mixed Methods Research for Improved Scientific Study 165
Christopher Boachie, Central University College, Ghana

The purpose of the chapter is to review the role of quantitative methods in corporate research and the methods for analysing quantitative data. The study used a secondary data on quantitative research methods and a survey of published articles on schools, businesses and non- profit organizations. The key findings show that exploratory data can be analyzed using graphs and charts and hypothesis testing can be employed to test statements made. Impacts of one variable on another and the relationships between variables can be explained using correlation and regression analysis. The implications are that the value of a quantitative analysis arises when it is possible to identify features that occur frequently across the many participatory discussions aimed at studying a particular research theme.

Chapter 10
How Marketers Conduct Mixed Methods Research: Incorporating the Exploratory Sequential Design with the Hierarchy of Effects Model 187
Roger Baran, DePaul University, USA

The complimentary nature of qualitative and quantitative research methods are examined with respect to a study assessing the market's view of a training and development institute in the Middle East. The qualitative portion consisted of focus groups conducted with seven distinct market segments served by the institute. The results proved insightful with respect to uncovering and understanding differences of opinion among the seven groups; however, taken alone, the qualitative research would have been very misleading with respect to the institute's standing in the Middle East.

Chapter 11
Making Sense of All the Words: Analyzing Qualitative Data 197
Janice E. Jones, Cardinal Stritch University, USA
A. J. Metz, University of Utah, USA

This chapter provides an introduction to the process of qualitative analysis and to use step by step examples to provide an idea of how the process of qualitative analysis actually works. Crabtree and Miller, 1992, note that there are many different strategies for analysis, in fact, they suggest there are as many strategies as there are qualitative researchers. This chapter is intended to give the researcher a place to begin and to inspire a deeper dive into this rewarding form of data analysis. Stake, (1995) writes that qualitative data analysis is "a matter of giving meaning to first impressions as well as to final compilations. Analysis essentially means taking something apart. We take our impressions, our observations, apart... we need to take the new impression apart, giving meaning to the parts"(p. 71). While qualitative data analysis can be time consuming the rewards that come from immersion in the data far outweigh the time spent doing so.

Section 6
Data Analysis: Examples of Research Studies using Mixed Methods

In this section three studies are used as innovative examples of how researcher use mixed methods to provide holistic interpretations of phenomena. Mixed methods researchers need to be experienced in three types of data analysis strategies. This chapter provides examples of innovative mixed methods research studies combining the use of inductive and deductive logics. The strength of a mixed methods approach is that it provides ample opportunities for researchers to be creative and eclectic in their approach, examining an issue from various angles using a myriad of data collection tools and data analysis techniques.

Chapter 12
Examining Online Communication: A Method for the Quantitative Analysis of Qualitative Data 213
Michael G. Hughes, HumRRO, USA
Jennifer A. Griffith, Alfred University, USA
Cristina Byrne, University of Oklahoma, USA
Darin S. Nei, Hogan Assessment Systems, USA
Lauren Harkrider Beechly, IBM, USA
Thomas A. Zeni, East Central University, USA
Amanda Shipman, Kenexa, USA
Shane Connelly, University of Oklahoma, USA
Michael D. Mumford, University of Oklahoma, USA

Methods of individual communication continue to expand through online media. Given the dynamic nature of online communications, traditional methods for studying communications may not suffice. A hybridized content analytic approach that combines qualitative and quantitative methods offers a unique methodological tool to researchers who seek to better understand computer-mediated communications and the psychological characteristics of those who communicate online by evaluating qualitative information using quantitative methods. This means of measurement allows researchers to statistically evaluate whether investigated phenomena are occurring in combination with the richness that qualitative assessment provides. As with any approach to computer-mediated communication, various ethical considerations must be borne in mind, and, thus, are discussed in concert with this hybridized approach to content analysis.

Chapter 13
Mentoring and Support for the edTPA: A Mixed Methods Program Review of edTPA Support Practices 236
Randa Suleiman, Cardinal Stritch University, USA
Clavon Byrd, Cardinal Stritch University, USA

The research question was: How effective is the current edTPA mentoring and support program for teacher candidates? This research utilized mixed method interactive program evaluation. An online survey collected teacher candidates' perceptions of mentoring and support for edTPA. The survey questions were organized around four constructs: Preparedness, support from instructor, support from university supervisors, and support from cooperating teacher. With N = 46, a comprehensive data analysis was conducted that identified areas of strength and need of the program. As a result, the researchers developed an edTPA mentoring and support program model.

Chapter 14
Morphological Ontology Design Engineering: A Methodology to Model Ill-Structured Problems 262
Joey Jansen van Vuuren, Council for Scientific and Industrial Research, South Africa
Louise Leenen, Council for Scientific and Industrial Research, South Africa
Marthie M. Grobler, Council for Scientific and Industrial Research, South Africa
Ka Fai Peter Chan, Council for Scientific and Industrial Research, South Africa
Zubeida C. Khan, Council for Scientific and Industrial Research, South Africa

In the Social-technical domain scientists are often confronted with a class of problems that are termed messy, ill-structured or wicked. These problems address complex issues that not well-defined, contain unresolvable uncertainties, and are characterized by a lack of common agreement on problem definition. This chapter proposes a new mixed methods research technique, Morphological Ontology Design Engineering (MODE), which can be applied to develop models for ill-structured problems. MODE combines three different research methodologies into a single, methodology. MODE draws from research paradigms that include exploratory and descriptive research approaches to develop models. General morphological analysis offers a systematic method to extract meaningful information from domain experts, while ontology based representation is used to logically represent domain knowledge. The design science methodology guides the entire process. MODE is applied to a case study where an ontological model is developed to support the implementation of a South African national cybersecurity policy.

Section 7
Conducting Research from Start to Finish

The numerous steps to be considered when conducting a research study can be daunting for an inexperienced researcher. Similarly, teaching these skills to students requires considerable thought as to how to best provide an integrated approach to teaching both strands and then combining this orientation. This section provides a step by step linear approach to the research process. The mixed methods researcher needs to manage numerous competencies starting with the philosophical assumptions underlying the use of mixed methods. The researcher must gather considerable data from each component and make meaningful inferences. In addition, a mixed methods study needs to integrate, link and connect the two strands of research in order to provide a comprehensive understanding of the issue under investigation.

Chapter 15
Creating and Implementing a Mixed Method Research Study 293
Mette L. Baran, Cardinal Stritch University, USA
Janice E. Jones, Cardinal Stritch University, USA

This chapter serves as a guideline for outlining the core characteristics of mixed methods research (MMR) and the various steps researchers undertake in order to conduct a research study. The purpose is to create a worksheet assisting the researcher step by step from beginning to end following the seven steps to conducting research. While the focus is on MMR the steps are similar for any type of research methodology. It is important to note that MMR is not a limiting form of research. Researchers need a MMR question and a mixed methods purpose statement for the research project. This chapter will also help explain why mixed method research is one of the best approaches in answering a research question. Finally, the chapter includes a suggestion to the importance of adding a visual diagram of the MMR into the research project and into the final report.

Compilation of References 303

About the Contributors 326

Index 333

Foreword

The paradigms question has been laid to rest after being heavily debated these last three decades. While we have moved beyond the perennial discussion about the one best method to conduct research, we are still left with how to best teach novice researchers to become proficient in mixed methods. With this book, we feel a dream has become an opportunity to offer a step by step outline of how to conduct a quality mixed methods study for novice researchers. As professors, teaching our masters and doctoral students research methodology, we feel there is a void in comprehensive teaching materials that can take students in the social and behavioral sciences through a quality Mixed Methods Research (MMR) project from start to finish. While there are numerous definitions of MMR, we feel that Creswell's (2009) definition of the research approach "as involving philosophical assumptions, the use of qualitative and quantitative approaches, and the mixing of both approaches in a study" is precise and encompassing (p. 4).

The evolution of mixed methods research and its subsequent acceptance as a valid approach, as well as its contribution to the field, warrants our attention and requires us to focus on the logistics of teaching mixed methods. Now solidly established as the third research community or the third methodological movement (Teddlie & Tashakkori, 2003), the mixed methods approach provides researchers with an opportunity to gain a holistic understanding of a research problem. The proposition of our experience is that students, albeit new to the research process, need to be successful in learning both quantitative and qualitative approaches. While some research questions can only be answered using quantitative data and others are only suited for qualitative data, one can't argue that some studies will provide more comprehensive results from a mixed methods approach. The old adage we tell our students, "the research question drives everything," is still very true. We note that many traditional doctoral programs still place students in either qualitative or quantitative paradigms for their initial research orientation. We believe, as Gorard and Taylor (2004) urge, that novice researchers can become imprisoned within one of these purported corners, limiting their potential to become consumers and generators of research. Hence we concur with Teddlie and Tashakkori (2009) that it is "essential for social and behavioral scientists today to be methodologically bilingual" (p. 32). We would like to thank our students past and current for the opportunity to learn with them and to hone our own research skills of MMR as we are emerged in this journey together.

WHY CHOOSE MIXED METHODS RESEARCH

Why choose Mixed Methods Research (MMR)? MMR helps to improve communication across disciplines and among practitioners leading to better outcomes. MMR appears to be a form of research that will greatly benefit practitioners as well as experienced researchers who want to expand their research methodologies. Research that has been presented primarily by quantitative research can benefit from the voice of the participants and the added value of the different perspective that qualitative research can provide. The purpose of MMR is to draw from the positive aspects of both research paradigms to better answer the research question. MMR has become a widely accepted and well established technique used in many areas of research in order to enhance the proficiency and reliability of research result. This book contributes to the understanding of MMR technique. In writing this book we have attempting to break down the steps or decisions researchers need to consider and master in order to produce a quality study. It is our experience that the thought of conducting a study feels overwhelming to many, and in our own teaching we address these concerns and offer to break down the research process into manageable linear phases. Hence the outline of the chapters follows this logical process.

The intent of this book is to create a practical, easy to read and understand, user- friendly book that will reach a wide audience. The main target is the novice researcher, for example master or doctoral level students committed to writing their thesis, dissertation, or scholarly articles. Typically these students have very limited knowledge of research. The majority of books on research are highly complex and target the advanced researchers. This book will help to fill the void for beginning researchers who are committed to conducting high quality research while at the same time being novice researchers. MMR is often used in schools, businesses and non- profit organizations as they strive to address and resolve questions that will impact their organizations. We hope that the book will be an excellent instructional aid for a professor who wishes to utilize this book in a research methods course. The final chapter, Creating and Implementing a Mixed Methods Research Study, is a complete tool taking students through the research process step by step from start to finish.

Mette L. Baran
Cardinal Stritch University, USA

Janice E. Jones
Cardinal Stritch University, USA
Milwaukee, October 2015

REFERENCES

Creswell, J. W. (2009). *Research design: Qualitative, quantitative, and mixed methods approaches.* Thousand Oaks, CA: Sage.

Gorard, S., & Taylor, C. (2004). *Combining methods in educational and social research.* Buckingham, UK: Open University Press.

Teddlie, C., & Tashakkori, A. (2003). Major issues and controversies in the use of mixed methods in the social and behavioral sciences. In A. Tashakkori & C. Teddlie (Eds.), *Handbook of mixed methods in social and behavioral research* (pp. 3–50). Thousand Oaks, CA: Sage.

Teddlie, C., & Tashakkori, A. (2009). *Foundations of mixed methods research: Integrating quantitative and qualitative approaches in the social and behavioral sciences.* Thousand Oaks, CA: Sage.

Preface

The intent of this book is to create a practical, easy to read and understand, user-friendly book that will reach a wide audience of inexperienced researchers. The main target is the novice researcher; for example, master or doctoral level students committed to writing their thesis, dissertation, or scholarly articles. Typically, these students have very limited knowledge of research. The majority of books on research are highly complex and target the advanced researchers. This book will help to fill the void for beginning researchers who are committed to conducting high quality research while at the same time being new to the research task. Mixed methods research (MMR) is often used in schools, businesses and non-profit organizations as it is appropriate in addressing and resolving questions that they face.

Graduate students and faculty can used the book as a reference and/or textbook for research courses. Graduate students, while committed to conducting high quality research, often have a limited knowledge of conducting research for their thesis or dissertation. This book takes novice students through the entire research process by offering step by step guidance. This book is also intended for a broader audience in social and human sciences, representing many disciplines, as MMR is rapidly gaining in popularity due to the added richness in findings it affords the researcher. Faculty often ponder how to best train students in MMR. It can be a challenge since students come from a wide variety of backgrounds and arrive with various skill levels and research interests. Hence faculty may find the book useful. We also encourage all graduate students to become proficient in this third research paradigm.

THE ISSUE AT HAND

Conducting empirical research to complete either a master's thesis or dissertation is a graduation requirement at most higher educational institutions today. Even accomplished students may be faced with the responsibility of conducting research for the first time, requiring them to make decisions regarding methodology and design, followed by the daunting tasks of data analysis and interpretation. Similarly, course instructors are faced with selecting teaching materials that provide the resources students need in order to learn both quantitative and qualitative methods. This in itself can be challenging since students often like to preselect either a quantitative or qualitative track. Likewise, their advisors may be more experienced in one or the other and uncomfortable training their students in both. We need to break this trend as many issues today are highly complex and the MMR approach may be the best way to gain a complete understanding of the issue under investigation. This book demonstrates how MMR designs can address a wide array of research questions and help navigate the inexperienced researchers through the complicated mix of research decisions that must be made.

The contributors to this book are all experienced MM researchers and, combined, they walk the reader through the entire research process beginning with the need to establish an epistemological foundation to formulating the research questions, selecting the appropriate research design, collecting data, and interpreting both quantitative and qualitative findings. The final chapter includes a practical step by step outline for conducting a research study from start to finish.

ORGANIZATION OF THE BOOK

The book is organized into 15 chapters spread over seven sections which follows a natural research project outline. A brief description of each of the chapters follows:

Chapter 1, The Nature of Research Methodologies: Terms and Usage within Quantitative, Qualitative, and Mixed Methods

Mixed methods research is, generally speaking, an approach to knowledge (theory to practice) that attempts to consider multiple viewpoints, perspectives, and positions. Before the advent of mixed methods, many studies used multiple methods to achieve the benefits of triangulation without restricting themselves to any paradigm or methodological category. Today, the primary philosophy of mixed methods research is that of pragmatism. This chapter covers the history and the foundation of research methodologies and explains the purpose of research within various methodologies. This chapter also reviews and discusses key research terminology.

Chapter 2, Moving from Tension to Texture: The Paradigmatic Roots of Mixed Methods Research

Much like a jigsaw puzzle box-top guides one while connecting the pieces, an individual's research paradigm operates as a conscious or subconscious influence in conducting a research project. This chapter starts by making the argument for the critical role of research paradigms before moving into a thorough investigation of the paradigmatic origins of the qualitative-quantitative "debate." While mixed methods research is often seen as the mediator in the dispute, the authors then articulate four broad ways in which mixed-methods research addresses the paradigm divide at the heart of qualitative and quantitative research. The result is paradigmatically complex, but offers researchers flexibility as they seek to address their research question.

Chapter 3, Mixed Method Research: A Concept

Mixed methods has emerged as the third research community in the social and behavioral sciences during the past decades, joining quantitative and qualitative methods of scholarly inquiry. In recent years, mixed methods research, research paradigm, methodology, and action research have encouraged the combined use of quantitative and qualitative research to answer complex questions. Mixed methods research integrates both methods, the quantitative and the qualitative, to present research findings within a single-system process. The chapter introduces the concept of action research and how mixed methods is applied to action research and its processes. A plan of action for attaining better research results through mixed methods research is also put forth.

Chapter 4, It Is All in the Design: Creating the Foundations of a Mixed Methods Research Study

Various designs are used in research, all with specific advantages and disadvantages, and it is important to be clear about the role and purpose of each. Researchers need to know where the design fits into the whole research process from framing the research purpose and question(s), data collection and analysis, to reporting the findings. Researchers can decide to use a quantitative, qualitative, or a mixed methods design. This chapter introduces the various approaches aligned with each design. Which one the researcher selects, depends on the objective of the study and the nature of the phenomenon.

Chapter 5, Designs of Mixed Methods Research

Mixed methods research is becoming an increasingly popular approach in the fields of sociology, psychology, education and health sciences. Calls for the integration of quantitative and qualitative research methods have been advanced in these fields. A key feature of mixed methods research is its methodological pluralism, which frequently results in research which provides broader perspectives than those offered by mono-method designs. The central premise of mixed methods is that the use of quantitative and qualitative approaches, in combination, provides a better understanding of research problems and complex phenomena than either approach alone. This chapter surveys the common designs of mixed methods research and examines the main characteristics of each in terms of purpose, strength, and issues. Suggestions are also put forth regarding the application of these designs.

Chapter 6, Sampling in Research

Research is aimed at discovery and interpretation of facts, revision of accepted theories or laws in the light of new facts, or practical application of such new or revised theories or laws. Research is often conducted using a sample of the population of interest rather than conducting a census. There is no guarantee, however, that any sample will be precisely representative of the population from which it comes. This chapter is designed to equip researchers with knowledge of the general issues involved with sampling. The chapter covers the purpose of sampling and how to reduce sampling and non-sampling errors. Highlighted are the types of sample designs and how to implement them—along with the advantages and disadvantages of each. Also covered are guides to determine sample size.

Chapter 7, Analyzing Qualitative Data: Visualizing Lived Experiences through Poems and Photography

Use of photography and poetry offer a way for participants to express lived experiences through a visual and written means of self-expression. These forms of data collection can provide a rich, thick description of those often lying on the periphery of society. Traditional means of qualitative research such as interview and observation can at times create a barrier between the researcher and the participants because of the face-to-face interaction. Participants may be uncomfortable expressing authentic feelings during a formal interview process. By offering participants the opportunity to personally select descriptive photographs and articulate expression in their own voice through poetry, an authentic expression occurs.

Chapter 8, Analyzing Quantitative Data

The main purpose of this chapter is to present a conceptual and practical overview of some of the basic and advanced statistical tools for analyzing quantitative data. Analyzing quantitative data involves two broad analytical methods that serve two main purposes, which are descriptive and inferential statistical methods. The chapter covers both descriptive and inferential quantitative methods. It introduces some of the descriptive statistical methods such as mean, median, mode, variance, standard deviation, and graphical methods (e.g., histograms). It also covers inferential statistical methods such as correlation, simple regression, multiple regression, t-test for two independent samples, t-test for two dependent samples, and analysis of variance (ANOVA).

Chapter 9, Analyzing Quantitative Data in Mixed Methods Research for Improved Scientific Study

The purpose of the chapter is to review the role of quantitative methods in corporate research and the methods for analyzing quantitative data. The study used secondary data and a survey of published articles on schools, businesses and non- profit organizations. The key findings show that exploratory data can be analyzed using graphs and charts, and hypothesis testing can be employed to test statements made. Impacts of one variable on another and the relationships between variables are explained using correlation and regression analysis. The implications are that the value of a quantitative analysis arises when it is possible to identify features that occur frequently across the many participatory discussions aimed at studying a particular research theme.

Chapter 10, How Marketers Conduct Mixed Methods Research: Incorporating the Exploratory Sequential Design with the Hierarchy of Effects Model

The complimentary nature of qualitative and quantitative research methods are examined with respect to a study assessing the market's view of a training and development institute in the Middle East. The qualitative portion consisted of focus groups conducted with seven distinct market segments served by the institute. The results proved insightful with respect to uncovering and understanding differences of opinion among the seven groups; however, taken alone, the qualitative research would have been very misleading with respect to the institute's standing in the Middle East.

Chapter 11, Analyzing Qualitative Data

This chapter provides an introduction to the process of qualitative analysis and to use step by step examples to provide an idea of how the process of qualitative analysis actually works. Crabtree and Miller, 1992, note that there are many different strategies for analysis, in fact, they suggest there are as many strategies as there are qualitative researchers. This chapter is intended to give the researcher a place to begin and to inspire a deeper dive into this rewarding form of data analysis. While qualitative data analysis can be time consuming the rewards that come from immersion in the data far outweigh the time spent doing so.

Chapter 12, Examining Online Communities: A Method for Quantitative Analysis of Qualitative Data

Methods of individual communication continue to expand through online media. Given the dynamic nature of online communications, traditional methods for studying communications may not suffice. A hybridized content analytic approach that combines qualitative and quantitative methods offers a unique methodological tool for researchers who seek to better understand computer-mediated communications and the psychological characteristics of those who communicate online by evaluating qualitative information using quantitative methods. This means of measurement allows researchers to statistically evaluate whether investigated phenomena are occurring in combination with the richness that qualitative assessment provides. As with any approach to computer-mediated communication, various ethical considerations must be kept in mind and these are discussed in concert with this hybridized approach to content analysis.

Chapter 13, Mentoring and Support for the edTPA: A Mixed Methods Program Review of edTPA Support Practices

The research question is: How effective is the current edTPA mentoring and support program for teacher candidates? This research utilized mixed method interactive program evaluation. An online survey collected teacher candidates' perceptions of mentoring and support for edTPA. The survey questions were organized around four constructs: preparedness, support from instructor, support from university supervisors, and support from cooperating teacher. With $N = 46$, a comprehensive data analysis was conducted that identified areas of strength and need of the program. As a result, the researchers developed an edTPA mentoring and support program model.

Chapter 14, Morphological Ontology Design Engineering: A Methodology to Model Ill-Structured Problems

In the social-technical domain, scientists are often confronted with a class of problems that are termed "messy", "ill-structured" or "wicked". These problems address complex issues that not well-defined, contain unresolvable uncertainties, and are characterized by a lack of common agreement on problem definition. This chapter proposes a new mixed methods research technique, Morphological Ontology Design Engineering (MODE), which can be applied to develop models for ill-structured problems. MODE combines three different research methodologies into a single, methodology. MODE draws from research paradigms that include exploratory and descriptive research approaches to develop models. General morphological analysis offers a systematic method to extract meaningful information from domain experts, while ontology-based representation is used to logically represent domain knowledge. The design science methodology guides the entire process. MODE is applied to a case study where an ontological model is developed to support the implementation of a South African national cybersecurity policy.

Chapter 15, Creating and Implementing a Mixed Methods Research Study

This chapter serves as a guideline for outlining the core characteristics of mixed methods research (MMR) and the various steps researchers undertake in order to conduct a research study. The purpose is to create a worksheet assisting the researcher step by step from beginning to end following the seven steps to conducting research. While the focus is on MMR, the steps are similar for any type of research methodology. It is important to note that MMR is not a limiting form of research. Researchers need a mixed method research question and a mixed methods purpose statement for the research project. This chapter will also help explain why mixed methods research is one of the best approaches in answering a research question. Finally the chapter includes a suggestion of the importance of adding a visual diagram of the mixed methods research project into the implementation and final report.

Acknowledgment

There are numerous people who contributed to this book. First of all, we would like to thank our Master of Educational Leadership students and our Doctoral students in the Leadership for the Advancement of Learning and Service at Cardinal Stritch University in Milwaukee, WI. We would like to dedicate this book to all of our past, current, and future students who cross our paths and forever engage and inspire us in the inquiry of social research. You motivate us daily to become better teachers and research advisors and we thank you for the opportunity to learn along with you.

We would also like to thank our assistant editors for their time and attention to detail, we could not have completed this book without your support. Dr. Roger Baran, Evan Diehnelt, Dr. Eric Dimmitt, and Dr. Glady van Harpen, we will forever be grateful.

Finally, we must thank all of the contributors to this book for their commitment and dedication to the exciting field of mixed methods research. We are truly honored to call ourselves your colleagues.

Section 1

Research Paradigms

Research paradigms address the philosophical dimensions of social sciences. A research paradigm is a set of fundamental assumptions and beliefs as to how the world is perceived which then serves as a thinking framework that guides the behavior of the researcher. This chapter explains the different points of view on paradigms. The historical background of mixed methods research, including the debate among the three methodology communities, is described. In addition, the nature of mixed methods and its rational is discussed.

Chapter 1
The Nature of Research Methodologies:
Terms and Usage within Quantitative, Qualitative, and Mixed Methods

Ben Tran
Alliant International University, USA

ABSTRACT

Mixed methods research is, generally speaking, an approach to knowledge (theory to practice) that attempts to consider multiple viewpoints, perspectives, positions, and standpoints. As such, before the advent of mixed methods, many studies used multiple methods to achieve the benefits of triangulation without restricting themselves to any paradigmatic membership or methodological category. Today, the primary philosophy of mixed research is that of pragmatism. This chapter will cover the history and the foundation of research methodologies and explain the purpose of research within various methodologies. This chapter will also explain the various terminologies used within research and research design as well as the meaning of these terminologies. This chapter will not cover statistics, however, mixed methods, methodology, research, and paradigm, statistical research methodology will be touched upon.

INTRODUCTION

Debates about singular or universal truths or approaches to viewing the world (Socrates, Plato), versus multiple or relative truths (the Sophists such as Protagoras and Gorgias), versus balances or mixtures of the extremes (Aristotle's "golden mean" or principle of balance, moderate skepticism, Cicero, Sextus Empiricus), go back, at least, to ancient Western philosophy, and the spirit of these debates lives today in the different views of the three major approaches to social research. According to Plato, Protagoras said that "man is the measure of all things," and in many ways the history of Western philosophy still is debating Protagoras and the other Sophists[1]. This debate continues to affect how we view knowledge, what we look for, what we expect to find, and how we believe we are to go about finding and justifying *knowledge*. Mixed research is between the extremes Plato (quantitative research) and the Sophists (quali-

DOI: 10.4018/978-1-5225-0007-0.ch001

tative research), with mixed research attempting to respect fully the wisdom of both of these viewpoints while also seeking a workable middle solution for many (research) problems of interest.

Today, the primary philosophy of mixed research is that of pragmatism. Mixed methods research is, generally speaking, an approach to knowledge (theory to practice) that attempts to consider multiple viewpoints, perspectives, positions, and standpoints. This chapter will cover the history and the foundation of research methodologies and explain the purpose of research within various methodologies. This chapter will also explain the various terminologies used within research and research design as well as the meaning of these terminologies. This chapter will not cover statistics, however, through the discussion of mixed methods, methodology, research, and paradigm, statistical research methodology will be touched upon.

BRIEF HISTORY OF RESEARCH METHODOLOGIES

Before the advent of mixed methods, many studies used multiple methods to achieve the benefits of triangulation (Galton & Wilcocks, 1983) without restricting themselves to any paradigmatic membership or methodological category (Tashakkori & Teddlie, 2003a). However, there was still awareness of the difficulties present in producing results based on multiple data types. Thus, during the last 50 years, writers have used different names, making it difficult to locate articles that might relate to mixed methods research. Mixed methods has been called *multitrait/multimethod research* (Campbell & Fiske, 1959), which recognizes the collection of several quantitative methods in a single investigation; *integrated* or *combined*, in the sense that two forms of data are blended together (Jason & Onwuegbuzie, 2004, p. 17; Steckler, McLeroy, Goodman, Bird, & McCormick, 1992); and *quantitative and qualitative methods* (Fielding & Fielding, 1986), which acknowledges that the approach is actually a combination of methods. It has been called *hybrids* (Ragin, Nagel, & White, 2004); *methodological triangulation* (Morse, 1991a), which recognizes the convergence of quantitative and qualitative data; *combined research* (Creswell, 1994); and *mixed methodology*, which acknowledges that it is both a method and a philosophical worldview (Tashakkori & Teddlie, 1998). It has also been called the *third methodological movement* following the developments of first quantitative research and then qualitative research (Tashakkori &Teddlie, 2002, p. 5), the *third research paradigm* (Johnson & Onwuegbuzie, 2004, p. 15), and *a new star in social science sky* (Mayring, 2007, p. 1).

Nevertheless, the beginning of mixed methods is cited by some (Creswell & Plano-Clark, 2007, p. 5; Johnson, Onwuegbuzie, & Turner, 2007) to Campbell and Fiske (1959) who used multiple quantitative measures in a single study and referred to this as *multitrait* of *multimethod* research. These numerical beginnings served to demonstrate how by juxtaposing the results of multiple methods, different facets of a phenomenon can be identified—a concept later formalized by Webb, Campbell, Schwartz, and Sechrest (1966) as *triangulation*. Triangulation is seen to increase validity when multiple findings wither confirm or confound each other. A second argument for triangulation is that "all methods have inherent biases and limitations, so use of only one method to assess a given phenomenon will inevitably yield biased and limited results" (Greene, Caracelli, & Graham, 1989, p. 256). In accord, triangulation is often cited as having methodological superiority over single methods (Tran, 2014a).

Mixed research, in its recent history in the social and behavioral or human sciences, started with researchers and methodologists who believed qualitative *and* quantitative viewpoints and methods were useful as they addressed their research questions (Johnson et al., 2007). For the first 60 years or so of the 20th century, mixed research can be seen in the work of cultural anthropologists and, especially, the fieldwork sociologists (Gans, 1963; Hollongshead, 1949; Jahoda, Lazarsfeld, & Zeisel, 1931/2003; Lynd & Lynd, 1929/1959). However, the mixed methods label would not be coined scientists to see how they blended qualitative and quantitative data as they studied their communities. Although mixed methods research is not new, it is a new movement, or discourse, or research paradigm that has arisen in response to the current of *quantitative research* and *qualitative research.* In the history of ideas, new antitheses and syntheses continually develop in response to current theses. Mixed research is a synthesis that includes ideas from qualitative and quantitative research.

In the social science methodological literature, Campbell and Fiske's (1959) article sometimes is viewed as formalizing the practice of using multiple research methods. In this 1959 article, Campbell and Fiske introduced the idea of triangulation, referring to *multiple operationalism*, in which more than one method is used as part of a validation process that ensures that the explained variance is the result of the underlying phenomenon or trait and not of the method. It was argued that the convergence of findings stemming from two or more methods "enhances our beliefs that the results are valid and not a methodological artifact" (Bouchard, 1976, p. 268). Today, the most frequently used name is *mixed methods research*, a name associated with the *Handbook of Mixed Methods in Social and Behavioral Research* (Tashakkori & Teddlie, 2003a). Although the term *mixed methods* may not be familiar to a large number of social behavioral, and human science scholars, its frequent use will encourage researchers to see this approach as a distinct methodology and method, used by an increasingly larger scholarly community.

The idea of multiple operationalism is more of a measurement and construct validation technique, in its original formulation, than it is a full research methodology. Furthermore, early researchers' idea of multiple operationalism follows more closely what today is called *multimethod research*, in contrast to what currently is called *mixed methods research.* However, Campbell and Fisk (1959) are rightfully credited as being the first to show explicitly how to use multiple research methods for validation purposes. The ideas of Campbell and Fiske (1959) were extended further by Webb, Campbell, Schwartz, and Sechrest (1966), who defined multiple operationalism as representing the use of multiple measures that "are hypothesized to share in the theoretically relevant components but have different patterns of irrelevant components" (p. 3). Thus, Webb et al. are credited with being the first to coin the term *triangulation.* This type of triangulation is referred to as between-method triangulation or across-method triangulation (Tran, 2014a). With that said, Cook (1985) is credited for coining the term critical multiplism (also see Houts, Cook, & Shadish, 1986) to refer to the ideas that research questions can be examined from different perspectives and it is often useful to combine different methods with different biases.

THE USAGE OF RESEARCH METHODOLOGIES

In the usage of research methodologies, it was Denzin (1978, p. 291) who first outlined how to triangulate methods. Denzin defined triangulation as "the combination of methodologies in the study of the same phenomenon." Denzin outlined the following four types of triangulation: (1) data triangulation, (2) investigator triangulation, (3) theory triangulation, and (4) methodological triangulation (Tran, 2014a). According to Tran, Denzin also distinguished *within-methods* triangulation, which refers to the use of either multiple quantitative or multiple qualitative approaches, from between-methods triangulation, which involves the use of both quantitative and qualitative approaches. Denzin surmised that within-methods triangulation had limited values because essentially only one paradigm is being used such that any inherent weakness stemming from the paradigmatic approach used will prevail regardless of the specific research design used within a methodological paradigm (Tran, 2014a).

Denzin (1978) recommended the use of between-method triangulation, contending that by utilizing mixed methods, (1) "the bias inherent in any particular data source, investigators, and particularly method will be canceled out when used in conjunction with other data sources, investigators, and methods" and (2) "the result will be a convergence upon the truth about some social phenomenon" (p. 14). According to Denzin, three outcomes arise from triangulation: convergence, inconsistency, and contradiction. Whichever of these outcomes prevail, the researcher can construct superior explanations of the observed social phenomena. As such, although acknowledging that triangulation may not be suitable for all research purposes, Jick (1979) noted the following advantages of triangulation: (1) it allows researchers to be more confident of their results, (2) it stimulates the development of creative ways of collecting data, (3) it can lead to thicker, richer data, (4) it can lead to the synthesis or integration of theories, (5) it can uncover contradictions, and (5) by virtue of its comprehensiveness, it may serve as the litmus test for competing theories (Tran, 2014a). Morse (1991b) on the other hand, outlined two types of methodological triangulation: simultaneous or sequential. According to Morse (1991b), on the one hand, simultaneous triangulation represents the simultaneous use of qualitative and quantitative methods in which there is limited interaction between the two sources of data during data collection stage, but the findings complement one another at the data interpretation stage. On the other hand, sequential triangulation is utilized when the results of one approach are necessary for planning the next method.

While Denzin (1978), Jick (1979), and others were promoting triangulation, Sieber (1973) provided a list of reasons to combine quantitative and qualitative research. Sieber (1973) outlined how such a combination can be effective at the research design, data collection, and data analysis stages of the research process. For example, at the research design stage, on the one hand, quantitative data can assist the qualitative component by identifying representative sample members, as well as outlying cases. On the other hand, at the research design stage, qualitative data can assist the quantitative component of a study by helping with conceptual and instrument development. At the data collection stage, on the one hand, quantitative data can play a role in providing baseline information and helping to avoid *elite bias*. Furthermore, at the data collection stage, qualitative data can facilitative the data collection process. During the data analysis stage, quantitative data can facilitative the assessment of generalizability of the qualitative data, and shed new light on qualitative findings. Alternatively, during the data analysis stage, qualitative data can play an important role by interpreting, clarifying, describing, and validating quantitative results, as well as through grounding and modifying (Tran, 2014a).

Rossman and Wilson (1985) identified three reasons for combining quantitative and qualitative research. First, combinations are used to enable confirmation or corroboration of each other through triangulation. Second, combinations are used to enable or to develop analysis in order to provide richer data. Third, combinations are used to initiate new modes of thinking by attending to paradoxes that emerge from the two data sources. By examining published research, Greene, Caracelli, and Graham (1989) inductively identified the following five broad purposes of rationales of mixed methodological studies: (1) triangulation, (2) complementarity, (3) development, (4) initiation, and (5) expansion. Sechrest and Sidana (1995) listed four reasons for methodological pluralism: (1) for verification purposes, (2) to provide some basis for estimating possible error in the underlying measures, (3) to facilitate the monitoring of data collected, and (4) to probe a data set to determine its meaning. Also, Dzurec and Abraham (1993) identified the following six *pursuits* that link qualitative and quantitative research: (1) the pursuit of mastery over self and the world, (2) the pursuit of understanding through recomposition,

(3) the pursuit of complexity reduction to enhance understanding, (4) the pursuit of innovation, (5) the pursuit of meaningfulness, and (6) the pursuit of truthfulness (pp. 76-77).

Most recently, Collins, Onwuegbuzie, and Sutton (2006) identified four rationales for conducting mixed research: participant enrichment, instrument fidelity, treatment integrity, and significance enhancement. Meanwhile, in recent years, some of the strongest supporters of qualitative research, such as Denzin, Lincoln, and Guba, have, at times, made statements that appear to give credence to mixed methods research. For example, Lincoln and Guba (1985) acknowledged that "indeed, there are many opportunities for the naturalistic investigator to utilize quantitative data—probably more than are appreciated" (pp. 198-199). Also, Guba and Lincoln (1989) stated that "the information may be quantitative or qualitative. Responsive evaluation does not rule out quantitative modes, as is mistakenly believed by many, but deals with whatever information is responsive to the unresolved claim, concern, or issue" (p. 174). Furthermore, Guba and Lincoln (1994) noted that, "both qualitative and quantitative may be used appropriately with any research paradigm" (p. 105). Similarly, Guba and Lincoln (2005) reiterated that "within each paradigm, mixed methodologies may make perfectly good sense" (p. 200). Guba and Lincoln (2005) also declared, "as we tried to make it clear, the argument arising in the social sciences was not about method, although many critics of the new naturalistic, ethnographic, phenomenological, and/or case study approaches assumed it was" (p. 200).

In sum, the 20th century started with some use of what later came to be called mixed research, but social and psychological research quickly became primarily quantitative. Partially in reaction, many qualitative currents developed throughout the century, coalescing into a qualitative research paradigm in the 1980s and 1990s (Guba, 1990). In reaction to the polarization between quantitative and qualitative research, another intellectual movement occurred, and it has come to be called mixed methods research. The research community is in a three methodological or research paradigm world, with quantitative, qualitative, and mixed methods research all thriving and coexisting. In contrast to Thomas Kuhn's (1962) expectation for single paradigms characterizing *normal science*, the research community suggest that a three-paradigm *methodological* world might be healthy because each approach has its strengths and weaknesses and times and places of need.

TERMINOLOGIES WITHIN THE PRACTICE OF RESEARCH

In the research community, the process of research, is generally defined as the procedures used in research that involve introducing a problem, narrowing the research problem into purpose statements, research questions, and hypotheses, collecting and analyzing data to address these questions and hypotheses, and using a writing structure that best fits the problem and the methods. As such, in the research community, terms such as *investigator* (often associated with quantitative research) and *inquirer* (often associated with qualitative research) are used interchangeably.

Thus, a *methodology* refers to the philosophical framework and the fundamental assumptions of research (van Manen, 1990). *Research design* refers to the plan of action that links the philosophical assumptions to specific methods (Creswell, 2003; Crotty, 1998). Experimental research, survey research, ethnography, and mixed methods are all research designs. Methods, on the other hand, are more specific. *Methods* are techniques of data collection and analysis, such as a quantitative standardized instrument or a qualitative theme analysis of text data (Creswell, 2003; van Manen, 1990). Some mixed methods writers, like Tashakkori and Teddlie (1998), consider this form of research a methodology and focus on the philosophical assumptions. To call a process a *methodology* introduces a complexity to the process of research. Other mixed methods writers, like Creswell, Plano-Clark, Guttmann, and Hanson (2003), Greene, Caraceli, and Graham (1989), and Onwuegbuzies and Teddlie (2003), emphasize the techniques or methods of collecting and analyzing data. To call mixed methods research a *method* is clean and concise and resonates with many researchers (Elliott, 2005).

Design decisions depend on the purposes of the study, the nature of the problem, and the alternatives appropriate for its investigation. Once the purposes have been specified, the study should have explicit scope and direction, and attention can be focused on a delimited target area. The nature of the problem then plays the major role in determining what approaches are suitable. Design alternatives can be organized into nine functional categories based on these problem characteristics (Isaac & Michael, 1995, pp. 45-47):

1. **Historical:** To construct the past objectively and accurately, often in relation to the tenability of an hypothesis,
2. **Descriptive:** To describe systematically a situation or area of interest factually and accurately,
3. **Developmental:** To investigate patterns and sequences of growth and/or change as a function of time,
4. **Case or Field:** To study intensively the background, current status, and environmental interactions of a given social unit: an individual, a group, an institution, or a community,
5. **Correlational:** To investigate the extent to which variations in one factor correspond with variations in one or more other factors based on correlation coefficients,
6. **Causal-Comparative (or Ex-Post Facto):** To investigate possible cause-and-effect relationships by observing some existing consequence and searching back through the data for plausible causal factors,
7. **True Experimental:** To investigate possible cause-and-effect relationships by exposing one or more experimental groups to one or more treatment conditions and comparing the results to one or more control groups not receiving the treatment (random assignment being essential),
8. **Quasi-Experimental:** To approximate the conditions of the true experiment in a setting which does not allow the control and/or manipulation of all relevant variables. The researcher must clearly understand what compromises exist in the internal and external validity of the design and proceed within these limitations, and
9. **Action:** To develop new skills or new approaches and to solve problems with direct application to the classroom or other applied setting.

With that said, within all research design, the validity and generalizability of research findings derived are affected by two factors: internal validity and external validity. Internal validity is concerned with controlling as many extraneous or irrelevant variables as possible so that alternative hypotheses regarding factors other than X that could have contributed to the change in the dependent variable can be minimized or eliminated. The following factors can constitute a threat to the realization of the internal validity of an experiment: contemporary history, maturation process, pretesting procedures, measuring instruments, and statistical regression (Campbell & Stanley, 1966; Van Dalen, 1979). External validity is the generalizability or representativeness of the experimental findings. When examining the external validity of a design, the experimenter checks the following threats to representativeness: interaction effects of selection biases and X, reactive or interaction effect of pretesting, reactive effects of experimental procedures, and multiple-treatment interference (Campbell & Stanley, 1966; Van Dalen, 1979).

STATISTICAL USAGE WITHIN THE PRACTICE OF RESEARCH

Statistics is a branch of mathematics used to summarize, analyze, and interpret what we observe—to make sense or meaning of our observations. Thus, statistics, for the purpose of this chapter, can be called that body of analytical and computational methods by which characteristics of a population are inferred through observations made in a representative sample from that population. Since scientists rarely observe entire populations, sampling and statistical inference are essential. Although, the objective of statistical methods is to make the process of scientific research as efficient and productive as possible, many scientists and engineers have inadequate training in experimental design and in the proper selection of statistical analyses for experimentally acquired data.

As such, any experimental or observational investigation is motivated by a general problem that can be tackled by answering specific questions. Associated with the general problem will be a population, and often, experiments aim at comparing two subpopulations and determining if there is a (significant) difference between them. With that said, in the field of research (methodologies), there are two types of statistical data: quantitative and qualitative. Hence, statistics is the science of collecting and organizing data and then drawing conclusions based on data. There are essentially three types of statistics with quantitative data: descriptive, multivariate, and inferential.

Quantitative

Quantitative research options have been predetermined and a large number of respondents are involved. By definition, measurement must be objective, quantitative and statistically valid. Simply put, it is about numbers, and objective hard data. The sample size for a survey is calculated by statisticians using formulas to determine how large a sample size will be needed from a given population in order to achieve findings with an acceptable degree of accuracy. Generally, researchers seek sample sizes which yield findings with at least a 95% confidence interval, plus/minus a margin error of 5% points. However, many surveys are designed to produce a smaller margin of error. As such, quantitative data includes closed-ended information such as that found on attitude, behavior, or performance instruments. The collection of this kind of data might also involve using a closed-ended checklist, on which the researcher checks the behaviors seen. Sometimes quantitative information is found in documents such as census records or attendance records. The analysis consists of statistically analyzing scores collected on instruments, checklists, or public documents to answer research questions or to test hypotheses. Hence, quantitative data are divided into two main categories, descriptive and continuous.

Descriptive Statistics

Descriptive statistics are tabular, graphical, and numerical methods by which essential features of a sample can be described. Although these same methods can be used to describe entire populations, they are more often applied to samples in order to capture population characteristics by inference. As such, descriptive statistics are used to describe the basic features of the data in a study, such that they provide simple summaries about the sample and the measures. Together with simple graphics analysis, they form the basis of virtually every quantitative analysis of data. Descriptive statistics are typically distinguished from inferential statistics. With descriptive statistics you are simply describing what is or what the data shows. In other words, descriptive statistics are used to present quantitative descriptions in a manageable form: (1) measures of central tendency and (2) measures of spread.

Multivariate

Multivariate statistics is a form of statistics encompassing the simultaneous observation and analysis of more than one outcome variable. The application of multivariate statistics is multivariate analysis. Multivariate statistics concerns understanding the different aims and background of each of the different forms of multivariate analysis, and how they relate to each other. The practical implementation of multivariate statistics to a particular problem may involve several types of univariate and multivariate analyses in order to understand the relationships between variables and their relevance to the actual problem being studied. In other words, multivariate statistics includes all statistical techniques for analyzing two or more variables of interest, or if you like, two or more dependent variables. Univariate statistics analysis involves the examination across cases of one variable at a time. There are three major characteristics of a single variable: (1) distribution, (2) central tendency, and (3) dispersion. In other words, univariate statistics includes all statistical techniques for analyzing a single variable of interest, and this single variable of interest is the dependent variable. Examples of univariate methods are: (1) t-tests, (2) ANOVA, and (3) multiple regression.

Inferential Statistics

Inferential statistics are used to reach conclusions that extend beyond the immediate data alone. As such, inferential statistics are techniques that allow us to use samples to make generalizations about the populations from which the samples were drawn. It is, therefore, important that the sample accurately represents the population. The process of achieving this is called *sampling*. Inferential statistics arise out of the fact that sampling naturally incurs sampling error and thus a sample is not expected to perfectly represent the population. The methods of inferential statistics are: (1) the estimation of parameter(s) and (2) testing of statistical hypotheses.

Qualitative

Qualitative research is collecting, analyzing, and interpreting data by observing what people do and say. Whereas, quantitative research refers to counts and measures of things, qualitative research refers to the meanings, concepts, definitions, characteristics, metaphors, symbols, and descriptions of things. Qualitative research is much more subjective than quantitative research and uses very different methods of collecting information, mainly individual, in-depth interviews, and focus groups. The nature of this type of research is exploratory and open-ended. Small numbers of people are interviewed in-depth and/or a relatively small number of focus groups are conducted. Participants are asked to respond to general questions and the interviewers or group moderator probes and explores their responses to identify and define people's perceptions, opinions, and feelings about the topic or ideas being discussed and to determine the degree of agreement that exists in the group. The quality of the finding from qualitative research is directly dependent upon the skills, experience and sensitive of the interviewer or group moderator.

This type of research is often less costly than surveys and is extremely effectively in acquiring information about people's communications needs and their responses to and views about specific communications. Qualitative data consists of open-ended information that the researcher gathers through interviews with participants. The general, open-ended questions asked during these interviews allow the participants to supply answers in their own words. Also, qualitative data may be collected by observing participants or sites of research, gathering documents from a private or public source, or collecting audiovisual materials such as videotapes or artifacts. The analysis of the qualitative data typically follows the path of aggregating the words or images into categories of information and presenting the diversity of ideas gathered during data collection. Hence, qualitative data are divided into two main categories, nominal and ordinal. The following are qualitative research approaches:

Phenomenology

Phenomenology, the study of conscious experience, can be traced back to early 20th-century philosophers such as Husserl, Sartre, and Merleau-Ponty. Many of the ideas embodied in the works of these early phenomenologists were later adopted in the behavioral and social sciences by notable scholars such as psychologist Amedeo Giorgi (1970) and social scientist Alfred Schütz (1967). In contemporary social science, the term is used more broadly to denote the study of individuals' perceptions, feelings, and life experiences (Smith, Flowers, & Larkin, 2009). Phenomenology is a commonly employed approach in clinical psychology, and in this context it is associated with a unique set of methods and procedures (Moustakes, 1994).

Many of the ideas within the phonological field are embedded within qualitative inquiry in general and much qualitative research is phenomenological in nature in that it attempts to understand individuals' life experiences have for them. With that said, of course, not all qualitative research has phenomenological underpinnings. In some cases, the topic of study might be social structures or cultural processes that transcend individual experience, such as we might find in ethnographic research. But even so, data on such topics are often collected through interviews with individuals and hence through their experiential lens.

Ethnography

Ethnography literally means "to write about a group of people". Its roots are grounded in the field of anthropology and the practice of in situ research, where a researcher is immersed within the community s/he is study for extended periods of time. Early 20th-century anthropologists such as Bronislaw Malinowksi and Franz Boas pioneered traditional ethnography, which historically has focused on the cultural dimensions of life and behavior, such as shared practices and belief systems. A hallmark feature of the ethnographic approach is a holistic perspective, based on the premise that human behavior and culture are complicated phenomena and are composed of, and influenced by, a multitude of factors. These might include historical precedents, the physical context in which people live and work, the social structures in which individuals are embedded, and the symbolic environment in which they act.

Traditionally, ethnographic research has involved a researcher's total and prolonged immersion within a study community, often for a year or longer. With the luxury of time, proximity to the field site, and the ability to coordinate data collection in an integrated and inductive manner, research can be more fluid. Thus, another strength of the ethnographic approach is the naturalistic, in situ manner in which it is carried out and its emphasis on understanding the *emic* (insider/local) perspective. Observing individual and group behavior in it natural context and participating in that context can generate insights that other forms of research cannot. Not surprisingly, participant observation has historically been an integral component of ethnographic inquiry.

Ethnography has evolved significantly since its formal emergence in the early 20th century. Many disciplines outside of anthropology now utilize an ethnographic approach, and its adoption in applied research has grown as well. But commensurate with these changes is a dilution of the method. Contemporary use of ethnography outside of anthropology, particularly in applied research, is generally not as immersive as traditional ethnography. Most researchers cannot afford—in time or funds—to spend a year or more in the field. Extended fieldwork is still considered by many anthropology departments a rite of passage for doctoral students, but outside of this context, modern ethnography is much shorter in duration than its traditional predecessor.

Another change, at least in applied fields, is the move toward team research. Ethnography used to be a lonely enterprise. An ethnographer would travel by himself or herself to some exotic and remote place and live with the local people for a year or more. In such a research context, the entire research process—design, data collection, analysis, and write-up—was embodied within individuals. Nowadays, field research is often a collaborative enterprise, with teams consisting of multiple members from professional disciplines who represent various organizations.

Inductive Thematic Analysis and Grounded Theory

Inductive thematic analysis is probably the most common qualitative data analysis method employed in the social, behavioral, and health sciences. The process consist of reading through textual data, identifying themes in the data, coding those themes, and then interpreting the structure and content of the themes (Guest, MacQueen, & Namey, 2012). Grounded theory, on the other hand, is a type of inductive thematic analysis. Developed by Glaser and Strauss (1967), grounded theory is a set of iterative techniques designed to identify categories and concepts within text that are then linked into formal theoretical models (Cordin & Strauss, 2008). Charmaz (2006) describes grounded theory as a set of methods that "consist of systematic, yet flexible guidelines for collecting and analyzing qualitative data to construct theories *grounded* in the data themselves" (p. 2).

The process entails systematically reviewing units of text as they are collected, creating emergent codes for those units, and writing memos that expand on created codes and the relationships between codes. This process is repeated until data collection is completed. A defining feature of grounded theory is the "constant comparison method." Done properly, grounded theory requires that all segments of text are systematically compared and contrasted with each other. Theoretical models are created and continuously revised as data are progressively collected and analyzed. The exhaustive comparison between small units of text, such as lines or words, is often not a part of many applied inductive thematic analyses because it is extremely time consuming, especially for larger datasets (Guest et al., 2012). Another difference is that the output of an applied thematic analysis is not necessarily a theoretical model but often recommendations for program and policy. In terms of data collection, one of the primary distinctions between general inductive thematic analyses and grounded theory is that the latter requires an iterative research design in which data collection and analyses are emerged and sample sizes are not predetermined. Sampling and data collection procedures in an applied thematic analysis context can be iterative, but these can also be predetermined and temporally separate from analysis.

Case Study Approaches

A qualitative case study examines a phenomenon within its real-life context. Data are collected on or about a single individual, group, or event. In some cases, several cases or events may be studied. The primary purpose of a case study is to understand something that is unique to the case(s). Knowledge from the study is then used to apply to other cases and contexts. Qualitative case study methods often involve several in-depth interviews over a period of time with each case. Interviews explore the unique aspects of the case in great detail, more so than would be typical for a phenomenological interview. Implications of a case study approach for qualitative data collection and analysis are several: (1) participants and/or cases, by definition, should be selected for their unique properties, (2) inquiry in these types of studies focuses largely on their larger population, and (3) the overall idea is to tease out what makes them so different and why.

Discourse and Conversation Analysis

Both discourse and conversation analysis approaches stem from the ethnomethodological tradition (Garfinkel, 1967, 2002), which is the study of the ways in which people produce recognizable social orders and processes. Thus, ethnomethodology and its subgenres view text as a window into broader social and cultural processes. Referring back to Ryan and Bernard's typology, both of these approaches tend to examine text as an object of analysis. Analysis can be quite detailed, looking at the specific structure of discourse and interaction between two or more speakers to understand how shared meanings are socially constructed. Both approaches study *naturally* occurring language, as opposed to text resulting from more *artificial* contexts, such as formal interviews, and aim to extract social and cultural meanings and phenomena from the discourse studied.

While conversation and discourse analysis are similar in a number of ways, there are some key differences. Discourse analysis (DA) is generally broader in what it studies, utilizing pretty much any naturally occurring text (existing) written texts, lectures, documents, and so forth. Conversation analysis (CA) is a subset of discourse analysis. Its scope is narrower and confined to natural conversations between two or more people. Another difference is that discourse analysis emphasizes how humans construct meaning through speech and text, and its object of analysis typically goes beyond individual sentences. Conversation analysis, on the other hand, tends to be more granular, looking at elements such as grammatical structures and concentrating on smaller units of text, such as phrases and sentences.

Narrative Analysis

As with all of the above techniques, narrative analysis is based on the study of discourse and the textual representation of discourse. What distinguishes it from CA or DA is the *type* of discourse or text it deals with—narrative. Narratives, in this context, refer to stories that represent a sequence of events. They can be generated during the data collection process, such as through in-depth interviews or focus groups, they can be incidentally captured during participant observation, or they can be embedded in written forms, including diaries, letters, the Internet, or literary works. Narratives are analyzed in numerous ways and narrative analysis itself is represented within a broad range of academic traditions—sociology, anthropology, literature, psychology, health sciences, and cultural studies. Narrative analysis can be used for a wide range of purposes. Some of the more common purpose include formative research for a subsequent study, comparative analysis between groups, understanding social or historical phenomena, or diagnosing psychological or medical conditions. The underlying principle of a narrative inquiry is that narratives are the source of data used, and their analysis opens a gateway to better understanding of a given research topic.

Mixed Methods Approaches

Research studies are becoming increasingly diverse and inclusive of both quantitative and qualitative methods—that is, they are mixing methods to address specific objectives. The basic premise behind using a mixed methods research design is that *the combined* of both approaches provides a better understanding of a research problem than either approach could alone. Creswell and Plano-Clark (2011) argue that integrating methodological approaches strengths the overall research design, as the strengths of one approach offset the weaknesses of the other, and can provide more comprehensive and convincing evidence than mono-method studies. Another more practical benefit is that mixed method research can encourage interdisciplinary collaboration and the use of multiple paradigms.

There are more than a dozen mixed methods research typologies in the literature, each emphasizing different angles. For the most part, however, typologies include at the very least two basic dimensions—timing of data integration and purpose of integration (Guest et al., 2012). Timing of integration refers to how quantitative and qualitative datasets are used chronologically and analytically with respect to each other. The two most commonly used terms in this regard are *sequential* and *concurrent* designs (Creswell & Plano-Clark, 2007; Morgan, 1998; Morse, 1991b). Sequential designs are those in which integration occurs across chronological phases of a study and where data analysis procedures for one dataset type inform another type of dataset.

Mixing the Data: Quantitative and Qualitative

The mixing of data is a unique aspect of the research community when it comes to definition. By mixing the datasets, the researcher provides a better understanding of the problem than if either dataset had been used alone. There are three ways in which mixing occurs: merging or converging the two datasets by actually bringing them together, connecting the two datasets by having one build on the other, or embedding one dataset within the other so that one type of data provides a supportive role for the other dataset. Thus, it is not enough to simply collect and analyze quantitative and qualitative data, they need to be *mixed* in some way so that together they form a more complete picture of the problem then they so when standing alone.

MIXED METHODS

Mixed methods research has been established as a third methodological movement over the past twenty years, complementing the existing traditions of quantitative and qualitative movements (Tashakkori & Teddlie, 2003a; Teddlie & Tashakkori, 2009). This development has been accompanied by a search for an appropriate paradigm to provide a legitimation for the use of mixed methods comparable to the paradigms that have been widely accepted as justifying the use of quantitative and qualitative methods separately. The term *mixed methods* has come to be used to refer to the use of two or more methods in a research project yielding both qualitative and quantitative data (Creswell & Palno-Clark, 2007; Greene, 2007; Teddlie & Tashakkori, 2009). Multimethods do not have the same paradigmatic problem as do mixed methods since they can adopt the paradigm appropriate to the single type of data being collected.

The paradigm problem for mixed methods arises because of the so called paradigm wars of the 1970s and 1980s where the positivist paradigm of quantitative research came under attack from social scientists supporting qualitative research and proposing constructivism, or variants thereof, as an alternative paradigm (Reichhardt & Rallis, 1994). What has consequently been seen as a problem for mixed methods researchers is finding a rationale for combining qualitative and quantitative data in the face of seemingly incompatible paradigms underpinning them. Indeed it has been claimed that mixed methods are not possible due to the incompatibility of the paradigms underlying them (Guba & Lincoln, 1994). To deal with this problem a range of alternative approaches have been developed (Creswell & Plano-Clark, 2007; Tashakkori & Teddlie, 2003a).

These approaches, according to Hall (2013), can be classified into three basic categories: a-paradigm stance, multiple paradigm approach, and the single paradigm approach. The first of these simply ignores paradigmatic issues altogether, the second asserts that alternative paradigms are not incompatible and can be used in the one research project, and the third claims that both quantitative and qualitative research can be accommodated under a single paradigm. A paradigm may be viewed as a set of *basic beliefs* that deals with ultimate or first principles. It represents a *worldview* that defines, for its holder, the nature of the *world*, the individual's place in it, and the range of possible relationships to that world and its parts. The beliefs are basic in the sense that they must be accepted simply on faith because there is no way to establish their ultimate truthfulness. If there were, the philosophical debates reflected in these pages would have been resolved millennia ago (Guba & Lincoln, 1994).

Paradigm Issues

A (research) paradigm is a common body of beliefs, assumptions, and rules that govern research (Kuhn, 1970). It is a basic beliefs system that represents a worldview of individuals, their positions in it and a range of possible relationship to it (Guba & Lincoln, 1994, p. 105). It is an overall conceptual framework that guides the researchers (Sobh & Perry, 2006). A paradigm plays an important role in the mixed methods literature for reasons outlined above. The term has been adapted from Kuhn (1962, 1970) but as Morgan (2007) points out it has been given at least four different meanings in this literature. These have been identified as a worldview, an epistemological stance, as shared beliefs among a community of researchers and as model examples of research. Although Morgan (2007) argues that the third of these is closest to what Kuhn defined as a paradigm, Morgan does acknowledge that the second meaning, namely a paradigm as an epistemological stance has been the most commonly used meaning in discussions of social science methodology. It is, however, the first of these meanings that has been adopted by some of the major writers in the field. Teddlie and Tashakkori (2009) define a paradigm as "a worldview, together with the various philosophical assumptions associated with that point of view" (p. 84). Likewise, Creswell and Plano-Clark (2007) refer to a paradigm as a "worldview" (p. 21). Similarly, Greene (2007) uses the term *mental model*, in much the same way as a worldview.

According to these authors, a worldview consists of stances adopted on each of the elements (Creswell & Plano-Clark, 2007), or dimensions of contrast (Teddlie & Tashakkori, 2009) comprising ontology, epistemology, axiology, and methodology. Using these dimensions, Creswell and Plano-Clark (2007) identify four worldviews and Teddlie and Tashakkori (2009) identify five, and the only difference being the separation of positivism and postpositivism by Teddlie and Tashakkori but not by Creswell and Plano-Clark. There are so few differences between positivism and postpositivism that treating them as distinct worldviews is hardly warranted. Indeed, postpositivism modifies some of the excesses of positivism such as the claim that research must be value free, so that it can be regarded as the successor of positivism.

The four commonly agreed worldviews are: (1) postpositivism, (2) constructivism, (3) transformative, and (4) pragmatism. Of these, only the transformative and pragmatism worldviews, are seen to be compatible with mixed methods research. Harrits (2011), on the other hand, has proposed two further paradigms for mixed methods research, which he refers to as *nested analysis* and *praxeological knowledge*. Harrits (2011) use of the term paradigm does not, however, correspond with the *worldview* used adopted by the authors discussed above, but rather, corresponds more closely with Morgan's (2007) fourth use, namely as model examples of research. Under this usage of the term, there would be very many paradigms in mixed methods research alone, giving the term a much more specific meaning than that generally used in the literature on mixed methods.

Paradigm Options in Mixed Methods Research

There are three possible positions mixed methods researchers can take to adopting a paradigm to underpin their research. These are the a-paradigmatic stance, the multiple paradigm stance, and the single paradigm stance (Teddlie & Tashakkori, 2003) have been reduced to three by grouping several [Teddlie and Tashakkori's (2003) original six stances] under the multiple paradigm stance. The a-paradigmatic stance sidesteps the paradigm issue by ignoring it. Teddlie and Tashakkori (2003) cite Patton's (1990) claim that methodology is independent of the epistemology that gave rise to it. This can be true only in a very general sense in that, empiricism mandates some kind of observational method, but it does not specify any particular method. It is, however, at the point of interpretation of the information gained by using a research method that epistemology plays a role.

According to Hall (2013), the paradigmatic stance adopted by a researcher may not be made explicit. Indeed, in most cases, researchers get on with the research without regard to their paradigmatic position, which is left implicit. This does not mean that they do not have one, only that they do not articulate it in their research papers. This means that the a-paradigmatic stance supported by Patton cannot be sustained as a viable approach to justifying mixed methods research since no research is paradigm free. The multiple paradigm stance claims that researchers can draw on more than one paradigm in their research. This takes three forms outlined by Teddlie and Tashakkori (2003), namely the complementary strengths thesis, the dialectical thesis and the multiple paradigm thesis.

The complementary strengths thesis keeps the methods separate so as to draw on the strength of each (Morse, 2003). The dialectical thesis (Greene, 2007; Greene & Caracelli, 2003) claims that insights can be gained from mixing *mental models* where a *mental model* is "the set of assumptions, understandings, predispositions, and values and beliefs with which all social inquiries approach their work" (Greene, 2007, p. 12). The multiple paradigm thesis (Cresswell & Plano-Clark, 2007) contends that the mixed-methods design determines the appropriateness of paradigm choice. That is, that some paradigms are appropriate for some designs but not others. The researcher needs to then choose the paradigm most suited to the research design being implemented. A problem with these approaches is that it is not made clear which paradigms are to be mixed and how the mixing is to be done. This is particularly so with paradigms for which there has been a history of antagonism and for which claims of incompatibility have been made.

The third paradigmatic positions that can be taken by mixed methods researchers is the single paradigm approach. In this stance, researchers adopt a single paradigm that encompasses both qualitative and quantitative research methods. Two such paradigms have been identified as contenders for this approach, namely, pragmatism and the transformative approach. The former has been advocated by a number of mixed methods researchers (Johnson & Onwuegbuzie, 2004; Maxcy, 2003; Morgan, 2007) and the latter by Mertens (2003). This approach at least in principle overcomes the problem inherent in the multiple paradigm approach of the difficulties involved in attempting to integrate paradigms based on fundamentally different assumptions.

Single Paradigm Options for mixed Methods Research

Pragmatism has gained considerable support as a stance for mixed methods researchers (Feilzer, 2010; Johnson & Onwuegbuzie, 2004; Maxcy, 2003; Morgan, 2007). It is oriented "toward solving practical problems in the real world" (Feilzer, 2010) rather than on assumptions about the nature of knowledge (p. 8). It is derived from the writings of Peirce, Dewey, and James in the 19th and early 20th centuries and Rorty in the late 20th century. According to Hall (2013), the early formulations of pragmatism by James and Dewey, were criticized by Russell (1910, 1945) on a number of grounds, but particularly on the difficulty of determining what *works*. While this criticism was leveled at James' pragmatic theory of truth it is relevant to the use of pragmatism in mixed methods research because it assumes that the usefulness of any particular mixed methods design can be known in advance of it being used. The choice of a mixed methods research design is based on a number of considerations including the research questions and the purpose of the research. The question of whether a mixed methods design *works* or not can only be decided once the research product is completed and the findings interpreted. Pragmatism does not enter into the choice of mixed methods nor justify its use.

The transformative-emancipatory paradigm proposed by Mertens (2003) as a paradigm for mixed methods research places "central importance on the lives and experiences of marginalized groups such as women, ethnic/racial minorities, members of the gay and lesbian communities, people with disabilities, and those who are poor" (pp. 139-140). This focus of the paradigm limits its application to only a small range of social scientific research. As Tashakkori and Teddlie (2003b) point out "the transformative-emancipatory orientation might be better conceptualized as a purpose of a research project" (p. 680). However, neither of these paradigms can encompass the wide range of mixed methods research currently employed. As Bergman (2011) point s out "… it is time to bring in a second generation of theoretical considerations about the shape and reasons for mixed methods research" (p. 101). What is needed is a paradigm that does not limit the range of topics to be researched, nor the methods that can legitimately be used to conduct research, and can accommodate the mixing of qualitative and quantitative methods in use.

A candidate for such a paradigm is the realist perspective used in the evaluation field by Pawson and Tilly (1997), by Henry, Julnes, and Mark (1998), and extended to other areas by Sayer (2000). Whilst realism is an ontological position usually associated with positivism and postpositivism it is by no means confined to these positions as these applications have shown. Pawson and Tilly (1997) developed what they called a scientific realist approach to evaluation in which mixed methods play a prominent role in the conduct of evaluation. Henry et al. (1998) developed an emergent realist paradigm for evaluation in which they argue that the objectives of their approaches "will often best be served by a combination of quantitative and qualitative methods" (p. 19). Sayer (2000) uses Bhaskar's (1975) approach of critical realism to develop a paradigm for social science. This approach, according to Sayer, is compatible with a wide range of research methods including both qualitative and quantitative. What these approaches have in common is a version of realism that recognizes the complexity of social phenomena by enabling a role for values and interpretive meaning while at the same time accepting explanation as a legitimate goal of social research (Hall, 2013).

FUTURE RESEARCH

Research paradigms address the philosophical dimensions of social sciences. A research paradigm is a set of fundamental assumptions and beliefs as to how the world is perceived which then serves as a thinking framework that guides the behavior of the researcher (Jonker & Pennink, 2010). Although the philosophical backgrounds usually remain implicit in most research, they affect the practice of research. Some writers (e.g. Berry & Otley, 2004; Creswell, 2009; Neuman, 2011; Saunders, Lewis, & Thornhill, 2009) emphases that it is important to initially question the research paradigm to be applied in conducting research because it substantially influences how one undertake a social study from the way of framing and understanding social phenomena. According to Arbnor and Bjerke (2008), Guba (1990), Guba and Lincoln (1994), and Wahyuni (2012), the two main philosophical dimensions to distinguish existing research paradigms are ontology and epistemology (Kalof, Dan, & Dietz, 2008; Laughlin, 1995; Saunders, Lewis, & Thornhill, 2009), whereas only Arbnor and Bjerke (2008), Guba (1990), and Guba and Lincoln (1994) claim that is a *third paradigm*: methodology.

In terms of social research, ontologically one can perceive that the existence of reality is external and independent of social actors and their interpretations of it, termed objectivist (Saunders et al., 2009) or realist (Neuman, 2011). On the other hand, subjectivist, or nominalist adopted theory, believes that reality is dependent on social actors and assumes that individuals contribute to social phenomena. As such, research purpose and research questions are the suggested starting points to develop a research design because they provide important clues about the substance that a researcher is aiming to assess (Berry & Otley, 2004; Saunders et al., 2009; Yin, 2012). Conducting a research study should be started off by considering how the researcher views the observed social phenomena, which leads to the dominant research paradigm to be applied. The choice of a research paradigm leads to a relevant research methodology. The research design then needs to be developed to link research methodology and a set of research methods in order to enable the drawing of logical and valid inferences (Wahyuni, 2012). It is worth noting that the research purpose and research questions are the fundamental basis on which to craft a research design.

Hence, the rigor of qualitative research has been subject to continuous discussion in the literature (e.g. Boeije, 2010; Denzin & Lincoln, 2005). Qualitative research has been criticized as lacking generalizability by its counterpart, the quantitative mainstream. The quantitative tradition believes that research should reply heavily and validity to ensure its replicability and generalizability. Reliability refers to the consistency of measures whereas validity concerns with the extent to which it reflects the social phenomena being observed. Measurement consistency facilitates the replication or the repeatability of a study. The degree to which the results can be generalized to a larger population, which is called generalizability, has been the major point of the criticism of qualitative research (Wahyuni, 2012).

However, these traditional concepts of reliability and validity do not fit perfectly into the qualitative research landscape. Following Parker's (2012) stance, qualitative research operates in a completely different domain with different missions and agendas. In a similar vein, some social scientists (Bryman, 2012; Kalof, Dan, & Dietz, 2008) explain that reliability and validity per se cannot be practically used as criteria to assess qualitative research. Some alternative terms have been used to sensitize reliability and validity to the specific nature of qualitative research. There are four criteria of research trustworthiness developed by Lincoln and Guba (1985) and Guba and Lincoln (1989) which have been widely cited in the social science research method literature (e.g. Bryman, 2012; Kalof, Dan, & Dietz, 2008) to evaluate the quality of qualitative research: *credibility* which parallels internal validity, *transferability* which resembles external validity, *dependability* which parallels reliability, and *confirmability* which resembles objectivity (Wahyuni, 2012). In addition, with the continuous changes of our society and rapid growth of technologies, researchers also need to take into consideration and reevaluate the various new trends of data, data collection methodologies, and the meaning of quantitative and qualitative data.

Hence, when it comes to new trends of data, trends such as blogs, webinars, virtual intercepts, and virtual reality, design decisions depend on the purposes of the study, the nature of the problem, and the alternatives appropriate for its investigation. Once the purposes have been specified, the study should have explicit scope and direction, and attention can be focused on a delimited target area. Design alternatives can be organized into nine functional categories based on these problem characteristics (Isaac & Michael, 1995, pp. 45-47): historical, descriptive, developmental, case or field, correlational, causal-comparative, true experimental, quasi-experimental, and action. Traditional qualitative research methods, like study groups, polls and observational studies, are going digital and expanding the ability for researchers and businesses to target participants and collect information. The following are four types of qualitative trends: virtual intercepts, virtual reality (Bryson, N. Y.), blogs, and webinars.

Blogs

A blog is a personal internal journal, a discussion, or an informational site published on the World Wide Web and consisting of discrete entries typically displayed in reverse chronological order.

Webinars. Web conferencing may be used as an umbrella term for various types of online collaborative services including web seminars (or webinars), webcasts, and peer-level web meetings. It may also be used in a more narrow sense to refer only to the peer-level web meeting context, in an attempt to disambiguate it from the other types of collaborative sessions.

Virtual Intercepts

A web intercept (survey) is a survey which is used by websites to get feedback from users and visitors about the website. Such surveys are used to collect information regarding the website and how helpful it is for the users. A web intercept survey is an easy and quick way to intercept a website's performance consisting of a few main questions which are related to the topics directly and are usually objective in nature. Researchers can intercept respondents online from social media or surveys and route them into a one-on-one interview.

Virtual Reality

A technology-based Tavistock Method (Tran, 2014b) for research that harnesses virtual and augmented reality to enhance the research experience—for participant and client alike. Virtual reality captures new data, creates new experiences, and gives new insight by creating amazing reproductions of the physical world. With virtual reality technology, "where" becomes "anywhere". Researchers can bring more people together—in virtual environments—than ever before. Researchers can bring people together from all across the country, or the world, and capture fresh data and experience it all "through the eyes" of the participants.

CONCLUSION

Mixed methods research is, generally speaking, an approach to knowledge (theory to practice) that attempts to consider multiple viewpoints, perspectives, positions, and standpoints. As such, before the advent of mixed methods, many studies used multiple methods to achieve the benefits of triangulation (Galton & Wilcocks, 1983) without restricting themselves to any paradigmatic membership or methodological category (Tashakkori & Teddlie, 2003a). However, there was still awareness of the difficulties present in producing results based on multiple data types. Thus, during the last 50 years, writers have used different names, making it difficult to locate articles that might relate to mixed methods research.

The combined use of quantitative and qualitative research methods is commonly known as mixed methods. It is expected that a mixed methods approach can overcome the drawbacks of using a single method by incorporating the relative advantages of both quantitative and quality methods (Harrigan, Ramsey, & Ibbotson, 2012; Harrison & Reilly, 2011; Tashakkori & Teddlie, 2003). Many researchers have used mixed methods (Arora, 2011; Bazeley, 2004; Davism Golicic, & Boerstler, 2011; Harrigan et al., 2012; Koller, 2008; Parry, Kupiec-Teahan, & Rowley, 2012), and this approach has become the third methodological movement (Cameron & Miller, 2007). However, one of the main issues in mixed methods is that the combination of quantitative and qualitative methods tends to pose inconsistencies in research paradigms that underpin mixed methods research (Bazeley, 2004; Harrison & Reilly, 2011; Hewege, 2011; Hunt, 1992; Perry, Riege, & Brown, 1998; Sobh & Perry, 2006). With that said, when it comes to the nature of research methodologies, there are four criteria of research trustworthiness developed by Lincoln and Guba (1985) and Guba and Lincoln (1989) which have been widely cited in the social science research method literature (e.g. Bryman, 2012; Kalof, Dan, & Dietz, 2008) to evaluate the quality of qualitative research: *credibility* which parallels internal validity, *transferability* which resembles external validity, *dependability* which parallels reliability, and *confirmability* which resembles objectivity (Wahyuni, 2012). In addition, with the continuous changes of our society and rapid growth of technologies, researchers also need to take into consideration and reevaluate the various new trends of data, data collection methodologies, and the meaning of quantitative and qualitative data.

REFERENCES

Arbnor, I., & Bjerke, B. (2008). *Methodology for creating business knowledge*. London: Sage Publications.

Arora, R. (2011). Bundling or unbundling frequently purchased products: A mixed method approach. *Journal of Consumer Marketing*, *28*(1), 67–75. doi:10.1108/07363761111101967

Bazeley, P. (2004). Issues in mixing qualitative and quantitative approaches to research. In R. Buber, J. Gadner, & L. Richards (Eds.), *Applying qualitative methods to marketing management research* (pp. 141–156). Palgrave Macmillan.

Bergman, M. M. (2011). The politics, fashions, and conventions of research methods. *Journal of Mixed Methods Research*, *5*(2), 99–102. doi:10.1177/1558689811408318

Berry, A. J., & Otley, D. T. (2004). Case-based research in accounting. In C. Humprey & B. Lee (Eds.), *The real life guide to accounting research: A behind-the-scenes view of using qualitative research methods* (pp. 231–256). Oxford, UK: Elsevier. doi:10.1016/B978-008043972-3/50016-5

Bhaskar, R. (1975). *A realist theory of science*. Leeds, UK: Leeds Books.

Boeije, H. (2010). *Analysis in qualitative research*. London: Sage Publications.

Bouchard, T. J. Jr. (1976). Unobtrusive measures: An inventory of uses. *Sociological Methods & Research*, *4*(3), 267–300. doi:10.1177/004912417600400301

Bryman, A. (2012). *Social research methods* (4th ed.). New York: Oxford University Press.

Bryson, J. (n.d.). *Current and emerging trends in qualitative market research*. Retrieved on March 4, 2015. Available at http://qualblog.com/current-and-emerging-trends-in-qualitative-market-research/

Cameron, R., & Miller, P. (2007). Mixed methods research: Phoenix of the paradigm wars. In R. Chapman (Ed.), *Managing our intellectual and social capital: Proceedings of the 21st ANZAM Conference*. Sydney, Australia: Promaco.

Campbell, D. T., & Fiske, D. W. (1959). Convergent and discriminant validation by the multitrait-multimethod matrix. *Psychological Bulletin*, *56*(2), 81–105. doi:10.1037/h0046016 PMID:13634291

Campbell, D. T., & Stanley, J. C. (1966). *Experimental and quasi-experimental designs for research*. Chicago, IL: Rand McNally & Co.

Charmaz, K. (2006). *Grounded theory: A practical guide through qualitative analysis*. Thousand Oaks, CA: Sage Publications.

Collins, K. M. T., Onwuegbuzie, A. J., & Sutton, I. L. (2006). A model incorporating the rationale and purpose for conducting mixed methods research in special education and beyond. *Learning Disabilities (Weston, Mass.)*, *4*(1), 67–100.

Cook, T. D. (1985). Postpositivist critical multiplism. In L. Shotland & M. M. Mark (Eds.), *Social science and social policy* (pp. 21–62). Beverly Hills, CA: Sage Publications.

Corbin, J., & Strauss, A. (2008). *Basics of qualitative research: Techniques and procedures for developing grounded theory*. Thousand Oaks, CA: Sage Publications.

Creswell, J. W. (1994). *Research design: Qualitative and quantitative approaches*. Thousand Oaks, CA: Sage Publications.

Creswell, J. W. (2003). *Research design: Qualitative, quantitative, and mixed methods approaches* (2nd ed.). Thousand Oaks, CA: Sage Publications.

Creswell, J. W. (2009). *Research design: Qualitative, quantitative, and mixed methods approaches* (3rd ed.). Thousand Oaks, CA: Sage Publications.

Creswell, J. W., & Plano-Clark, V. L. (2007). *Designing and conducting mixed methods research*. Thousand Oaks, CA: Sage Publications.

Creswell, J. W., & Plano-Clark, V. L. (2011). *Designing and conducting mixed methods research*. Thousand Oaks, CA: Sage Publications.

Creswell, J. W., Plano-Clark, V. L., Guttmann, M. L., & Hanson, E. E. (2003). Advanced mixed methods research design. In A. Tashakkori & C. Teddlie (Eds.), *Handbook of mixed methods in social and behavioral research* (pp. 209–240). Thousand Oaks, CA: Sage Publications.

Crotty, M. (1998). *The foundations of social research*. Thousand Oaks, CA: Sage Publications.

Davis, D. F., Golicic, S. L., & Boerstler, C. N. (2011). Benefits and challenges of conducting multiple methods research in marketing. *Journal of the Academy of Marketing Science*, *39*(3), 467–479. doi:10.1007/s11747-010-0204-7

Denzin, N., & Lincoln, Y. (2005). *The sage handbook of qualitative research* (3rd ed.). London: Sage Publications.

Denzin, N. K. (1978). *The research act: A theoretical introduction to sociological methods*. New York: Praeger.

Dzurec, L. C., & Abraham, J. L. (1993). The nature of inquiry: Linking quantitative and qualitative research. *ANS. Advances in Nursing Science*, *16*(1), 73–79. doi:10.1097/00012272-199309000-00009 PMID:8311427

Elliott, J. (2005). *Using narrative in social research: Qualitative and quantitative approaches*. Thousand Oaks, CA: Sage Publications.

Feilzer, M. Y. (2010). Doing mixed methods research pragmatically: Implications for the rediscovery of pragmatism as a research paradigm. *Journal of Mixed Methods Research*, *4*(1), 6–16. doi:10.1177/1558689809349691

Fielding, N. G., & Fielding, J. L. (1986). *Linking data: Sage university paper series on qualitative research methods*. Beverly Hills, CA: Sage Publications. doi:10.4135/9781412984775

Galton, M., & Wilcocks, J. (1983). *Moving from the primary school*. London: Routledge and Kegan Paul.

Gans, H. J. (1963). *Urban villagers: Group life and class in the life of Italian-Americans*. New York: Free Press.

Garfinkel, H. (1967). *Studies in ethnomethodology*. Englewood Cliffs, NJ: Prentice-Hall.

Garfinkel, H. (2002). *Ethnomethodology's program: Working out Durkheim's aphorism*. Lanham, MD: Rowman & Littlefield.

Giorgi, A. (1970). *Psychology as a human science*. New York, NY: Harper & Row.

Glaser, B., & Strauss, A. (1967). *The discovery of grounded theory: Strategies for qualitative research*. New Brunswick, NJ: Aldine Transaction.

Green, J. C., Caracelli, V. J., & Graham, W. F. (1989). Toward a conceptual framework for mixed-method evaluation designs. *Educational Evaluation and Policy Analysis*, *11*(3), 255–274. doi:10.3102/01623737011003255

Greene, J. C. (2007). *Mixed methods in social inquiry*. San Francisco, CA: John Wiley & Sons.

Greene, J. C., & Caracelli, V. J. (2003). Making paradigmatic sense of mixed methods practice. In A. Tashakkori & C. Teddlie (Eds.), *Handbook of mixed methods in social and behavioral research* (pp. 91–110). Thousand Oaks, CA: Sage Publications.

Guba, E. G. (1990). *The paradigm dialog*. Newbury Park, CA: Sage Publications.

Guba, E. G., & Lincoln, Y. S. (1989). *Fourth generation evaluation*. Newbury Park, CA: Sage Publications.

Guba, E. G., & Lincoln, Y. S. (1989). *Fourth generation evaluation*. Newbury Park, CA: Sage Publications.

Guba, E. G., & Lincoln, Y. S. (1994). Competing paradigms in qualitative research. In N. K. Dezin & Y. S. Lincoln (Eds.), *Handbook of qualitative research* (pp. 105–117). Thousand Oaks, CA: Sage Publications.

Guba, E. G., & Lincoln, Y. S. (2005). Paradigmatic controversies, contradictions, and emerging confluences. In N. K. Denzin & Y. S. Lincoln (Eds.), *The Sage handbook of qualitative research* (3rd ed., pp. 191–215). Thousand Oaks, CA: Sage Publications.

Guest, G., MacQueen, K., & Namey, E. (2012). *Applied thematic analysis*. Thousand Oaks, CA: Sage Publications. doi:10.4135/9781483384436

Hall, R. F. (2013). Mixed methods: In search of a paradigm. In T. Le & Q. Le (Eds.), *Conducting research in a changing and challenging world* (pp. 71–78). Nova Science Publishers, Inc.

Harrigan, P., Ramsey, E., & Ibbotson, P. (2012). Exploring and explaining SME marketing: Investigating e-CRM using a mixed methods approach. *Journal of Strategic Marketing*, *20*(2), 127–163. doi:10.1080/0965254X.2011.606911

Harrison, R. L., & Reilly, T. M. (2011). Mixed methods designs in marketing research. *Qualitative Market Research: An International Journal*, *14*(1), 7–26. doi:10.1108/13522751111099300

Harrits, G. S. (2011). More than method?: A discussion of paradigm differences within mixed methods research. *Journal of Mixed Methods Research*, *5*(2), 150–166. doi:10.1177/1558689811402506

Henry, G. T., Julnes, G., & Mark, M. M. (1998). *Realist evaluation: An emerging theory in support of practice: New directions for evaluation, No. 78*. San Francisco, CA: Jossey-Bass Publishers.

Hewege, C. R. (2011). *Resolving structure-egency dichotomy in management research: Case for adaptive theory research methodology*. Paper presented at the 24th Annual Australian and New Zealand Academy of Management Conference, North Terrace, Australia.

Hollongshead, A. B. (1949). *Elmtown's youth*. New York: John Wiley.

Houts, A. C., Cook, T. D., & Shadish, W. R. (1986). The person-situation debate: A critical multiplist perspective. *Journal of Personality*, *54*(1), 52–105. doi:10.1111/j.1467-6494.1986.tb00390.x

Hunt, S. D. (1992). For reason and realism in marketing. *Journal of Marketing*, *56*(2), 89–102. doi:10.2307/1252045

Isaac, S., & Michael, W. B. (1995). *Handbook in research and evaluation: A collection of principles, methods, and strategies useful in the planning, design, and evaluation of studies in education and the behavioral sciences* (3rd ed.). San Diego, CA: Educational and Industrial Testing Services.

Jahoda, M., Lazarsfeld, P. F., & Zeisel, H. (2003). *Marienthal: The sociography of an unemployed community*. New Brunswick, NJ: Transaction Publishers. (Original work published 1931)

Jick, T. D. (1979). Mixing qualitative and quantitative methods: Triangulation in action. *Administrative Science Quarterly*, *24*(4), 602–611. doi:10.2307/2392366

Johnson, R. B., & Onwuegbuzie, A. J. (2004). Mixed methods research: A research paradigm whose time has come. *Educational Researcher*, *33*(7), 14–26. doi:10.3102/0013189X033007014

Johnson, R. B., Onwuegbuzie, A. J., & Turner, L. A. (2007). Toward a definition of mixed methods research. *Journal of Mixed Methods Research*, *1*(2), 112–133. doi:10.1177/1558689806298224

Jonker, J., & Pennink, B. (2010). *The essence of research methodology: A consice guide for master and PhD students in management science*. Heidelberg, Germany: Springers.

Kalof, L., Dan, A., & Dietz, T. (2008). *Essential of social research*. New York: McGraw-Hill.

Koller, M. (2008). A future research agenda for mixed-method-design in business research. *International Journal of Business Research*, *8*(4), 50–60.

Kuhn, T. S. (1962). *The structure of scientific revolutions*. Chicago: University of Chicago Press.

Kuhn, T. S. (1970). *The structure of scientific revolution* (2nd ed.). Chicago: University of Chicago Press.

Laughlin, R. (1995). Empirical research in accounting: Alternative approaches and a case for "middle-range" thinking. *Accounting, Auditing & Accountability Journal*, *8*(1), 63–87. doi:10.1108/09513579510146707

Lincoln, Y. S., & Guba, E. G. (1985). *Naturalistic inquiry*. Beverly Hills, CA: Sage Publications.

Lynd, R. S., & Lynd, H. M. (1929/1959). *Middletown: A study in modern American culture*. Orlando, FL: Harcourt Brace.

Maxcy, S. J. (2003). Pragmatic threads in mixed methods research in the social sciences: The search for multiple modes of inquiry and the end of the philosophy of formalism. In A. Tashakkori & C. Teddlie (Eds.), *Handbook of mixed methods in social and behavioral research* (pp. 51–90). Thousand Oaks, CA: Sage Publications.

Mayring, P. (2007). Introduction: Arguments for mixed methodology. In P. Mayring, G. L. Huber, L. Gurtler, & M. Kiegelmann (Eds.), *Mixed methodology in psychological research* (pp. 1–4). Rotterdam, Taipei: Sense Publishers.

Mertens, D. M. (2003). Mixed methods and the politics of human research: The transformative-emancipatory perspective. In A. Tashakkori & C. Teddlie (Eds.), *Handbook of mixed methods in social and behavioral research* (pp. 135–166). Thousand Oaks, CA: Sage Publications.

Morese, J. M. (2003). Principles of mixed methods and multimethod research design. In A. Tashakkori & C. Teddlie (Eds.), *Handbook of mixed methods in social and behavioral research* (pp. 189–208). Thousand Oaks, CA: Sage Publications.

Morgan, D. L. (1998). Practical strategies for combining qualitative and quantitative methods: Applications to health research. *Qualitative Health Research*, *8*(3), 362–376. doi:10.1177/104973239800800307 PMID:10558337

Morgan, D. L. (2007). Paradigms lost and pragmatism regained. Methodological implications of combining qualitative and quantitative methods. *Journal of Mixed Methods Research*, *1*(1), 48–76. doi:10.1177/2345678906292462

Morse, J. M. (1991a). Strategies for sampling. In J. M. Morse (Ed.), *Qualitative nursing research: A contemporary dialogue* (pp. 127–145). Newbury Park, CA: Sage Publications. doi:10.4135/9781483349015.n16

Morse, J. M. (1991b). Approaches to qualitative-quantitative methodological triangulation. *Nursing Research*, *40*(2), 120–123. doi:10.1097/00006199-199103000-00014 PMID:2003072

Moustakes, C. (1994). *Phenomenological research methods*. Thousand Oaks, CA: Sage Publications.

Neuman, W. L. (2011). *Social research methods: Qualitative and quantitative approaches* (7th ed.). Boston, MA: Pearson/Allyn and Bacon.

Onwuegbuzie, A. J., & Teddlie, C. (2003). A framework for analyzing data in mixed methods research. In A. Tashakkori & C. Teddlie (Eds.), *Handbook of mixed methods in social and behavioral research* (pp. 351–383). Thousand Oaks, CA: Sage Publications.

Parker, L. D. (2012). Qualitative management accounting research: Assessing deliverables and relevance. *Critical Perspectives on Accounting*, *23*(1), 54–70. doi:10.1016/j.cpa.2011.06.002

Parry, S., Kupiec-Teahan, B., & Rowley, J. (2012). Exploring marketing and relationships in software SMEsA mixed methods approach. *Management Research Review*, *35*(1), 52–68. doi:10.1108/01409171211190805

Patton, M. Q. (1990). *Qualitative evaluation and research methods* (2nd ed.). Newbury Park, CA: Sage Publications.

Pawson, R., & Tilly, N. (1997). *Realistic evaluation*. London: Sage Publications.

Perry, C., Riege, A., & Brown, L. (1998). *Realism rules OK: Scientific paradigms in marketing research about networks*. Paper presented at the 1998 Annual Australian and New Zealand Academy of Management Conference, Dunedin, New Zealand.

Ragin, C., Nagel, J., & White, P. (2004). *Sociology program methodology, measurement & statistics program directorate for social, behavioral & economic sciences*. Presented at the Workshop on Scientific Foundations of Qualitative Research, Arlington, VA.

Reichhardt, C. S., & Rallis, S. F. (1994). Qualitative and quantitative inquiries are not incompatible: A call for a new partnership. In C. S. Reichhardt & S. F. Rallies (Eds.), *The qualitative-quantitative debate: New perspectives*. San Francisco, CA: Jossey Bass. doi:10.1002/ev.1670

Rossman, G. B., & Wilson, B. L. (1985). Numbers and words: Combining quantitative and qualitative methods in a single large-scale evaluation study. *Evaluation Review*, *9*(5), 627–643. doi:10.1177/0193841X8500900505

Russell, B. (1910). *Philosophical essays*. London.

Russell, B. (1945). *A history of western philosophy*. Forage Village, MA: Simon & Schuster.

Saunders, M., Lewis, P., & Thornhill, A. (2009). *Research methods for business students*. London: Pearson Education.

Sayer, A. (2000). *Realism and social science*. London: Sage Publications. doi:10.4135/9781446218730

Schütz, A. (1967). *The Phenomenology of the social world*. Evanston, IL: Northwestern University Press.

Sechrest, L., & Sidana, S. (1995). Quantitative and qualitative methods: Is there an alternative? *Evaluation and Program Planning*, *18*(1), 77–87. doi:10.1016/0149-7189(94)00051-X

Sieber, S. D. (1973). The integration of fieldwork and survey methods. *American Journal of Sociology*, *78*(6), 1335–1359. doi:10.1086/225467

Smith, J., Flowers, P., & Larkin, M. (2009). *Interpretive phenomenological analysis: Theory, method and research*. Thousand Oaks, CA: Sage Publications.

Sobh, R., & Perry, C. (2006). Research design and data analysis in realism research. *European Journal of Marketing*, *40*(11/12), 1194–1209. doi:10.1108/03090560610702777

Steckler, A., McLeroy, K. R., Goodman, R. M., Bird, S. T., & McCormick, L. (1992). Toward integrating qualitative and quantitative methods: An introduction. *Health Education Quarterly*, *19*(1), 1–8. doi:10.1177/109019819201900101 PMID:1568869

Tashakkori, A., & Teddlie, C. (1998). *Mixed methodology: Combining qualitative and quantitative approaches*. Thousand Oaks, CA: Sage Publications.

Tashakkori, A., & Teddlie, C. (2002). *Handbook of mixed methods in social & behavioral research*. Sage Publications, Inc.

Tashakkori, A., & Teddlie, C. (2003a). *Handbook of mixed methods in social and behavioral research*. Thousand Oaks, CA: Sage Publications.

Tashakkori, A., & Teddlie, C. (2003b). The past and future of mixed methods research: From data triangulation to mixed model designs. In A. Tashakkori & C. Teddlie (Eds.), *Handbook of mixed methods in social and behavioral research* (pp. 671–702). Thousand Oaks, CA: Sage Publications.

Teddlie, C., & Tashakkori, A. (2000). *Foundations of mixed methods research*. Thousand Oaks, CA: Sage Publications.

Teddlie, C., & Tashakkori, A. (2003). Major issues and controversies in the use of mixed methods in the social and behavioral sciences. In A. Tashakkori & C. Teddlie (Eds.), *Handbook of mixed methods in social and behavioral research* (pp. 3–50). Thousand Oaks, CA: Sage Publications.

Teddlie, C., & Tashakkori, A. (2009). *Foundations of mixed methods research.* Thousand Oaks, CA: Sage Publications.

Tran, B. (2014a). Triangulation in organizational research: Validating knowledge in human competence at work. In A. Takhar & A. Ghorbani (Eds.), *Market research methodologies: Multi-method and qualitative approaches* (pp. 93–117). Hersey, PA: Premier Reference Source/IGI Global.

Tran, B. (2014b). Rhetoric of play: Utilizing the gamer factor in selecting and training employees. In T. M. Connolly, L. Boyle, T. Hainey, G. Baxter, & P. Moreno-Ger (Eds.), Psychology, pedagogy and assessment in serious games (pp. 175-203). Hershey, PA: Information Science Reference/IGI Global.

Van Dalen, D. B. (1979). *Understanding educational research* (4th ed.). New York: McGraw-Hill.

van Manen, M. (1990). *Research lived experience: Human science for an action sensitive pedagogy.* London, Canada: State University of New York Press.

Wahyuni, D. (2012). The research design maze: Understanding paradigms, cases, methods and methodologies. *Journal of Applied Management Accounting Research, 10*(1), 69–80.

Webb, E. J., Campbell, D. T., Schwartz, R. D., & Sechrest, L. (1966). *Unobtrusive measures.* Chicago: Rand McNally.

Yin, R. K. (2012). *Applications for case study research* (3rd ed.). Thousand Oaks, CA: Sage Publications.

ADDITIONAL READING

Bloor, M. A., & Wood, F. A. (2006). *Keywords in qualitative methods: A vocabulary of research concepts. Limited.* London: Sage Publications.

Creswell, J. W. (2008). *Research design: Qualitative, quantitative, and mixed methods approaches.* Thousand Oaks, CA: Sage Publications.

Davidson, J., & di Gregorio, S. (2011). Qualitative research and technology: In the midst of a revolution. In N. K. Denzin and S. Lincoln (4th Ed.), Handbook of qualitative research (pp. 627-643). Thousand Oaks, CA: Sage Publications.

Deshpande, R. (1983). Paradigms lost: On theory and method in research in marketing. *Journal of Marketing*, *47*(4), 101–110. doi:10.2307/1251403

Eisenhardt, K. M. (1989). Building theories from case study research. *Academy of Management Review*, *14*(4), 532–550.

Flick, U. (2007). *Managing quality in qualitative research.* London: Sage Publications.

Flick, U. (2014). *An introduction to qualitative research* (5th ed.). London: Sage Publications.

Flick, U., Garms-Homolová, V., Herrmann, W. J., Kuck, J., & Röhnsch, G. (2012). "I can't prescribe something just because someone asks for it ...": Using mixed methods in the framework of triangulation. *Journal of Mixed Methods Research*, *6*(2), 97–110. doi:10.1177/1558689812437183

Given, L. M. (2008). The sage encyclopedia of qualitative research method: Volumes 1 & 2. Newbury Park, CA: Sage Publications. doi:10.4135/9781412963909

Gobo, G. (2012). Glocalizing methodology?: The encounter between local methodologies. *International Journal of Social Research Methodology*, *14*(6), 417–437. doi:10.1080/13645579.2011.611379

Hanson, D., & Grimmer, M. (2007). The mix of qualitative and quantitative research in major marketing journals, 1993-2002. *European Journal of Marketing*, *41*(1/2), 58–70. doi:10.1108/03090560710718111

Hennink, M. M., Hutter, I., & Bailey, A. (2011). *Qualitative research methods*. London: Sage Publications.

Hsiung, P. C. (2012). *The globalization of qualitative research: Challenging anglo-american domination and local hegemonic discourse* [27 paragraphs]. Forum Qualitative Sozialforschung/Forum: Qualitative. *Social Research*, *13*(1), 21. Available at http://nbn-resolving.de/urn:nbn:de0114-fqs1201216 Retrieved on September 20, 2014

Layder, D. (1994). *Understanding social theory*. London, England: Sage Publications.

Layder, D. (1998). *Sociological practice: Linking theory and social research*. London, England: Sage Publications.

Marsden, D., & Littler, D. (1996). Evaluating alternative research paradigms: A market-oriented framework. *Journal of Marketing Management*, *12*(7), 645–655. doi:10.1080/0267257X.1996.9964442

Maxwell, J. A., & Loomis, D. M. (2003). Mixed methods design: An alternative approach. In A. Tashakkori & C. Teddlie (Eds.), *Handbook of mixed methods in social and behavioral research* (pp. 241–272). Thousand Oaks, CA: Sage Publications.

Maxwell, S. E., & Delaney, H. D. (1994). *Designing experiments and analyzing data*. Mahwah, NJ: Lawrence Erlbaum.

McMullen, L. M. (2011). A discursive analysis of Teresa's protocol: Enhancing oneself, diminishing others. In F. J. Wertz, K. Charmaz, L. M. McMullen, R. Josselson, R. Anderson, & E. McSpadden (Eds.), *Five ways on doing qualitative analysis: Phenomenological psychology, grounded theory, discourse analysis, narrative research, and intuitive inquiry* (pp. 205–223). New York: Guilford.

Nuttall, P., Shankar, A., Beverland, M. B., & Hooper, C. S. (2011). Mapping the unarticulated potential of qualitative research. *Journal of Advertising Research*, *51*(1), 153–163. doi:10.2501/JAR-51-1-153-166

Onwuegbuzie, A. J., & Leech, N. L. (2010). Generalization practices in qualitative research: A mixed methods case study. *Quality & Quantity*, *44*(5), 881–892. doi:10.1007/s11135-009-9241-z

Ryan, A., & Gobo, G. (2011). Managing the decline of globalized methodology. *International Journal of Social Research Methodology*, *14*(6), 411–415. doi:10.1080/13645579.2011.611378

Schnettler, B., & Rebstein, B. (2012). International perspective on the future of qualitative research in Europe. *Qualitative Sociology Review*, *8*(2), 6–11.

KEY TERMS AND DEFINITIONS

Between-Methods Triangulation: Involves the use of both quantitative and qualitative approaches.

Mixed Methods Research: Is an approach to knowledge (theory to practice) that attempts to consider multiple viewpoints, perspectives, positions, and standpoints. Mixed methods research has been established as a third methodological movement over the past twenty years, complementing the existing traditions of quantitative and qualitative movements.

Mixing Data: There are three ways in which mixing occurs: merging or converging the two datasets by actually bringing them together, connecting the two datasets by having one build on the other, or embedding one dataset within the other so that one type of data provides a supportive role for the other dataset.

Paradigm (Research): Is a common body of beliefs, assumptions, and rules that govern research. It is a basic beliefs system that represents a worldview of individuals, their positions in it and a range of possible relationship to it.

Qualitative Data: Consists of open-ended information that the researcher gathers through interviews with participants. The general, open-ended questions asked during these interviews allow the participants to supply answers in their own words.

Quantitative Data: Quantitative data includes closed-ended information such as that found on attitude, behavior, or performance instruments.

Qualitative Research: Is much more subjective than quantitative research and uses very different methods of collecting information, mainly individual, in-depth interviews, and focus groups.

Quantitative Research: Refers to counts and measures of things, qualitative research refers to the meanings, concepts, definitions, characteristics, metaphors, symbols, and descriptions of things.

Triangulation: Is the combination of methodologies in the study of the same phenomenon. The four types of triangulation: (1) data triangulation, (2) investigator triangulation, (3) theory triangulation, and (4) methodological triangulation.

Within-Methods Triangulation: Refers to the use of either multiple quantitative or multiple qualitative approaches.

ENDNOTES

1 Here are the approximate dates of these classical philosophers' lives: Protagoras (c. 490-420 B.C.E.), Socrates (470-399 B.C.E.), Plato (429-347 B.C.E.), Aristotle (384-322 B.C.E.), Gorgias (483-375 B.C.E.), Cicero (c. 106-43 B.C.E.), and Sextus Empiricus (c. late first century C.E. to mid-2nd century).

Chapter 2
Moving from Tension to Texture:
The Paradigmatic Roots of Mixed Methods Research

Preston B. Cosgrove
Cardinal Stritch University, USA

Peter M. Jonas
Cardinal Stritch University, USA

ABSTRACT

Much like a jigsaw puzzle box top guides one in how to connect the pieces, an individual's research paradigm operates as a conscious or subconscious influence in conducting a research project. This chapter starts by making the argument for the critical role of research paradigms before moving into a thorough investigation of the paradigmatic origins of the qualitative-quantitative "debate." While mixed-methods research is often seen as the mediator in the dispute, the authors then articulate four broad ways in which mixed methods research addresses the paradigm divide at the heart of qualitative and quantitative research. The result is paradigmatically complex, but offers researchers flexibility as they seek to address their research question.

INTRODUCTION: WHY SHOULD I CARE ABOUT RESEARCH PARADIGMS?

Consider the question: What is the most important part of putting a puzzle together? The most obvious answers always come first – start with the corner pieces, followed by locating the edges, and then grouping pieces of similar color. While you may consider yourself in command of the puzzle, a neutral observer would note something quite different: that the placement of those corners, edges, and colors was dictated in part by the image on the cover of the box top. The importance of that box top was so implicit and inherent that most do not identify it as an essential answer to the initial question, even though it controls most of our actions. You may be thinking that a puzzle *could* still be completed without a

DOI: 10.4018/978-1-5225-0007-0.ch002

box top. A *simple* puzzle, perhaps. But what of advanced puzzles? The infamous Ravensburger puzzle with 32,000 pieces? 3D puzzles that do not fit within the confines of corners and edges? Or the set of "Impossibles" puzzles that are borderless, have irregular edges, and come with five extra pieces? Our perception of what the puzzle looks like will always influence our work to complete it.

Not significantly different from a puzzle box top, a research paradigm is a significant force in how one considers and conducts research. Before answering questions regarding potential data collection methods, or research designs (methodologies) to govern those methods, one must first consult the theoretical perspectives at hand (Crotty, 1998). The broader research *gestalts* are not a new phenomenon, and have their origins in the ancient debates among the Sophists, Socrates, Plato, and Aristotle about what constituted the pursuit and meaning of knowledge (Johnson & Gray, 2010). In this sense, "mixed methods" is not about the simple merger of words and numbers; it is the process of potentially balancing seemingly opposing paradigms, or attempting to fuse them to create a new, coherent idea. The "how" of that process is the focus of this chapter, with an emphasis on comprehending the nature of research paradigms, their role in the qualitative-quantitative debate, and how the researcher understands their function in a mixed-methods project. To pursue mixed-methods research without this foundation is akin to constructing a puzzle with no box top; and without that framing image to guide your hands as they move about the top of the table, you're lost. And there is nothing worse than an incomplete puzzle.

TENSION: THE QUALITATIVE-QUANTITATIVE DEBATE AND THE PARADIGM DIVIDE

\ten(t)-shən\ noun ~ a state of latent hostility or opposition between individuals or groups

An examination of the tension inherent in the qualitative and quantitative debate first requires an understanding of the two terms in question. Both forms of research follow from an empirical process involving the standard collection, analysis, and interpretation of data (Leedy & Ormrod, 2010). But to describe their distinctions as a simple comparison between words (qualitative) and numbers (quantitative) is to over-generalize their fundamentally different ways of approaching the research process. Qualitative research involves emergent and inductive research designs, processes, and analyses that focus on the interpretation and meaning of phenomenon and/or participant experience in a natural setting. Conversely, quantitative research is a deductive and theory-driven approach focusing on strict measurement and control of variables within large samples and using analysis to identify statistical links among the data (Creswell, 2014; Ponterotto, 2005).

It is important note the terminology used above. The brief descriptions articulate qualitative and quantitative *research*, not qualitative and quantitative *methods*. The distinction in terminology is critical, for as Morgan (2007) questions, is the problem "simply about how we use methods, [or does is it concern] basic issues about the nature of research" (p. 48). Biesta (2010) helps clarify by arguing that the research process or design is not innately qualitative or quantitative, but that the kinds of data obtained by a preceding method are the focus of these terms. The answer to the question then is that the crux of the dispute centers not on various methods and the corresponding types of data, but instead on the nature of social science research all together. In this context, qualitative and quantitative are clusters and categories of research designs and meanings; that is, they represent distinct approaches, or specifically paradigms (i.e. puzzle box tops) about the essence and nature of research.

The divide between paradigms was not instantaneous, but instead emerged gradually as a reaction to the dominant research archetype of positivism. (The extensive background on the historical foundations and formation of current paradigms is not revisited here, but has been documented by Johnson and Gray (2010), Teddlie and Johnson (2009a), and Teddlie and Johnson (2009b)). Positivism (or logical-positivism) represented a hyper-empiricist and deductive model of research with a belief that the natural and social world could be studied with the same approach: value-free, causal explanations, measurement of observation, and a single identifiable reality. By the mid-twentieth century, disagreements with the absolute nature of positivistic claims made it increasingly subject to critique. Phillips and Burbules (2000) explain that positivistic logic falls under the broader form of *foundationalist philosophy*, in which scholars had identified six main philosophical problems (pp. 14-25). Put simply, the attributes of positivism "[fall] short when applied to human behavior" (Mertens, 2010, p. 11). Or as Phillips and Burbules (2000) describe, ultimately, non-positivists of all stripes successfully argued that "scientists... are no less human, and no less biased and lacking in objectivity than anyone else; they work within frameworks that are just that: frameworks" (p. 1; see also Denzin, 2010). The positivist research paradigm never disappeared (and has witnessed a revival, particularly within educational research [Feuer, Towne, & Shavelson, 2002; Hammersley, 2002; Lather, 2004; Mosteller & Boruch, 2002]), but was no longer the dominant approach within the social sciences. With the reign of positivism gone, two alternative paradigms emerged. No surprise given its name, postpositivism was marked by an adherence to much of the positivistic ethos, but with modifications to many of its challenged tenets. The second approach arriving after postpositivism was constructivism (or interpretivism or naturalism), which sought to posit an understanding of research far from the positivist or postpositivist orientation.

Before a more detailed analysis of each approach, it is critical to realize how the very *understanding* of paradigm was also shifting. The historic 1962 publication of Thomas Kuhn's *The Structure of Scientific Revolutions* became responsible not only for the popularity of the term "paradigm", but also its adoption within the philosophy of science as a way to capture the various perspectives on research. Bergman (2010) remarks that, "according to Kuhn, scientific paradigms determine the kind of questions researchers ask, how these questions are to be understood, what data to collect, and how to interpret research results to derive answers to these questions" (p. 172). Despite Kuhn's (2012) admission that scholars had identified approximately two dozen ways in which 'paradigm' was used in his original text (p. 181), Morgan (2007) classified the four general versions of paradigm as understood by Kuhn: 1) Paradigms as worldviews; 2) Paradigms as epistemological stances; 3) Paradigms as shared beliefs in a research field; and 4) Paradigms as model examples (pp. 50-54). Kuhn (1962) favored the final two as appropriate understandings of paradigm, but as Morgan (2007) documents, researchers made "epistemological stances" the dominant paradigm model in the social sciences. The implication was significant, as Denscombe (2008) describes:

...the focus of attention on [the philosophical] version of research paradigm both caused and reflected a shift in the nature of [the qualitative-quantitative] debate from the level of practice or outcomes to that of metaphysics. Paradigms, in this sense, concern higher level belief systems and the way these link with research questions. ...the methodological debate [shifted] from the level of research practice to the level of theoretical principles. p. 275

No longer were qualitative and quantitative seen simply as different research approaches; rather, they were now nested in broader paradigms conceptualized by various philosophical orientations, and as a result, research distinctions gave way to paradigm dichotomies.

Philosophical Foundations

In order to understand the nature of post-positivistic and constructivistic paradigms, we must explore the embedded philosophical foundations. Answers to the following questions help identify the core characteristics of each research paradigm:

- **Epistemology:** What is the theory of knowledge and the researcher-participant relationship?
- **Ontology:** What is the nature and role of reality?
- **Axiology:** What is the nature and role of values?
- **Causality:** What is the role and possibility of causal relationships?
- **Generalizability:** What is the role and possibility generalizing knowledge beyond the sample?
- **Logic:** What is the nature and role of theory?
- **Method:** What is the means by which the researcher gathers knowledge?

Table 1 illustrates the ways in which post-positivistic and constructivistic paradigms address these foundational questions.

It is important to know that positivism/post-positivism and constructivism are not the only research paradigms. The table above could include multiple versions of critical theory (race, pedagogy, etc.), feminist ideology, and post-modernist, among others. The paradigms in question are highlighted because they are not only the foundational philosophies behind quantitative and qualitative research, they also represent the opposing ends of a spectrum.

Table 1. Traditional research paradigms and philosophical orientations

		Research Paradigm	
		Post-Positivim	**Constructivism**
Philosophical Orientation	Epistemology	Knowledge is objectively discovered with the researcher independent of the data	Knowledge is subjectively constructed with interaction between researcher and participants
	Ontology	A single reality that is knowable based upon agreed limits of probability	Multiple realities that are socially constructed and understood
	Axiology	The influence of values in research can be checked and controlled	All research is value-bound by the context and interpretations from the researcher and participant
	Possibility for Causalities	Research can lead to an identification of causes and effects, within a specified probability	Research can make credible inferences with explanations made in the natural context
	Possibility for Generalizations	Critical to seek conditions leading to broad, context-free generalizations	Less important, with an emphasis on specific, context-bound theories
	Logic	Primarily deductive	Primarily inductive
	Methods	Primarily quantitative	Primarily qualitative

Table adapted from work by Creswell (2011), Crotty (1998), Mertens (2010), Ponterotto (2005), Teddlie and Tashakkori (2009)

The qualitative-quantitative divide widened in part because of the zero-sum nature of the debate. With a theoretical/philosophical basis for paradigm, the division between the two was quickly labeled as an "incompatibility thesis" (Howe, 1988, p. 10). Incompatible could seem absolute, particularly given the arguments that research paradigms fall on a continuum (Teddlie & Tashakkori, 2009). But as Guba (1990) reminded us, answers to the kinds of questions above, "are the starting points or givens...[that] cannot be proven or disproven in any foundational sense" (p. 18). Biesta (2010) also discussed the problem of theoretically-driven understandings of paradigms by stating that they,

...bring under one heading a range of different ideas and assumptions that do not necessarily have to go together. This tends to make the notion of paradigm into a container concept and leads to a situation in which paradigms have to be embraced or rejected in a wholesale manner rather than letting the discussion focus on smaller elements. (p. 98)

The result has been the production of two independent box tops displaying very different pictures, with each believing they are the fulfillment of Dewey's 1966 proclamation that "we know some methods of inquiry are better than others" (p. 104).

TEXTURE: MIXED-METHODS RESEARCH AS A PARADIGMATIC BRIDGE

\teks-chər\ noun ~ something composed of closely interwoven elements

As other chapters in this volume describe, mixed-methods research is an alternative in the qualitative-quantitative debate. Researchers advocating this bridge refused to "reduce one's opponents to stereotyped paradigms," which, as Shadish (1995) claimed, "is an intellectually lazy solution to an intellectually important set of problems" (p. 49). While Teddlie and Tashakkori (2010) argue that there are nine broad characteristics most scholars would agree upon, mixed-methods research still lacks a common definition in several senses: when does the mixing occur in the research process? How much breadth does the mixing involve? Why is mixing being used? What is being mixed in the first place? Despite Johnson, Onwuegbuzie, and Turner's (2007, p. 123) composite definition, it is critical to understand that there is incongruity in the mixed-methods community about some fundamental components. As a result, it should not be surprising that even though mixed methods broadly helps address the qualitative-quantitative debate, there are various positions about just how that paradigmatic bridge takes shape. And the uniqueness of these stances hold implications for those considering the potential use of mixed-methods in a research project.

Bridge #1: Mixed Methods Research Has Its Own Distinct Paradigm

Scholars advocating this position believe that there are unique presuppositions underlying mixed methods research that do not align with post-positivism or constructivism. The most prevalent paradigm articulated is pragmatism. Historically derived from the philosophies of William James and John Dewey, pragmatism in the twentieth century moved through three periods of relevance (Maxcy, 2003), before Howe's 1988 argument for a pragmatic research approach in the social sciences signaled a rejection of the incompatibility thesis. As an alternative, the pragmatist paradigm "sidesteps the contentious issues of truth and reality, accepts, philosophically, that there are singular and multiple realities that are open to empirical inquiry and orients itself toward solving practical problems in the "real world"" (Feilzer, 2010, p. 8). The specifics of pragmatism as a distinctive paradigm are contingent on the author [see Biesta (2010); Johnson & Onwuegbuzie (2004); Teddlie & Tashakkori (2009)], but Morgan's (2007) analysis provides a general understanding of the pragmatic alternative's response to three paradigmatic dualisms: 1) an *abductive* form of logic that employs both deductive and inductive reasoning; 2) an *intersubjective* approach that rejects the near impossibility of either complete objectivity or subjectivity; and 3) *transferability* of results that moves away from the dichotomy of outcomes that are either generalizable or only context-specific. And despite claims that a pragmatic paradigm skews toward a post-positivist orientation (Giddings, 2006), it remains the dominant option for a single-paradigm approach.

But pragmatism is not the sole paradigmatic alternative for mixed methods research. Mertens' (2010) transformative paradigm has become a broadly recognize option that goes beyond the pragmatic nature described above. And as a distinct paradigm, it includes scholars whose research confronts directly social oppression in myriad ways: critical theorists, feminists, Marxists, etc. Mertens (2010) identifies four primary characteristics for the transformative paradigm: 1) it centers on the experiences of marginalized and oppressed groups; 2) it offers an analysis of how inequity is reflected in traditional demographic groups (gender, race, etc.); 3) it offers an examination of the effect of inquiry on social/political action; and 4) it's philosophical orientation is immersed in transformative theory and philosophy (p. 21). The transformative paradigm has also been adopted to address concerns that mixed-methods research lacks an advocacy purpose with an emphasis on social justice. Indeed, with its unique focus, the transformative worldview allows for a mixed-methods approach, but without losing the emancipatory inclinations traditionally located in the constructivist arena (Sweetman, Badiee, & Creswell, 2010).

The third alternative is the critical realist paradigm, although its adoption as a single mixed-methods approach is much less formal and recognized than the pragmatic and transformative. While Maxwell and Mittapalli (2010) do not think realism should constitute a distinct paradigmatic approach, they do argue that "as a philosophical perspective that validates and supports key aspects of both qualitative and quantitative approaches while identifying some specific limitations of each, realism can constitute a productive stance for mixed methods research and can facilitate a more effective collaboration between qualitative and quantitative researchers" (p. 146). Specifically, the critical realist is an ontological cousin of post-positivists, yet affirms the epistemological character of the constructivist (Krauss, 2005). Critical realism represents a definitive philosophical approach, but given that it has not been identified as a mixed methods paradigm with the same formality as pragmatism or transformative, readers should consult Maxwell and Mittapalli (2010) for a more comprehensive take.

Bridge #2: Multiple Paradigms Should be used in Mixed Methods Research

This position has been promoted primarily by Greene and Carcelli (1997), who label it as the dialectical perspective. In this approach, the differences among the various paradigms and philosophical orientations are recognized, and the researcher should ensure the integrity of any paradigm in question. That said, "[multiple paradigms] should be deliberately used both within and across studies toward a dialectical discovery of enhanced understandings, of new and revisioned perspectives and meanings" (p. 8). The dialectical approach rejects the pursuit for a solitary paradigm, and that mixed methods should take advantage of the benefits that unique research paradigms bring. Indeed, Greene (2007) has more recently stated that a mixed methods research project should involve "the planned and intentional incorporation" of diverse paradigms in the research inquiry (p. 30). Critical to this paradigmatic bridge is the ability to actively engage with and balance the tension created from the juxtaposition of contrasting ideas (Tashakorri & Teddlie. 2010, p. 811).

Bridge #3: Multiple Paradigms Can be used as it Supports the Mixed Methods Research Design

The third bridge directs the researcher back to the research design and the type of paradigm(s) required to conduct the study. Creswell and Plano Clark (2011) serve as the advocates for this position, which is ultimately driven by the nature of the research question and the specific design. If the mixed-methods design is concurrent or parallel, a single paradigm such as transformative or pragmatic could be utilized. But in a sequential study, Creswell (2009) argues that it only makes sense for the researcher's paradigm to shift (post-positivist to constructivist, or vice-versa) depending on the data collection method. This understanding acknowledges the *potential* role of the first bridge, in contrast to the dialectical interpretation, and it only disagrees with the second bridge in that the research question/design should drive the use of multiple paradigms, and not an indiscriminating "planned and intentional" merger (Greene, 2007, p. 30).

Bridge #4: Mixed Methods Research can Operate Free from Paradigmatic Constraints

Three different perspectives form the narrative for this bridge. Argued by Morgan (2007) and Denscombe (2008), the first questions the application of philosophical orientations as the primary understanding for the concept of paradigm. As observed earlier, the adoption of a more theoretical frame for paradigm led, in part, to the incompatibility between research traditions. Morgan (2007) in particular challenged the role of epistemological influence, questioning "to what extent to metaphysical assumptions guide our research?" (p. 63). Instead, he favored Kuhn's (1962) third conceptualization focused on the shared beliefs within a field of research; or as Creswell and Plano Clark (2011) describe, paradigms as understood by the "scholarly community" (p. 46). No longer restricted by notions that particular paradigmatic assumptions govern an entire discipline, researchers can acknowledge that, practically, the social sciences has tended to concentrate on "beliefs and practices…in specific subfields" (Morgan, 2007, p. 53). Denscombe (2008) expanded the idea by advocating for "communities of practice" to determine the nature of paradigmatic assumptions. Generally credited to the work of Lave and Wegner (1991), communities of practice allow for a more malleable notion of paradigm to fit the various discipline arenas, orientations, inquiry traditions out of which research operates, which makes mixed methods research a much less paradigmatic challenge (Denscombe, 2008, p. 278).

The second perspective accepts philosophical foundations for paradigms, but rejects the fidelity and purity of a singular constructivist or post-positivist model. Maxwell and Loomis' (2003) work informs this perspective best, building of the research of McGawley (1982), Pitman and Maxwell (1990), and Sayer (1992). There is recognition of general "philosophical and methodological resonances" among various paradigmatic components, but that ultimately, they are just that: a bundle of components (Maxwell & Loomis, 2003, p. 250). While scholars can be in agreement about the holistic nature of a research paradigm, there can still be sharp disagreement about how the specific components are operationalized and the corresponding effects. The result, Teddlie and Tashakorri (2009) observe, is that mixing qualitative and quantitative does not have the dangerous philosophical consequences one might believe, because if "the two research paradigms are not 'pure' to begin with, researchers lose little when they creatively mix them" (p. 98).

The third perspective is perhaps the most radical, but also the simplest to describe. As Patton (2002) argues,

> *While these intellectual, philosophical, and theoretical traditions have greatly influenced the debate... [there is no need to] swear vows of allegiance to any single epistemological perspective. ...Indeed, [researchers need not] be concerned about theory. ...in real world practice, methods can be separated from the epistemology out of which they have emerged. (p. 136)*

This stance has been labeled "a-paradigmatic" given the refusal to recognize the integrity of any philosophically-infused paradigm. Others, such as Morgan (2007), could theoretically fit here, but in those cases, the author is often questioning the utility or role of the traditionally-understood paradigm, not making an unequivocal statement similar to Patton.

CONCLUSION: THE RESEARCH QUESTION DRIVES EVERYTHING – EVEN MIXED METHODS RESEARCH

The preceding discussion has established the paradigmatic roots at the heart of mixed methods research. At this point it is critical to clarify that while a research box top dictates one's actions and decisions, the *research question* determines which paradigm the study resides in. As Bryman (2006) states, "the research question...relegates epistemological and ontological debates to the sidelines...[and] clears the path for research that combines quantitative and qualitative research" (p. 118). Or as Tashakkori and Teddlie (1998) famously described in their first book, the research question holds a "dictatorship" over everything else (p. 20). This is not to say the research question is a trump card *over* paradigms; rather that it *informs* and *drives* the type of paradigm, and corresponding methodology and methods, required to sufficiently answer the question (using one of the four bridges described above). As a result, mixed-methods research should not be chosen independent of one's research purpose because of an assumption that the approach will yield "better" conclusions. Instead, it should be selected "when the quantitative or qualitative approach, each by itself, is inadequate to best understand a research problem" (Creswell, 2014, p. 20). When this is the case, it is critical to frame the research question appropriately so that it both addressed the research problem and directs the researcher to a mixed-methods design, and Onwuegbuzie and Leech (2006) and Plano-Clark and Badiee (2010) offer in-depth guides with recommendations, considerations, and techniques. What is missing in many research studies is an understanding of this interconnectedness throughout the entire project; or as Crotty (1998) argues, the link among epistemology (paradigm), theoretical perspective, methodology, and methods (pp. 2-3). After all, putting a puzzle together does not have to be overly strenuous, but it does require coherent planning, execution, and clear insight into the power and implications of the box top picture itself.

REFERENCES

Biesta, G. (2010). Pragmatism and the philosophical foundations of mixed methods research. In A. Tashakkori & C. Teddlie (Eds.), *SAGE Handbook of Mixed Methods in Social & Behavioral Research* (2nd ed.; pp. 95–118). Thousand Oaks, CA: Sage. doi:10.4135/9781506335193.n4

Bryman, A. (2006). Paradigm peace and the implications for quality. *International Journal of Social Research Methodology Theory and Practice*, *9*(2), 111–126. doi:10.1080/13645570600595280

Creswell, J. W. (2009). Editorial: Mapping the field of mixed methods research. *Journal of Mixed Methods Research*, *3*(2), 95–108. doi:10.1177/1558689808330883

Creswell, J. W. (2014). *Research design: Qualitative, quantitative, and mixed methods approaches* (4th ed.). Los Angeles, CA: Sage.

Creswell, J. W., & Plano Clark, V. L. (2011). *Designing and conducting mixed methods research* (2nd ed.). Thousand Oaks, CA: Sage.

Crotty, M. (1998). *The foundations of social research: Meaning and perspective in the research process*. London: Sage Publications.

Denscombe, M. (2008). Communities of practice: A research paradigm for the mixed methods approach. *Journal of Mixed Methods Research*, *2*(3), 270–283. doi:10.1177/1558689808316807

Denzin, N. K. (2010). Moments, mixed methods, and paradigm dialogs. *Qualitative Inquiry*, *16*(6), 419–427. doi:10.1177/1077800410364608

Dewey, J. (1966). *Logic: They theory of inquiry*. New York: Henry Holt.

Feilzer, M. Y. (2010). Doing mixed methods research pragmatically: Implications for the rediscovery of pragmatism as a research paradigm. *Journal of Mixed Methods Research*, *4*(1), 6–16. doi:10.1177/1558689809349691

Feuer, M. J., Towne, L., & Shavelson, R. J. (2002). Scientific culture and educational research. *Educational Researcher*, *31*(8), 4–14. doi:10.3102/0013189X031008004

Giddings, L. S. (2006). Mixed-methods research: Positivism dressed in drag? *Journal of Research in Nursing*, *11*(3), 195–203. doi:10.1177/1744987106064635

Greene, J. C. (2007). *Mixing methods in social inquiry*. San Francisco: Jossey-Bass.

Greene, J. C., & Caracelli, V. J. (1997). Defining and describing the paradigm issue in mixed- method evaluation. In J. C. Green & V. J. Caracelli (Eds.), *Advances in Mixed-Method Evaluation: The Challenges and Benefits of Integrating Diverse Paradigms* (pp. 5–18). San Francisco: Jossey-Bass. doi:10.1002/ev.1068

Guba, E. C. (1990). The alternative paradigm dialogue. In E. C. Guba (Ed.), *The Paradigm Dialogue* (pp. 17–27). Newbury Park, CA: Sage.

Hammersley, M. (2002). *Educational research, policymaking and practice*. London: Sage.

Howe, K. R. (1988). Against the quantitative-qualitative incompatibility thesis or dogmas die hard. *Educational Researcher*, *17*(8), 10-16.

Johnson, B., & Gray, R. (2010). A history of philosophical and theoretical issues for mixed methods research. In A. Tashakkori & C. Teddlie (Eds.), *SAGE Handbook of Mixed Methods in Social & Behavioral Research* (2nd ed.; pp. 69–94). Thousand Oaks, CA: Sage. doi:10.4135/9781506335193.n3

Johnson, R. B., & Onwuegbuzie, A. J. (2004). Mixed methods research: A research paradigm whose time has come. *Educational Researcher, 33*(7), 14–26. doi:10.3102/0013189X033007014

Johnson, R. B., Onwuegbuzie, A. J., & Turner, L. A. (2007). Toward a definition of mixed methods research. *Journal of Mixed Methods Research, 1*(2), 112–133. doi:10.1177/1558689806298224

Krauss, S. E. (2005). Research paradigms and meaning making: A primer. *Qualitative Report, 10*(4), 758–770.

Kuhn, T. S. (1962). *The structure of scientific revolutions.* Chicago: The University of Chicago Press.

Kuhn, T. S. (2012). *The structure of scientific revolutions* (4th ed.). Chicago: The University of Chicago Press. doi:10.7208/chicago/9780226458144.001.0001

Lather, P. (2004). This *IS* your father's paradigm: Government intrusion and the case of qualitative research in education. *Qualitative Inquiry, 10*(1), 15–34. doi:10.1177/1077800403256154

Lave, J., & Wenger, E. (1991). *Situated learning: Legitimate peripheral participation.* Cambridge, UK: Cambridge University Press. doi:10.1017/CBO9780511815355

Leedy, P. D., & Ormrod, J. E. (2013). *Practical research: Planning and design* (10th ed.). Boston: Pearson.

Maxcy, S. J. (2003). Pragmatic threads in mixed methods research in the social sciences: The search for multiple modes of inquiry and the end of the philosophy of formalism. In A. Tashakkori & C. Teddlie (Eds.), *Handbook of Mixed Methods in Social & Behavioral Research* (pp. 51–90). Thousand Oaks, CA: Sage.

Maxwell, J. A., & Loomis, D. M. (2003). Mixed methods design: An alternative approach. In A. Tashakkori & C. Teddlie (Eds.), *Handbook of Mixed Methods in Social & Behavioral Research* (pp. 241–271). Thousand Oaks, CA: Sage.

Maxwell, J. A., & Mittapalli, K. (2010). Realism as a stance for mixed methods research. In A. Tashakkori & C. Teddlie (Eds.), *SAGE Handbook of Mixed Methods in Social & Behavioral Research* (2nd ed., pp. 145–162). Thousand Oaks, CA: Sage. doi:10.4135/9781506335193.n6

McGawley, J. (1982). *Thirty million theories of grammar.* Chicago: University of Chicago Press.

Mertens, D. M. (2010). *Research and evaluation in education and psychology: Integrating diversity with quantitative, qualitative, and mixed methods* (3rd ed.). Thousand Oaks, CA: Sage.

Morgan, D. L. (2007). Paradigms lost and pragmatism regained: Methodological implications of combining qualitative and quantitative methods. *Journal of Mixed Methods Research, 1*(1), 48–76. doi:10.1177/2345678906292462

Mosteller, F., & Boruch, R. F. (2002). *Evidence matters: Randomized trials in education research.* Washington, DC: Brookings Institution Press.

Onwuegbuzie, A. J., & Leech, N. L. (2006). Linking research questions to mixed methods data analysis procedures. *Qualitative Report, 11*(3), 474–498.

Patton, M. Q. (2002). *Qualitative research & evaluation methods* (3rd ed.). Thousand Oaks, CA: Sage.

Phillips, D. C., & Burbules, N. C. (2000). *Postpositivism and educational research.* Lanham, MA: Rowman & Littlefield Publishers, Inc.

Pitman, M. A., & Maxwell, J. A. (1990). Qualitative approaches to evaluation. In M. D. LeCompte, W. L. Milroy, & J. Preissle (Eds.), *The Handbook of Qualitative Research in Education* (pp. 729–770). San Diego, CA: Academic Press.

Plano-Clark, V. I., & Badiee, M. (2010). Research questions in mixed methods research. In A. Tashakkori & C. Teddlie (Eds.), *SAGE Handbook of Mixed Methods in Social & Behavioral Research* (2nd ed., pp. 275–300). Thousand Oaks, CA: Sage. doi:10.4135/9781506335193.n12

Ponterotto, J. G. (2005). Qualitative research in counseling psychology: A primer on research paradigms and philosophy of science. *Journal of Counseling Psychology, 52*(2), 126–136. doi:10.1037/0022-0167.52.2.126

Sayer, A. (1992). *Method in social science: A realist approach.* London: Routledge.

Shadish, W. R. (1995). The quantitative-qualitative debates: "Dekuhnifying" the conceptual context. *Evaluation and Program Planning, 18*(1), 47–49. doi:10.1016/0149-7189(94)00048-3

Sweetman, D., Badiee, M., & Creswell, J. W. (2010). Use of the transformative framework in mixed methods studies. *Qualitative Inquiry, 16*(6), 441–454. doi:10.1177/1077800410364610

Tashakorri, A., & Teddlie, C. (1998). *Mixed methods methodology: Combining the qualitative and quantitative approaches.* Thousand Oaks, CA: Sage.

Tashakorri, A., & Teddlie, C. (2010). Epilogue: Current developments and emerging trends in integrated research methodology. In A. Tashakkori & C. Teddlie (Eds.), *SAGE Handbook of Mixed Methods in Social & Behavioral Research* (2nd ed., pp. 803–825). Thousand Oaks, CA: Sage. doi:10.4135/9781506335193.n31

Teddlie, C., & Johnson, R. B. (2009a). Methodological thought before the 20th century. In C. Teddlie & A. Tashakkori (Eds.), *Foundations of mixed methods research: Integrating quantitative and quaitative approaches in the social and behavioral sciences* (pp. 40–61). Thousand Oaks, CA: Sage.

Teddlie, C., & Johnson, R. B. (2009b). Methodological thought since the 20th century. In C. Teddlie & A. Tashakkori (Eds.), *Foundations of mixed methods research: Integrating quantitative and qualitative approaches in the social and behavioral sciences* (pp. 62–82). Thousand Oaks, CA: Sage.

Teddlie, C., & Tashakkori, A. (2009). *Foundations of mixed methods research: Integrating quantitative and qualitative approaches in the social and behavioral sciences.* Thousand Oaks, CA: Sage.

Teddlie, C., & Tashakkori, A. (2010). Overview of contemporary issues in mixed methods research. In A. Tashakkori & C. Teddlie (Eds.), *SAGE Handbook of Mixed Methods in Social & Behavioral Research* (2nd ed.; pp. 1–41). Thousand Oaks, CA: Sage. doi:10.4135/9781506335193.n1

Chapter 3
Mixed Method Research: A Concept

Aroop Mukherjee
Universiti Putra Malaysia, Malaysia

Nitty Hirawaty Kamarulzaman
Universiti Putra Malaysia, Malaysia

ABSTRACT

Mixed methods have emerged as the third research community in the social and behavioural sciences during the past decades, joining quantitative and qualitative methods of scholarly inquiry. Mixed methods research, research paradigm, methodology, and action research have encouraged the combined use of quantitative and qualitative research to answer complex questions in recent years. Mixed methods research integrates both methods, the quantitative and the qualitative, to present research findings within a single system process. The chapter aims to provide an insight between mixed method research and action research, which includes the basic foundation of mixed method research and research paradigm. The chapter will discuss the concept of action research and how mixed method is applied to action research and its processes. A brief idea about the future plan of action required for mixed methods research to attain better research designs and processes is also discussed in the chapter.

INTRODUCTION

Mixed methods research is being articulated as the research practice and recognised as the third major research approach along with quantitative and qualitative approaches for research. The importance of mixed methods is reflected when there is no perfect or an essential workable definition to achieve a goal in the field and based on the understanding of the definition changes over time as an approach and continuously grows (Johnson, Onwuegbuzie & Turner, 2007). Mixed methods research is also known as mixed research, where researchers combine the quantitative and qualitative research approaches. Mixed methods research helps to build on the strength and tries to reduce the weakness by both the approaches such as quantitative and qualitative to draw inferences that can lead to increase that can lead to increased understanding on the researched topic (Hayes, Bonner & Douglas, 2013) Thus, mixed methods research is defined as:

DOI: 10.4018/978-1-5225-0007-0.ch003

Mixed methods research is the type of research in which a researcher or team of researchers combines elements of qualitative and quantitative research approaches for the broad purposes of breadth and depth of understanding and collaboration.

- Johnson et al., 2007:123

A mixed method researcher focuses on the combination of numeric and narrative data and analysis, whereas the quantitative researcher typically focuses on numerical data and analysis and the qualitative researcher focuses on narrative data. In other words, the mixed methods researcher explores and examines the problems and issues, which are based on the researcher's belief and are important to the scholarly community (Teddlie & Tashakkori, 2009). According to Howe (1988), mixed methods researchers accept ideas, which are compatible with both the quantitative and qualitative methods. Thus, mixed methods researcher works in engaging objective or subjective point of view with respect to the study.

FOUNDATION OF MIXED METHODS RESEARCH

Teddlie and Tashakkori (2003) revealed that the importance of mixed methods research has started from the traditional positivistic period 1900 to 1950. Many researchers have explored that a mixed method was set out in the early to mid-20th century but did not distinguish their work as it belongs in the area of mixed method research and nor they have critically reviewed that how the mixed method has been applied and under what circumstance that it is dissimilar from traditional research (Datta, 1994). However, actually the beginning of mixed methods was started in the late 1980s with several publications, which focused on describing and defining the importance of mixed methods. The majority of the publications from different countries were published on the same idea at the same time. Thus, researchers from management in the United Kingdom (Bryman, 2003), from evaluation in the United States (Greene, Caracelli & Grahman, 1989), from sociology in the United States (Brewer & Hunter, 1989), from nursing in Canada (Morse, 1991) and from education in the United States (Creswell, 1994) were developing the concepts in the late 1980s to early 1990.

Many of these researchers were giving thought of linking the methods and were discussing how it should mix the method and data and to provide a proper reason for this and give a name to this research. There have been discussions to provide the proper name for this form of inquiry. It was during the last 50 years the researchers have applied different names and, therefore, making it difficult to distinguish the specific character of research subjects that would ring as "mixed methods" research. It used to be called as an "integrated" or "combined" research by advancing the notion that the data are blended (Steckler, McLeroy, Goodman, Bird & McCornick, 1992) and by Fielding and Fielding (1986) as "quantitative" and "qualitative" methods. Ragin, Nagel and White (2004) have acknowledged mixed methods as "hybrid research" or "methodological triangulation" (Morse, 1991) which later Creswell (1994) identified as the conjunction of quantitative and qualitative data as "combined research". Accordingly, Tashakkori and Teddlie (1998) acknowledged that it is a combination of both methods and philosophical worldview and is called as "mixed methodology." Therefore, Bryman (2006) integrated the approaches together after several years and has discussed the possible research designs and the problems associated with mixing the methods. In 2009, Leech and Onwuegbuzie (2009) called "mixed research" as it reinforces them to be an idea that it is an approach rather than the methods and tries to face them as research as philosophical assumptions.

In today's scenario, it is more likely to be used as mixed methods research, which has been discussed in the *Handbook of Mixed Methods in Social & Behavioural Research* by Tashakkori and Teddlie (2003). The use of mixed methods research and designing the studies with a historical view is not only sufficient, but it also helps the researcher to know and understand mixed methods research and defend the use of this approach in the research while discussing the methods. Hence, Creswell and Plano Clark (2010) reviewed different stages of mixed methods, and have organized it into five, often overlapping, and periods in Table 1.

Table 1. Taxonomy of mixed methods and their contribution

Development Stage	Author (Year)	Contribution to Mixed Methods Research
Constructive period	Campbell and Fiske (1959)	Multiple quantitative methods use introduced
	Sieber (1973)	Interviews and surveys conjoined
	Jick (1979)	Triangulating qualitative and quantitative data methods discussed
	Cook and Reichardt (1979)	Collectively quantitative and qualitative data methods presented in ten (10) ways
Paradigm period	Rossman and Wilson (1985)	Methods combining – purists, situationists, and pragmatists discussed
	Bryman (1988)	Provided review of the debate establishing connections within the two traditions
	Reichardt and Rallis (1994)	Reconciled two traditions and discussed paradigm debate
	Greene and Caracelli (1997)	Recommended to proceed with the paradigm debate
Procedural development period	Greene, Caracelli, and Graham (1989)	Types of mixed methods design classification system identified
	Brewer and Hunter (1989)	The multi-method approach used in the process of research concentrated
	Morse (1991)	Notation system developed
	Creswell (1994)	Three types of mixed method designs identified
	Morgan (1998)	Typology for determining design to use developed
	Newman and Benz (1998)	Overview or procedures gave
	Tashakkori and Teddlie (1998)	Topical overview of mixed methods research introduced
	Bamberger (2000)	International policy for mixed methods research given
Introspective period	Tashakkori and Teddlie (2003a)	Comprehensive treatment of aspects of mixed methods research given
	Creswell (2003)	The quantitative, qualitative and mixed methods approach compared in the process of research
	Johnson and Onwuegbuzie (2004)	Mixed methods research a natural complement to traditional qualitative and quantitative research determined

Source: Adapted from Creswell and Plano Clark (2010)

The constructive period for the mixed methods was from 1950 until 1980. During this period, it was seen that there was an interest in using more than one method in the study. It can be termed as antecedents of mixed methods in today's scenario for mixed methods (Creswell & Plano Clark, 2010). Denzin (1978) used multiple methods, in general, and further Jick (1979) has been triangulating both quantitative and qualitative approaches.

The concept development was in the period from the 1970s and 1980s, when more of the qualitative researchers were adamant that different assumptions provided the foundations for quantitative and qualitative researchers (Creswell & Plano Clark, 2010). The concept development period is sometimes called as paradigm period. It generally involves different scholars who are in the argument that whether quantitative and qualitative data can be combined, as the qualitative data is linked with different philosophical assumptions. According to Smith (1983), the philosophical assumptions that are being commented by others and mixed methods research was incommensurable as mixed methods research asked for paradigms. Thus, currently the link between the methods of data collection and the large philosophical assumptions are not as tightly drawn as envisioned in 1990s (Creswell & Plano Clark, 2010). According to Tashakkori and Teddlie (2003a), mixed methods research embraces pragmatism as the best philosophical foundation for mixed methods research. Greene and Caracelli (1997) said to use different paradigms in mixed methods research, but to honour them explicit when they are used (Creswell & Plano Clark, 2010).

During the procedural development period, the researchers were focused on methods of data collection, data analysis, research designs, and the purpose of conduction the mixed methods research. Greene, Caracelli and Graham (1989) have studied around 57 evaluations mixed methods that have been classified into five types and talks about the design decisions. The work in the field of evaluation that have laid down the groundwork for mixed methods research design. The linking of mixed methods research, discussion has laid down the steps in the process of research. In 1988, the reasons for combining the quantitative and qualitative research were discussed by Bryman (1992). Tashakkori and Teddlie (1998) have mapped the contours of mixed methods procedures in policy research and in attending the issues as validity and inferences.

Many researchers have started to endorse mixed methods as a separate methodology, method or as a different approach to the study with respect to different disciplines in different parts of the country. The publication of the book, "Handbook of Mixed Methods in Social and Behavioural Research" by Tashakkori and Teddlie (2003a) has created awareness and use of mixed methods as the third method to conduct the study, which has combined the quantitative and qualitative methods of research.

In this introspective period, it is time to identify, interrogate, and critically view the underpinning theory of mixed method. Many researchers have identified the four methodological domains and have discussed the importance of considering the mixed method distinctive methodology. During this period, majority of the researchers have mapped with issues and priorities in the mixed method. Now, mixed methods research is being published widely in all areas. Several journals have started publishing the empirical as conceptual and also review papers on mixed methods.

As all researchers have a philosophical foundation, similarly mixed methods do have philosophy base that provides the foundation for conducting the research during the study. This helps to conduct the research and that too of inquiry. Therefore, to conduct mixed methods research it is better to articulate the philosophical assumption in the project rather just having knowledge or awareness of the philosophical. This will acknowledge the worldwide of the study that can relate to the elements of the mixed methods project.

RESEARCH PARADIGM

A different decision maker has conducted the research prototype, which is being popularised by Kuhn (1962) which have included in one's opinion as per the "nature of reality and knowledge" of its roots and institutions. Researchers embark on its task by making certain assumptions about how they memorize and what they should do while making the inquiry. Thus, these can be called as paradigms (Creswell, 2003). The paradigms are positioning them on the scale are positivism, post-positivism, constructivism/ interpertivist, emancipator/transformative, and pragmatism are presented in Table2 (Creswell, 2003).

According to Teddlie and Yu (2006), the French philosopher August has said that positivism holds truth and is represented by measurable, naturally occurring phenomenon. The phenomenon, which is not measured, does not exist, and the measurement is the proof of the existence (Crotty, 1998). A naturally occurring phenomenon broken down into measurable moments together reproduces truth. The use of deductive reasoning by positivism researchers to generate theory evolves specific hypothesis, which are tested. Therefore, present day researchers consider that positivism comprises many opposing misconceptions (Shadish, 1998) and associate positivism with quantitative research.

Post-positivism has been evolving since 1930s (Popper, 1959). Post-positivism asserts that truth is discovered and is well understood through objective, standardization, and deductive reasoning and has control within the research process (Teddlie & Yu, 2006). The post-positivism is being established by research design, statistical hypothesis and with possible explanations for the findings. The strength of post-positivism research is to be accurate, generalize, reliable, and replicable. According to Shadish, Cook, and Campbell (2002), the post-positivism research is focused on addressing the fundamental research and is considered well suited for confirmatory research. Positivism and post-positivism research is commonly considered well suited for explanatory research and is often associated with the quantitative approach where the specific question on "What" and "Where" of phenomenon is asked. This is also called as "scientific method" research. The post-positivism refers to the thinking process after the positivism, challenging the traditional notion of the absolute truth of knowledge (Creswell, 2003). Post-positivism reflects its probabilistic philosophy, which actually determines the effects or outcomes of the research. It is a process of reducing the ideas into small, discrete sets of ideas to test. The body of knowledge is created based on careful observation and measurement of the objective reality that exists in the world. Thus, in the post-positivism the scientific method accepts the approach that an individual begins with a theory, then collects the data that either supports the theory and make necessary changes to test the theory.

Table 2. Research paradigm

Post-Positivism Determination Reductionism Empirical observation and measurement Theory verification	**Constructivism/Interpretivist** Understanding Multiple participant meanings Social and historical construction Theory generation
Emancipator/Transformative Political Empowerment issue-oriented Collaborative Change-oriented	**Pragmatism** Consequences of actions Problem-centered Pluralistic Real-world practice-oriented

Source: Creswell (2003)

Constructivism is found to be anti-positivism. The terms of epistemology constructivism are linked to interpretivism. Constructivism looks for "culturally derived and historically situated interpretations of the social life-world" (Gray, 2009:21). Constructivism ascertains that there are multiple truths exists which is determined by individuals unique perspectives and experience. However, the truth is understood through critical subjective and inductive reasoning and research in usual situations where the researcher has accessed to the research participants (Smith & Heshusius, 1986). The emphasis is on the description and utilizing the researcher in garnering data and to do analysis. It prioritizes the inquiry by exploring the ontological, epistemological, and axiological and with specific methodologies that can be considered to have their own paradigms (Creswell, 1998). Constructivism research is well suited for exploratory research associated with the qualitative approach and is especially asked to uncover the question "How" and "Why". Constructivism does have five approaches to perform the research: symbolic interactionism, phenomenology, realism, hermeneutics, and naturalistic inquiry. Thus, constructivist researchers address the process of interaction among the mortals. Researchers have realized that their own background shapes their interpretation and set them in the situation that the researchers are acknowledging their interpretation flows from their own personal, ethnic and historical experience (Creswell, 2003).

Emancipators/Transformatives are influenced by political concerns and are often associated with the qualitative approach than quantitative approaches. During 1980s and 1990s, an individual has felted that post-positivism has imposed structural laws and theories that are not fit to marginalized individuals and groups to address the social issues. According to Neuman (2000 as cited in Welsh, 2007), the emancipators have drawn the full behaviour of Marx, Marcuse, Habermas, and Freire. In this type of inquiry, the inquiries can feel that constructivist stance is not far from an action on the agenda to help and marginalized peoples. Thus, the researchers need to receive an action agenda to reform that can alter the lifespan of the social club. During the research, specific issues need to be addressed such as social issues of the day, issues such as empowerment, inequality, oppression, and suppression (Creswell, 2003). Therefore, theoretical positions may be mixed with the philosophical assumptions that construct issues are being analyzed.

Pragmatism is a philosophical movement by those who claim that ideology or proposition is true, and the function is satisfactory. The ideology that creates true proposition will extend to establish the practical issues of accepting and while the unpractical ideas need to be eliminated. C.S. Peirce as a technique to help us find the solutions than a philosophy first coined the term "Pragmatism". William James is the most famous philosopher of Pragmatism who made it popular by his work entitled "Philosophical Conceptions and Practical Results" that was delivered at the University of California. James argued that the ideas and beliefs have value only when it works. According to James, the pragmatism is the techniques for clarifying concepts and hypotheses. The proposed techniques will have disagreements of the metaphysical approach that appear to be reusable and will be dissolved. The principle of pragmatism is not the first principle rather "the principle of peirces' was the first principle of pragmatism. It was later the pragmatism was further developed by two philosophers John Dewey and C. I. Lewis. Dewey has described the pragmatism as the systematic exploration of what is called as "the logic and ethics of scientific inquiry." Pragmatism is a method for attaining clarity of ideas within a normative conception of logic; within the norms for continuing, self-correcting inquiry directed towards truth. The possibility for scientific truth does not derive from indubitable foundations, but from the self-correcting process of interpretation (Halton, 2004).

Pragmatism is a practical approach in solving the problems and has a strong association with mixed methods research. Pragmatism is considered to be a bridge between methodology and paradigm, or it is a state between philosophy and methodology (Cameron, 2011). Methodological views the choice of methods that are driven by philosophical (ontology's and epistemological) bases. The major task that researchers need to embark on to put them in a position by itself paradigmatically. According to Neuman (2006), the paradigm is referred to as a general organizing framework for theory and research that includes basic assumption, key issues, models of quality research, and methods for seeking answers (Cameron, 2011). In a broad manner, the researchers contain ontology's, epistemologies, and the methodology that may be termed as a paradigm. It is important the paradigm is based upon a research proposal, and design which are fully realised and made explicit in the research itself (Maxwell, 2005:36; Neuman, 2006:81).

Essentially, all researches are interpretive in nature and are guided by researcher's beliefs and opinions on what researchers look and read. According to Creswell (1994:176), there are different school of thoughts in the paradigm debate and are addressed as "paradigm wars." The discussion between the "purists" and "situationalists" which assert paradigm as methods should be mixed, and other methods can be used for specific situations (Cameron, 2011). Hence, in opposition to the "purists" paradigm the pragmatism argued for using qualitative and quantitative research paradigms to advocate better and efficient use of both approaches (Cameron, 2011). According to Wicks and Freeman (1998), pragmatism is an alternative paradigm besides positivism and constructivism used in behavioural research.

As a pragmatist, the thinking has influenced the research in agribusiness in great extent, even though, the pragmatic thoughts and research have not fully acknowledged. Pragmatism is an adequate research paradigm for research design than positivism and constructivism (Lee & Nickerson, 2010). "Pragmatism is concerned with action and change and the interplay between knowledge and action" (Goldkuhl, 2012:136). Thus, growing interest and the importance to investigate pragmatism is one possible way of conducting action research and design research. The mixed methods research emphasizes on pragmatism, which opens the door to perform multiple methods with different perspectives and with different assumptions. In addition, data collection and analysis in the mixed method study results in different forms.

Its ontological, epistemological, axiological, rhetoric, and methodological (Creswell, 2003), are specifying the paradigm philosophical. These paradigms inform the researcher that it informs the origin of the research, its formation, its design and at the end the methodology. The philosophical researchers have claimed the five dimensions as: what is knowledge (ontology's), how we know about it (epistemology), what values go into it (axiology), how we write about it (rhetoric), and the process of studying it (methodology) (Lincoln & Guba (1985); Creswell, 2003).

The five dimensions represented by Lincoln and Guba (1985) are:

1. **Ontology:** The positivist researchers consider a single reality, whereas constructivists consider multiple constructed realities.
2. **Epistemology:** The positivist researchers considered the knower and known are independent, whereas the constructivist researchers considered the knower and known are combined.
3. **Axiology:** The positivist researchers consider that inquiry value-free, whereas the constructivist researchers consider the inquiry with value.
4. **Rhetoric:** The positivist researchers consider it is possible to distinguish the cause and effects, whereas the constructivist researchers consider it is impossible to distinguish the cause and effects.
5. **Methodology:** The positivist researchers consider that it is possible for time and context to be free and independent, whereas constructivist researchers consider the time and context together.

POSITIVISM, CONSTRUCTIVISM, EMANCIPTORY, AND PRAGMATISM (PCEP) RESEARCH APPROACH

Dewey (1938) has given an outline of paradigm of inquiry that can be underpinned the scientific approach. The four research paradigm approaches that can be associated with collecting and converting data for further analysis to draw conclusions are:

1. **Positivism/Quantitative Approach:** During this research approach, the researchers would have one or more hypotheses. These are the questions or assumptions that the researchers would like to address or predict about the possible relationships between the things that they would like to investigate. The main emphasis on the quantitative approach is in deductive reasoning, which tends to move from the general to the specific. Deductive approach move towards hypothesis testing to verify the principal to be confirmed and or modified. These hypotheses are tended to assert two or more concepts that attempt to explain the relationship between them. The concepts are abstract ideas that can build the building blocks of the research hypotheses and theories. This is sometimes referred to as a top-down approach.
2. **Constructivism/Qualitative Approach:** It is associated with a social paradigm, which emphasis on constructing the nature of reality. It is about recording, analysis, and attempt to uncover the deeper meaning and significance of human behaviour and experience. In this, researchers are interested in gaining and understanding the complex scenario of people's experience to generalize the larger groups. The emphasis is on the qualitative approach and tends to be inductive, for which the theory can be developmental or try to look for the pattern for the basic data collected. This involves the movement from specific to general and is sometimes called as bottom to up approach. Through the inductive approach, the researchers move towards discovering a binding principle that takes care to have conclusions based on the data. The inductive approach does not set out to corroborate or falsify a theory, whereas with the gathering data it attempts to establish patterns consistencies and meanings.
3. **Emancipator/Transformative Approach:** The researchers who adopt an advocacy/participatory approach feel that the approaches do not tackle and respond to the needs or situation of people from marginalized groups. The main aim of this category is to bring positive change in the research, and their approach is sometimes is described as emancipator. The researchers will have a political agenda and try to give the group that they are studying a voice so that the research directly or indirectly result in some kind of reform and is being studied in different stages so that to avoid become insignificant. During this approach, the researchers would adopt a less neutral position that is usually required in scientific research. As this type of research is not being used more for not having proper objectives and is not fully understood.
4. **Pragmatic/Mixed Methods:** The pragmatic approach uses the method, which suits the research problem and is not caught in the philosophical debates of the best approach. The researchers who use pragmatic are free to use any of the methods, techniques, and procedures associated with quantitative and qualitative researchers. However, every method has limitations, which can be addressed with the use of mixed methods. In this approach, the researchers use different techniques at the same time or one after the other. Thus, it is possible that to transform from qualitative to the quantitative approach or vice versa, and even though, the transformative from quantitative to the qualitative approach is not common. Hence, the mix methods approach has advantages of enabling

the triangulation, which is the common feature of mixed methods research. The mix methods approach uses data triangulation (different variety of data sources), investigator triangulation (different researchers), theory triangulation (multiple perspectives to interpret the results), and methodological triangulation (multiple methods to study a research problem).

RELATIONSHIP WITH PARADIGMS, RESEARCH APPROACH, METHODOLOGY, AND METHODS

Mackenzie and Knipe (2006) were able to review the methodology and method with the research paradigms, which is commonly used in Table 3. Somekh and Lewin (2005) defined the term methodology as "the collections of methods or rules by which a particular piece of research is undertaken" and "principles, theories and values that underpin a particular approach to research" (pp. 346). With this, it will be suggested that which research paradigm and research question would be used for which type of research data collection and analysis methods will be used most appropriately for the study (Mackenzie & Knipe, 2006). In this way, researchers can apply the data collection and analysis methods, the most appropriate for a particular research study, which is possible to apply any research paradigms to employ mixed methods rather than being restricted to any one method, which can increase the richness of the research project.

MIXED METHOD DESIGN AND PROCEDURE

There are numerous classifications done for mixed methods research from the past literature, and to classify into different designs and use different terminologies. According to Tashakkori and Teddlie (2003) in the book "*Handbook of Mixed Method in Social and Behavioural Research*" mixed method design has been classified and nearly 40 different types of classification according to the literature are found. Creswell and Plano Clark (2007) have summarized these 40 such different types of classification into 12 types in Table 4. Various types of classification will evolve the mixed methods research design and its use.

The majority of the researchers have founded using these four major types of mixed method designs such as Exploratory, Explanatory, Embedded, and Triangulation.

Mixed Methods Design Prototypes

Mixed methods design falls under two categories: typology-based and dynamic (Creswell, Klassen, Plano Clark & Smith, 2011). The design of mixed methods designs and the selection of the particular designs for the study purpose and questions is classified with typology-based approach, whereas the dynamic approach is focused on the design process that is considered and interrelated with multiple components of research design from the existing typology. This will help to make the key decision on choosing mixed methods design by key principles. Creswell and Plano Clark (2011) have listed the four key principles while designing the study in mixed methods. The key principles are 1) the level of interaction between the strands, 2) the relative priority of the strands, 3) the timing of the strands, and 4) the procedure for mixing the strands (Creswell, *et.al.* 2011: 64). A strand is a combination of component of

Table 3. Relationship between Paradigms, Methodology, and Methods

Paradigms	Research Approach	Methodology/ Strategies of Inquiry	Methods	Data Collection Tools
Positivist/Postpositivist	Quantitative	Experimental Quasi-experimental Correlational Reductionism Theory Verification Causal Comparative Determination Normative	Quantitative. "Although qualitative methods can be employed within this paradigm, quantitative methods tend to be predominant…"	Experimentations Quasi-experiments Tests Scales
Constructtivist/ Interpretivist	Qualitative	Naturalistic Phenomenological Hermeneutics Interpretivist Ethnographic Multiple participant meanings Social and Historical Construction Theory generation Symbolic interaction theory	Qualitative methods predominate, although quantitative method may also be utilized.	Interviews Observations Document reviews Visual data analysis
Emancipator/ Transformative	Qualitative	Critical theory Neo-Marxist Feminist Critical Rae theory Freirean Participatory Emancipatory Advocacy Grand Narrative Empowerment issue Oriented Change-oriented Interventionist Queer Theory Race Specific Political	Qualitative methods with quantitative and mixed methods	Diverse range of tools – particular need to avoid discriminations
Pragmatic	Mixed Methods (Quantitative and Qualitative)	Consequences of actions Problem-centred Pluralistic Real-world practice oriented Mixed models	Qualitative and / or Quantitative methods may be employed. Methods are matched to the specific questions and purpose of the research	May include tools from both positivist and constructivist paradigms.

Source: Adapted from Mackenzie and Knipe (2006); Creswell (2003)

a study that encompasses the basic process of conducting quantitative and qualitative research (Teddlie & Tashakkori, 2009).

The researcher starts thinking about the decision point and selects the design that reflects the interaction, priority, timing, and mixing. Thus, the researchers make the decision about using quantitative, qualitative, or mixed methods before the research is started or even can think of adding the second methods after the study has begun. Creswell and Plano Clark (2011) has listed out six major mixed methods designs, which are useful for researchers to design the framework carefully that suit to be studied and best match with the research problem. With the selection of typology-based design, the researchers are

Table 4. Taxonomy of Mixed Method Design

Author	Mixed Methods Design	Discipline
Greene, Caracelli, and Graham (1989)	Initiation Expansion Development Complementary Triangulation	Evaluation
Patton (1990)	Experiment design, qualitative data, and content analysis Experiment design, qualitative data, and statistical analysis Naturalistic inquiry, qualitative data, and statistical analysis Naturalistic inquiry, quantitative data, and statistical analysis	Evaluation
Morse (1991)	Simultaneous triangulation QUAL + qual QUAN + quan Sequential triangulation [INSERT FIGURE 001]QUAL quan [INSERT FIGURE 002]QUAN qual	Nursing
Steckler, Mcleroy, Goodman, Bird, and McCormick (1992)	Model 1: Qualitative methods to produce quantitative measures Model 2: Qualitative methods to explain quantitative findings Model 3: Quantitative methods embellish qualitative findings Model 4: Qualitative and quantitative methods used equally and parallel	Public health education
Greene and Caracelli (1997)	Component designs Triangulation Complementary Expansion Integrated designs Iterative Embedded or nested Holistic Transformative	Evaluation
Morgan (1998)	Complementary Designs Qualitative preliminary Quantitative preliminary Qualitative follow-up Qualitative follow-up	Health Research
Tashakkori and Teddlie (1998)	Mixed methods designs Equivalent status (sequential or parallel) Dominant-less dominant (sequential or parallel) Multilevel use Mixed method designs: 9. Confirmatory, qualitative data, statistical analysis, and inference 10. Confirmatory, qualitative data, qualitative analysis, and inference 11. Exploratory, quantitative data, statistical analysis, and inference 12. Exploratory, qualitative data, statistical analysis, and inference 13. Confirmatory, quantitative data, qualitative analysis, and inference 14. Exploratory, quantitative data, qualitative analysis, and inference 15. Parallel mixed model 16. Sequential mixed model	Educational research
Creswell (1999)	Convergence Model Sequential Model Instrument-building model	Educational policy
Sandelowski (2000)	Sequential Concurrent Iterative Sandwich	Nursing

Author	Mixed Methods Design	Discipline
Creswell, Plano Clark, Gutmann, and Hanson (2003)	Sequential explanatory Sequential exploratory Sequential transformative Concurrent triangulation Concurrent nested Concurrent transformative	Educational research
Tashakkori and Teddlie (2003b)	Multistrand designs Concurrent mixed designs Concurrent mixed methods design Concurrent mixed model design Sequential mixed designs Sequential mixed methods design Sequential mixed model design Multistrand conversion mixed designs Multistrand conversion mixed methods design Multistrand conversion mixed model design Fully integrated mixed model design	Social and behavioural research
Creswell, Fetters, and Ivankova (2004)	Instrument design model Triangulation design model Data transformation design model	Primary medical care

Source: Adopted from Creswell and Plano Clark (2007)

provided with the framework and guide to implement the research methods to ensure to have a high quality of research. The basic four types of mixed methods designs are convergent parallel design, the explanatory sequential design, the exploratory design, and the embedded design, which bring multiple design elements together; the transformative design and the multiphase design (Creswell, *et. al.*, 2011). The prototype of the six major mixed methods is represented in Figure 1 (a-f).

Procedure for Selecting Mixed Methods Design

The conception of appropriate mixed methods projects the research can utilize these seven steps as a guidebook to identify the best research design for the survey. These seven steps (Table 5) are built on the work of Creswell (2003), Morgan (1998), Teddlie, and Tasakkori (2009).

ACTION RESEARCH

Lewin first coined the term "action research" in 1947. Where Lewin means "a process through which theory building and research on practical problems should be combined" (Gray, 2009:313); and in the *Handbook of Action Research: Participative Inquiry and Practice* by Reason and Bradbury (2001) "action research" has been defined as "a participatory process concerned with developing practical knowing in the pursuit of worthwhile human purposes. "It seeks to bring together action and reflection, theory and practice, in participation with others, in the pursuit of practical solutions to issues of pressing concern to people" (Reason & Bradbury, 2001:4). Lewin (1947) has viewed the action research as a way of improving social behaviour and encouraging social change; but more or less the approach was similar

Figure 1. (a) The convergent parallel design (b) The explanatory sequential design (c) The exploratory sequential design (d) The embedded design (e) The transformative (f) The multiphase design
Source: Creswell and Plano Clark (2011)

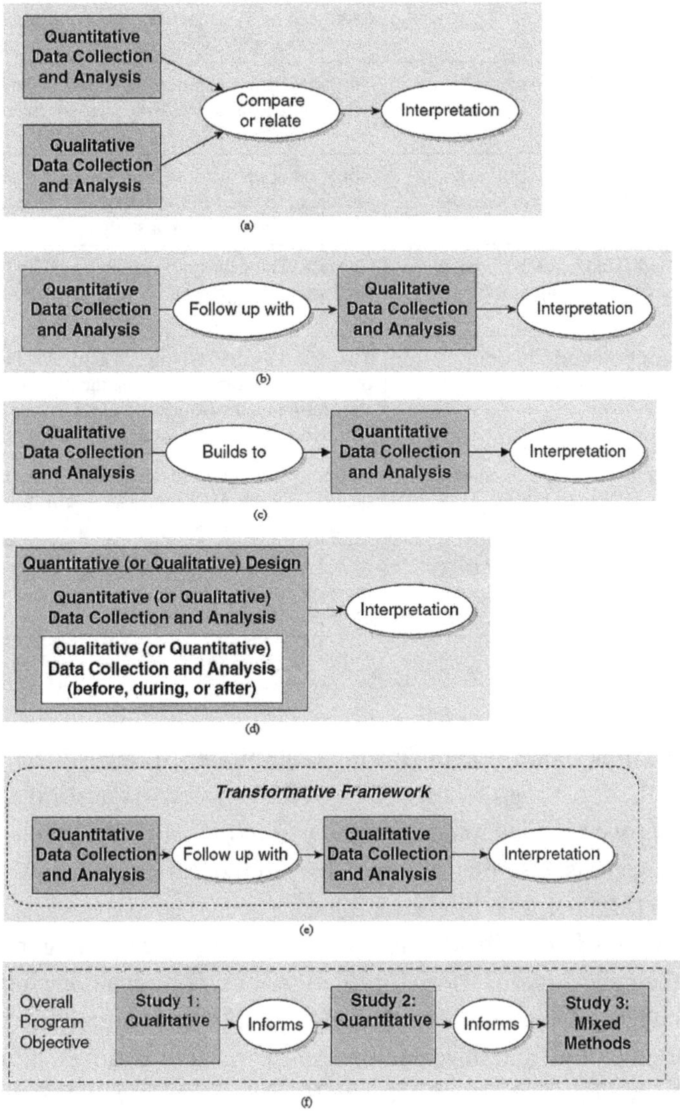

to the contemporary, traditional, scientific paradigm and is recognized as the value of experimentation and importance of creating knowledge. It was first formalized the action research from Stephen Corey in his book *Action Research to Improve School Practices* in 1953 (McNiff, 1988). Action research is seen the issues as only being understood and not through the study of a single variable, but as a holistic framework for the complex social system. According to Bowling (1997, cited in Gray, 2009) Lewin's concept of action research is a means of social engineering and how it will replace with the emphasis on raising awareness, empowerment and collaboration. Thus, action research is an approach that focuses on action and research simultaneously and in a particular manner (Coghlan & Brannick, 2014).

Table 5. Procedure for Selecting Mixed Method Designs

Step	Explanation
8. Do the research questions require one method or a mixed method design?	Research questions answered by either QUAL or QUAN data are addressed by a one method design. Research questions are required for both QUAL and QUAN to answer the require a mixed method design.
9. To note that a number of typologies of mixed methods research designs exist and they have knowledge to access information.	The original presentations of mixed methods designs can source the detailed data about the design and its characteristics upon accessing.
10. To choose the best available mixed methods research design apprehending that a figure may finally require to be generated for the study.	The most appropriate or best available research design to be selected instead of the best fit for the subject and study. Designs may be needed to combine or create for a study.
11. Responsive to the criteria highlighted by each of the mixed methods design typologies and its implications on the work.	The criteria for the typology proposed by Creswell (2003) are implemented with priority, stage of integration, and theoretical perspectives.
12. General criteria to be listed out before selecting the specific measurement that are important for the meditation.	The mixed method typologies general standards include a numeral of methods, number of stages, the implementation process, stage of integrating methods, priority of QUAL or QUAN, function of the research field, and theoretical perspectives.
13. To use the selected criteria for potential designs so as to select the best research design for the subject area.	Defining the research design that is almost coherent with the desired qualities on the chosen, measures will result in the best design for the study.
14. If there is one best design for a given field, a new mixed method, the design may need to be developed at the inception or during the development of the study.	Mixed methods studies may vary as the research progresses and yields more phases than were originally planned or includes phases that change in importance.

Source: Adapted from Graff (2013)

Action research is a form of inquiry that helps the practitioners to investigate and evaluate the work. The practitioners can think of, getting the answer from the respective question as "What should I do to improve?", "How do I improve it?", and many more questions (McNiff & Whitehead, 2006). Some of the social science research questions with that of action research questions to show the difference between them are listed in Table 6 (McNiff & Whitehead, 2006).

The action research is classified as insider action research in which managers are involved in the action research project in their own organization (Coghlan, 2011). The advantage of adopting insider action research is that the managers have a complete knowledge about the organisation that is being studied along with the organisation culture and have personal networks, but at times, it becomes difficult for an insider to cross-departmental hierarchical or network boundaries (Gray, 2009). The other approach can

Table 6. Social Science Research Question vs. Action Research Question

Social Science Questions	Action Research Question
How practitioner's motivation and employee retention are related?	What should be the quality of employees experience in school, so that they decide to stay long?
Does worker productivity be influenced by management style?	What should I do to improve my management style to promote productivity?
Does a different seating arrangement will increases audience participation?	What should I do to encourage greater audience participation through different seating arrangements?

Source: Adapted from McNiff and Whitehead (2006)

be termed as external action research, in which the researcher may be independent of the professional context and work within it and alongside professional practitioners to achieve change. Action research is a process of collaboration for bringing the change. Gummesson (2000) has classified the action research into societal action research and management action research. The societal action research is concerned with macro-social, political, and economic issues whereas the management action research focuses on a company as a business. Thus, action research will help to contribute the knowledge and understand the theoretical perspectives for conducting the research. An additional aspect of action research, the management action research is to interact and have collaboration between the researcher and the organization. Hence, it can be seen that the action research has a wide range of approaches to do research both in terms of relationships between researcher and participants and is the focus of the research itself.

Lewin (1947) presented action research as a cyclic process of four stages of reflecting, planning, acting, and observing which includes action research methodological steps as: identifying a problem, reconnaissance of fact-finding, planning action, acting, evaluating action, and monitoring action. The past literature on action research has identified the outcomes that are associated to conduct action research the few outcomes are:

A. Improve organizational effectiveness
B. Improve decision-making skills
C. Increase in understanding about dynamism
D. Create confidence and ability to sustain change
E. Need to increase awareness and link between the goals and practices
F. Help in understanding the theory and demystifies the research
G. Identify and verify the methods work
H. Make accountable for decision-making and help in creating awareness
I. Ready to accept challenges and willingness to accept the research findings and act on it
J. Strengths the connection between pure and applied research
K. Encourage critically and responsive stakeholders in research

Indeed with such type of comprehensive approach, collecting and analysing of quantitative and qualitative data that certifies the practical evaluation of the vast collection of cognitive, social, and interpersonal factors that can bear on the individuals or groups for their thoughts, behaviours and opinion. According to Stringer (2007), the aim of action research is not to present the full answer to the problem, but to create the link between the truth and realities held by the legitimate groups or an individual. Different people because of their previous experience and culture will do the interpretation of the answer in different ways. Therefore, the main aim of action research is to bring people together with different views and perceptions into one platform so that the group can formulate a joint solution.

PURPOSE OF USING ACTION RESEARCH

The purpose of action research is to create a special kind of knowledge (Mcniff & Whitehead, 2006). It has always been understood that taking action will improve the personal and social situations. It is being seen as a potential for promoting more productive and for sustainable organization development. Generally, researchers explain the conceptual analyzes and explanations of action research and its possible uses. The action research becomes more real when ideas are linked with action. The ideas need to be communicated as in the form of theories by explaining what they are doing, why they are doing it and what they hope to achieve (Mcniff & Whitehead, 2006). Thus, the purpose of action research is to generate living theories about how learning has improved practice and is informing new practices. Thus, the main objectives are to transform the practices into theories.

MIXED ACTION RESEARCH (MAR)

When action research is studied using mixed methods research, then it is called as mixed action research (MAR). The mixed method in the action research will provide an accurate methodological foundation that leads to solve the issues in more systematic approach for the assessment of the problem, the development, evaluation and monitoring of the natural process in the process of getting meaningful knowledge and change. With the systematic approach, the collection of data, analyzing and integrating the quantitative and qualitative methods would generate more consistent and have credible conclusions about the study (Ivankova, 2013). Mixed methods action research always starts with the diagnosis phase and is an iterative approach that implies to the study, but the researchers do not plan in the linear manner and generally skip certain phases of the action research cycle. The systematic way, of conducting action research, is to go beyond to solve the problem. It makes sure to consider the problem in a complex way, so that it leads to create a new set of research questions or issues. Applying mixed method in action research would help the researchers to secure more systematic approach to action, provide more comprehensive solutions for change, and promote sustainability. While keeping in the mind when there is an involvement of stakeholders in the action research process, the concept is termed as participatory mixed action research (PMAR).

MIXED ACTION RESEARCH PROCESS

The characteristics of action research are classified into eight steps in Figure 2 with the following elements: 1) Identification of the issues for investigation, 2) garnering and reviewing literature, 3) preparing research questions and hypothesis, 4) developing a research plan of action, 5) garnering data, 6) analysing data, 7) providing the findings, and 8) taking action based on findings. These eight steps are carried out in the cyclic process in line with Lewin (1947). During the process, the action research is emphasised on the need for flexibility in researching the problem, its issues and to have a possible solution that will handle the issue in a critical manner. The process of action research uses the cyclic model as to emphasize the importance of research results leading to possible solutions (Stringer, 1999; McNiff, 1988).

Figure 2. Steps in Action Research
Source: Stringer (1999); McNiff (1988)

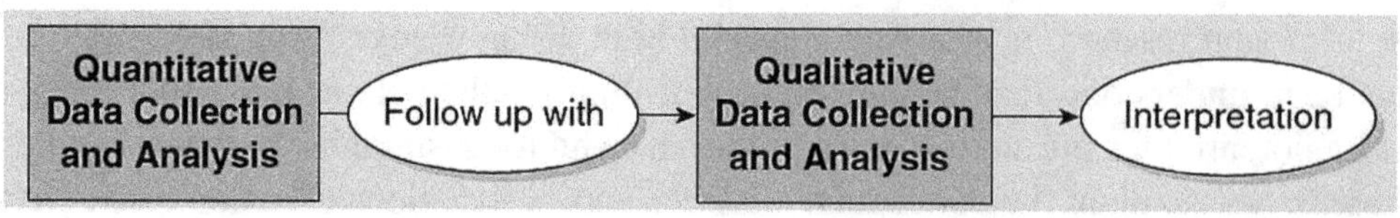

Identify the Issues to Investigate

Generally, the stakeholders are pressing issues that would like to address. The step of action research is to identify the issues to investigate the problems and should address the issues of the time in order to decide which of the goal is most prominent. According to Newman, Ridenour, Newman and DeMarco (2003), the basic goals for identifying the issues are relevant, predicting, add to knowledge, measure changes, understand complex phenomena, test new ideas, generate new ideas, and examine the past. To perform the standard research, Johnson & Christensen (2004) have emphasised on the five phases of action research, which are exploration, description, explanation, prediction, and influence. Hence, the deductive or inductive approach of research can be linked to one or more of the five research objectives.

Gathering and Reviewing the Relevant Literature

The literature review involves the evaluation of documents, which contain information related to the research problems. Schwartz, Slate, and Onwuegbuzie (1999) have stated that the purposes of the literature review are: -

1. To multifold to clarify the research problems
2. To determine the importance of the research problem
3. To formulate research questions
4. To provide the necessary knowledge for the development of a conceptual framework
5. To relate the proposed study
6. Helps to identify the variables
7. Help in identifying the limitations
8. To identify the strategies, methodology and methods to implement
9. Help to find the research gaps
10. To help in facilitating the interpretation and compare the future results with previous research.

To identify the complexity of the research is found by the depth and breadth of the literature review. The comprehensive literature review will help to conduct the action research studies, and if the research problem is unique, then the literature review process will be minimized.

Formulating Research Questions and Hypothesis

Once the research problem has been identified, and review of the literature is being done, then the researcher's job is to formulate the research questions. A good research question will have some specific qualities, which provide the direction to the research process. The research question should be stated clearly and is feasible and meaningful. The benefit of developing the mixed method action research questions will help the researchers explore the process to a great extent. With the formulation of research questions and create, the research objective will help to create the testable hypothesis within a realistic time frame and have expectations concerning the relationships between independent and dependent variables. There are two types of hypothesis namely inductive and deductive hypotheses. According to Johnson and Christensen (2004), the inductive hypothesis is generalizations based on the observation that are not related to any large research, whereas the deductive hypothesis is derived to satisfy the theory that can test and provides the empirical evidence to support and expands or contradicts the given theory.

Developing a Research Plan of Action

The plan for collecting the data is done once the research question is formulated. The idea of planning the research will ensure that the mixed action research study is planned properly so as to maximize the quality of research. The research plan is to find the target sample and also find out which sample is selected. It should be noted that the sample size should be consistent with the research goals, objectives, and questions. The research plan also includes the decision of involving the instruments. Data collection strategies include standardizing instruments and self-developed instruments. Hence, all procedures should be planned in a systematic manner and to have permission to replicate the study.

Collecting the Data

Johnsons and Turners (2003) have identified six specific data collection strategies in mixed methods research:

1. Mixture of open ended and closed ended items on one or more questionnaires
2. Mixture of depth and breadth interviewing
3. Mixture of focus group strategies
4. Mixture of standardized open and closed ended pre-designed tests
5. Mixture of standardized and exploratory observations
6. A mixture of non-numeric and numeric documents consists of archived data based on open and closed-ended items (pp.299)

Analyzing the Data

The seven stages of mixed method data are being analysed and summarized by Onwuegbuzie and Teddlie (2003) such as:

1. **Data Reduction:** Refers to reducing the dimensionality of the qualitative data
2. **Data Display:** Involves in pictorially display
3. **Data Transformation:** Performs quantification
4. **Data Correlations:** Involves the quantitative data to correlate with qualitized data
5. **Data Consolidation:** Combines quantitative and qualitative data to create new or consolidate variables
6. **Data Comparison:** Compares data from the qualitative and quantitative data sources
7. **Data Integration:** Quantitative and qualitative data are integrated into either a coherent whole or two separate sets (pp. 375).

Interpreting the Findings

Interpreting of the findings is to validate the data. The validation of data involves assessing of trustworthiness, credibility, dependability, validity, applicability, reliability, conformability, and transferability of data collected.

Taking Action

The researcher will be in the position to take action based upon findings and make decisions. The decisions are based on discipline and administrative needs and involve communicating the findings. The communication can be a formal or informal way of discussion that involves mixed action research. The findings should be effectively and efficiently as possible communicated to the respective audience.

ADVANTAGES OF APPLYING MIXED METHOD IN ACTION RESEARCH

The integration of mixed method and action research is not new, as many researchers have found the importance of using a mixed method and action research more effectively and efficiently for solving the current or life's problems. The utilization of using mixed methods in the various areas of research has widely been accepted in the last decade which has produced a sound methodology and framework for action research that are more rigorous and more consistent. According to McNiff and Whitehead (2011), the call for improvement for data driven and evidence based research would require action research to conduct the research and generate scientifically sound and effective plan of action. The action research goes by solving the issues. The combination of action research will assist the stakeholders in helping in the decision-making process. This will lead to understanding the different issues and able to formulate new research question, which is actually a cyclic process. With this, the research question will be solved and provides the solution in a systematic manner and create the background to change. Johnson and Turner (2003) have stated that the combination of quantitative and qualitative will enhance the solution credibility and will improvise the process by better performance using action research. Mixed method is an established research approach that has built the meaningful integration between quantitative and qualitative and has provided better ways of learning about the issues and creates new knowledge base along with action research. Thus, the combinations of mixed method and action research have provided a scientific methodology framework.

CHALLENGES IN USING MIXED METHOD

The challenges that mixed method has faced is mainly from the manipulation or reduction of data. The combination of both the approaches will require more resources and the issues of sample size in order to make the studies using the mixed method approach (Driscoll, Appiah-Yeboah, Salib & Rupert, 2007). Liber (2009) has identified generalizing of data, compiling, and analysing of data to relate to the research questions, which require more time and money to perform in an effective and efficient way.

BENEFITS OF MIXED METHODS DESIGN

Greene *et al.* (1989) and Creswell (2003) have reviewed and evaluated the studies and have identified five main purposes of using mixed methods in the study such as triangulation, complementary, development, initiation, and expansion. From the past literature, it is found that around 70-80% of the primary purposes of the study are either complementary or expansion however, many have used triangulation because the use of those in the study will ensure the inherent bias of one is overcome by the strengths of the others. Thus, the use of mixed methods would strengthen the validity of the findings (Creswell, 2003). Table 7 shows the purpose of mixed methods, evaluation strategy.

- **Triangulation:** It means a combination of quantitative and qualitative methods. It helps to overcome the weakness of one method with the other one, which operate side by side. Triangulation, which combines quantitative and qualitative methods, can focus on a single case in which same people can complete a questionnaire and also interviewed (Gray, 2009).
- **Embedded:** In the embedded form of mixed method, the quantitative and qualitative methods are combined to measure the overlapping with different elements of the phenomenon. This is a contrast with triangulation, which use different methods to assess the same conceptual phenomenon. Thus, the two measures assess similar as well as different aspects of the aspirations concept (Gray, 2009).
- **Sequential:** As the name suggests the results of one method are used to inform the development of the second method. In this, it can be exploratory or explanatory method of mixed method. In this, the purposive sample could then be followed up by an in-depth interview.
- **Initiation:** It uses mixed methods to uncover the paradoxes, new perspectives, and contradictions, which is the contrast to triangulation (Gray, 2009). The focus of this type method is to generate new insights which lead to the reframing the research questions.
- **Expansion:** It uses mixed methods to broaden and widen the range of the study (Gray, 2009). It helps to assess the study outcomes from the quantitative methods; whereas qualitative methods help to explore the perspectives of participating and make the group decision takes place.

Table 7. Summary on Purpose of Mixed Methods, Evaluation Design Benefits

Purpose	Rationale
Triangulation Seeks the convergence, corroboration, correspondence of results from different methods.	To increase the validity of the constructs by counteracting or maximizing the heterogeneity of irrelevant sources of variance attributable to inherent method bias, inquirer bias or biases in inquiry context.
Embedded Seeks elaboration, enhancement. Illustration, clarification of results of one method with the results from the other method.	To increase the meaningfulness and validity of the constructs by capitalizing on inherent method strengths and counteracting inherent method biases.
Sequential Seeks to use the results of one method to help or inform the other method.	To increase the validity of the constructs and inquiry results by capitalizing on inherent method strengths.
Initiation Seeks the discovery of paradox and contradiction, new perspectives, the recasting of questions or the results from one method with questions or results from the other method.	To increase the breadth and depth of inquiry results and interpretations by analyzing them from the different perspectives of different methods and paradigms.
Expansion Seeks to extend the breadth and range of inquiry by using different methods from different inquiry components.	To increase the scope of inquiry by selecting the methods, most appropriate for multiple inquiry components.

Source: Adapted from Gray (2009)

FUTURE PLAN FOR MIXED METHODS RESEARCH

Mixed methods research has passed the difficult phase of getting recognition from the researchers, scholars, and industrials players in a significant manner and has shown advancement greatly in the recent years. Creswell and Plano Clark (2007) have identified the area of development required, such as mixed method design with associated procedure issues, mixed method as it is viewed by the world, applications of mixed method research and its implication. Table 8 represents the overview of the topic and answers need to be determined.

Table 8. Future Research Topics and Answered Need to Determine

Topic	Questions need to be answered
Sampling Procedure	• What should be the sample size for the quantitative and qualitative of the mixed methods study? • Is the sample for both quantitative and qualitative be of similar sizes? • Should the quantitative data collection, use random sampling to make the sample more comparable to qualitative sample? • When should different types of purposive sampling be used? • Is purposive sampling is useful for both the quantitative and qualitative approach?
Procedures for integrating information	• To converge quantitative and qualitative data, what options exist beyond a comparison of the discussion and matrices? • What are the possibilities for matrices? • How can qualitative data be embedded into a correlational design? • How can quantitative data be embedded into an ethnography, ground theory, or another qualitative design? • Are some data transformation procedures more credible (and useful) than others? • Are there good examples of transformation or quantitative data into qualitative data? • What innovation designs might be used that have not already been discussed in the literature?
Software needs	• What are the current features available in software that can assist mixed methods researchers? • What new features need to be contrived to help implement the mixed methods designs and their variations?
Writing up mixed methods research	• How do the structures for writing up mixed methods research need to be accommodated to meet guidelines for funding proposals?
The philosophical foundations and worldviews of mixed methods research	• What components of an emancipator worldviews go into mixed methods study? • What are some models of mixed methods studies that use emancipator worldviews, and how are they planned? • What are some models of mixed methods studies that use pragmatism as a worldview, and how are they planned?
Validity and inference issues	• Does the validity vary by type of design? • Do different types of designs lead to different implications?
Applications of mixed methods research	• What chance occur when mixed methods research is utilized in a field or area? • Are there any common applications across different approaches? • What nuances or changes take place when mixed methods research is given in different cultural contexts? • In what fields are mixed methods research is more accepted than others?
Value of mixed methods research	• What value is added by using mixed methods research? • How do researchers and practitioners see the value of mixed method? • Is the value worth the time and resources required to conduct mixed methods research?

Source: Adopted from Creswell and Plano Clark (2007)

CONCLUSION

Mixed methods research has gained increasing acceptance in the last decades with the complexity of understanding the methodology and implementing in the previous years. The design provides a comprehensive knowledge of understanding the framework for collecting, analyzing, mixing, interpreting, and reporting in different approaches. The four principles will help the researchers to design their mixed method study. However, still there is a need for further enhancement of the design and methodology and mixing action research for future study to solve the practical issues. The new ideas will help to improvise the design structure and its methodology. According to Creswell (2003), the future plan which has been summarized at the end of the chapter will serve as the beginning of the new research. Some of the issues are still in the nascent stage and require further clarification to enhance in-depth knowledge of mixed method needed for further study.

REFERENCES

Brewer, J., & Hunter, A. (1989). *Multi-method research: A synthesis of styles*. London: Sage Publications, Inc.

Bryman, A. (1992). Quantitative and qualitative research: further reflections on their integration. *Mixing methods: Qualitative and quantitative research*, 57-78.

Bryman, A. (2003). *Quantity and quality in social research*. London: Routledge.

Bryman, A. (2006). Paradigm peace and the implications for quality. *International Journal of Social Research Methodology*, *9*(2), 111–126. doi:10.1080/13645570600595280

Cameron, R. (2011). Mixed methods research: The Five Ps framework. *The Electronic Journal of Business Research Methods*, *9*(2), 96–108.

Coghlan, D. (2011). Action research: Exploring perspectives on a philosophy of practical knowing. *The Academy of Management Annals*, *5*(1), 53–87. doi:10.1080/19416520.2011.571520

Coghlan, D., & Brannick, T. (2014). *Doing action research in your own organization*. London: Sage.

Creswell, J. W. (1994). *Research design: Qualitative and quantitative approaches*. Thousand Oaks, CA: Sage.

Creswell, J. W. (1994). *Research Design: Qualitative and Quantitative Approaches*. London: Sage.

Creswell, J. W. (1998). *Qualitative inquiry and research design: Choosing among five*. New York: Sage.

Creswell, J. W. (2003). *Research design: Qualitative, quantitative, and mixed methods approaches* (2nd ed.). Thousand Oaks, CA: Sage.

Creswell, J. W., Klassen, A. C., Plano Clark, V. L., & Smith, K. C. (2011). *Best practices for mixed methods research in the health sciences*. Bethesda, MD: National Institutes of Health. doi:10.1037/e566732013-001

Creswell, J. W., & Plano Clark, V. L. (2007). *Designing and conducting mixed methods research*. Thousand Oaks, CA: Sage.

Creswell, J. W., & Plano Clark, V. L. (2010). *Understanding research: A consumer's guide*. Thousand Oaks, CA: Sage.

Crotty, M. (1998). *The foundations of social research: Meaning and perspective in the research process*. Thousand Oaks, CA: Sage.

Datta, L.-E. (1994). Paradigm wars: A basis for peaceful coexistence and beyond. *New Directions for Program Evaluation*, *61*, 53-70.

Denzin, N. K. (1978). *The research act: A theoretical introduction to sociological methods*. New York: McGraw.

Dewey, J. (1938). *Logic: The theory of inquiry*. New York: Holt, Rinehart and Winston.

Driscoll, D. L., Appiah-Yeboah, A., Salib, P., & Rupert, D. J. (2007). Merging qualitative and quantitative data in mixed methods research: How to and why not. *Ecological and Environmental Anthropology*, *3*(1), 19–28.

Fielding, N. G., & Fielding, J. L. (1986). *Linking data: the articulation of qualitative and quantitative methods in social research*. Beverly Hills, CA: Sage. doi:10.4135/9781412984775

Goldkuhl, G. (2012). Pragmatism vs interpretivism in qualitative information systems research. *European Journal of Information Systems*, *21*(2), 135–146. doi:10.1057/ejis.2011.54

Graff, J. C. (2013). Mixed Methods Research. In Evidence-Based Practice: An Integrative Approach to Research, Administration and Practice (pp. 44-64). Academic Press.

Gray, D. E. (2009). *Doing research in the real world*. London: Sage.

Greene, J. C., & Caracelli, V. J. (1997). *Advances in mixed-method evaluation: The challenges and benefits of integrating diverse paradigms*. Jossey-Bass Publishers.

Greene, J. C., Caracelli, V. J., & Graham, W. F. (1989). Toward a conceptual framework for mixed-method evaluation designs. *Educational Evaluation and Policy Analysis*, *11*(3), 255–274. doi:10.3102/01623737011003255

Gummesson, E. (2000). *Qualitative methods in management research*. Thousand Oaks, CA: Sage.

Halton, E. (2004). In G. Ritzer (Ed.), *Pragmatism. Encyclopedia of Social Theory*. Thousand Oaks, CA: Sage.

Hayes, B., Bonner, A., & Douglas, C. (2013). An introduction to mixed methods research for nephrology nurses. *Ren Soc Australasia J*, *9*(1), 8–14.

Howe, K. R. (1988). Against the quantitative-qualitative incompatibility thesis or, Dogmas die hard. *Educational Researcher*, *17*, 10–16.

Ivankova, N. V. (2013). Implementing quality criteria in designing and conducting a sequential QUAN→ QUAL mixed methods study of student engagement with learning applied research methods online. *Journal of Mixed Methods Research*, 1–27. doi:10.1177/1558689813487945

Jick, T. D. (1979). Mixing qualitative and quantitative methods: Triangulation in action. *Administrative Science Quarterly*, *24*(4), 602–611. doi:10.2307/2392366

Johnson, R. B., Onwuegbuzie, A. J., & Turner, L. A. (2007). Toward a definition of mixed methods research. *Journal of Mixed Methods Research*, *1*(2), 112–133. doi:10.1177/1558689806298224

Johnson, R. B., & Turner, L. A. (2003). Data collection strategies in mixed methods research. In A. Tashakkori & C. Teddlie (Eds.), *Handbook of mixed methods in social and behavioral research* (pp. 297–319). Thousand Oaks, CA: Sage.

Kuhn, T. S. (1962). *The structure of scientific revolutions*. Chicago University.

Lee, A., & Nickerson, J. (2010). Theory as a case of design: Lessons for design from the philosophy of science. In *Proc. of the 43rd Hawaii International Conference on System Sciences.* doi:10.1109/HICSS.2010.484

Leech, N. L., & Onwuegbuzie, A. J. (2009). A typology of mixed methods research designs. *Quality & Quantity*, *43*(2), 265–275. doi:10.1007/s11135-007-9105-3

Lewin, K. (1947). Group decision and social change. In T. M. Newcombe & E. L. Hartley (Eds.), *Readings in Social Psychology*. New York: Holt.

Lieber, E., (2009). Mixing qualitative and quantitative methods: Insights into design and analysis issues. *Journal of Ethnographic and Qualitative Research*, (3), 218-227.

Lincoln, Y. S., & Guba, E. G. (1985). *Naturalistic inquiry.* Técnicas.

Mackenzie, N., & Knipe, S. (2006). Research dilemmas: Paradigms, methods and methodology. *Issues in Educational Research*, *16*(2), 193–205.

Maxwell, J. A. (2005). *Qualitative research design: An interactive approach* (2nd ed.). Thousand Oaks, CA: Sage.

McNiff, J. (1988). *Action Research: Principles and Practice*. Basingstoke, UK: Macmillan. doi:10.4324/9780203305676

McNiff, J., & Whitehead, J. (2006). *All you need to know about action research*. London: Sage.

McNiff, J., & Whitehead, J. (2011). *All You Need to Know About Action Research*. London: Sage.

Morgan, D. L. (1998). Practical strategies for combining qualitative and quantitative methods: Applications to health research. *Qualitative Health Research*, *3*(3), 362–376. doi:10.1177/104973239800800307 PMID:10558337

Morse, J. M. (1991). Strategies for sampling. In J. M. Morse (Ed.), *Qualitative nursing research: A contemporary dialogue* (pp. 127–145). Newbury Park, CA: Sage. doi:10.4135/9781483349015.n16

Neuman, S. (2006). Speak up. *Scholastic Early Childhood Today, 20*(4), 12–13.

Neuman. (2000). *Social research methods: qualitative and quantitative approaches* (4th ed.). Boston: Allyn & Bacon.

Newman, I., Ridenour, C. S., & Newman, C. D., Jr. (2003). A typology of research purposes and its relationship to mixed methods. In Handbook of Mixed Methods in Social and Behavioural Research, (pp. 167-188). Academic Press.

Popper, K. R. (1959). *The logic of scientific discovery*. London: Hutchinson.

Ragin, C. C., Nagel, J., & White, P. (2004). *Workshop on scientific foundations of qualitative research*. National Science Foundation.

Reason, P., & Bradbury, H. (Eds.). (2001). *Handbook of action research: Participative inquiry and practice*. London: Sage.

Schwartz, R., Slate, J., & Onwuegbuzie, A. J. (1999). Empowering teachers: Actingupon action research. *GATEways to Teacher Education, 11*(2), 44–59.

Shadish, W. R. (1998). Evaluation theory is who we are. *The American Journal of Evaluation, 19*(1), 1–19. doi:10.1177/109821409801900102

Shadish, W. R., Cook, T. D., & Campbell, D. T. (2002). *Experimental and quasi-experimental designs for generalized causal inference*. Wadsworth Cengage learning.

Smith, J. K. (1983). Quantitative versus qualitative research: An attempt to clarify the issue. *Educational Researcher, 12*(3), 6–13. doi:10.3102/0013189X012003006

Smith, J. K., & Heshusius, L. (1986). Closing down the conversation: The end of the quantitative-qualitative debate among educational inquirers. *Educational Researcher, 15*(1), 4–12. doi:10.3102/0013189X015001004

Somekh, B., & Lewin, C. (2005). *Research methods in social science*. London: Sage.

Steckler, A., McLeroy, K. R., Goodman, R. M., Bird, S. T., & McCormick, L. (1992). Toward integrating qualitative and quantitative methods: An introduction. *Health Education Quarterly, 19*(1), 1–8. doi:10.1177/109019819201900101 PMID:1568869

Stringer, E. T. (1999). *Action research* (2nd ed.). London: Sage.

Stringer, E. T. (2007). *Action research* (3rd ed.). London: Sage.

Tashakkori, A., & Teddlie, C. (1998). *Mixed methodology: Combining qualitative and quantitative approaches*. Sage.

Tashakkori, A., & Teddlie, C. (Eds.). (2003a). *Handbook of mixed methods in social and behavioral research*. Thousand Oaks, CA: Sage.

Teddlie, C., & Tashakkori, A. (2003). Major issues and controversies in the use of mixed methods in the social and behavioral sciences. In A. Tashakkori & C. Teddlie (Eds.), *Handbook of mixed methods in social & behavioral research* (pp. 3–50). Thousand Oaks, CA: Sage.

Teddlie, C., & Tashakkori, A. (Eds.). (2009). *Foundations of mixed methods research: Integrating quantitative and qualitative approaches in the social and behavioral sciences*. Sage Publications Inc.

Teddlie, C., & Yu, F. (2006). Mixed methods sampling: A typology with examples. *Journal of Mixed Methods Research*, *1*(1), 77–100. doi:10.1177/2345678906292430

Welsh, W. (2007). *Transformation changes everything: Exploring organizational leadership challenges in an "aspiring - to" transformative environment*. (Doctoral dissertation, Regent University). Available from Dissertations & Theses: Full Text. ProQuest

Wicks, A. C., & Freeman, R. E. (1998). Organization studies and the new pragmatism: Positivism, anti-positivism, and the search for ethics. *Organization Science*, *9*(2), 123–140. doi:10.1287/orsc.9.2.123

KEY TERMS AND DEFINITIONS

Action Research: Research in improving and initiating to solve an immediate problem.

Mixed Action Research: To handle a practical research problem from any perspectives.

Mixed Method Design: To use a multi-method model in the research.

Mixed Method Research: To handle research questions from any perspectives.

Pragmatism: The process of action to solve problems in a specific situation instead of ideas and theories.

Research Approach: The research style f dealing the situation or problem.

Research Paradigm: A specific pattern of model to perform research.

Section 2

Designing a Mixed Methods Research Study

Researchers need to make numerous methodical decisions including the research design which is the blueprint for the study. This chapter outlines the design decisions that need to be made. The design is chosen based on which strategy is best suited to answer the research question(s). The three basic mixed methods designs are discussed: Parallel Convergent, Sequential (Explanatory or Exploratory), and Embedded Design. Additional decisions need to be made about the implementation of the study including the sequencing of the data collection, weight of the two methods (quantitative and qualitative), and when and how to converge the data.

Chapter 4
It is All in the Design: Creating the Foundations of a Mixed Methods Research Study

Mette L. Baran
Cardinal Stritch University, USA

ABSTRACT

This chapter introduces the various design choices researchers need to decide on prior to conducting the study. In the first section of this chapter a detailed description of research design is followed by an explanation that the type of information that is collected is based on whether the research question is descriptive, explanatory, or exploratory. The major strategic implementation methods for quantitative, qualitative, and mixed methods are then discussed. The three strategies for mixed methods research Parallel Convergent, Sequential, and Embedded Design are presented in detail along with the rationale for their use. Finally, in the last section, the strands or sequencing of the data collection phase of the study is explained.

INTRODUCTION

After the method for the research is chosen: qualitative, quantitative or mixed methods, the researcher needs to determine the design. Various designs are used in research and researchers need to determine which design bet fits the purpose of the study. Researchers need to know where the design fits into the whole research process from framing the research purpose and question(s), data collection and analysis, to finally reporting the findings. There are various designs which are used in research, all with specific advantages and disadvantages. Which one the researcher selects, depends on the objective of the study and the nature of the phenomenon (Hartley & Muhit, 2003). Researchers can decide to use a quantitative, qualitative, or a mixed methods design. This chapter introduces the various approaches aligned with each design.

DOI: 10.4018/978-1-5225-0007-0.ch004

What Is Research Design?

The research design is a framework or blueprint which gives structure and direction to show how all of the major parts of the research project work together to try to address the research question (Malhotra, 2004). It is the logical structure of an inquiry grounded in the research purpose and research question(s) (Mertens, 2005). Furthermore, Creswell (2007) refers to designs as "procedures for collecting, analyzing, interpreting, and reporting data" (p. 58). It details the procedures necessary for obtaining the information needed to structure or solve the research question(s) (Malhotra, 2004). This implies that the researcher needs to decide on the design before the study can begin.

The research design does not imply or dictate any particular method of collecting data or any particular type of data. How the data are collected is irrelevant to the logic of the design. Any research design can, in principle, use any type of data collection method and utilize either quantitative or qualitative data as research design is different from the method by which data are collected. The main purpose of research design is to reduce the ambiguity of research evidence providing a step by step approach to the entire research plan reducing the possibility for errors to be made.

There are numerous research designs ranging from simple in mature to overly complex; however, for the purpose of this chapter, the number of methodologies are limited to the three major typologies in order not to overwhelm beginning researchers. Mixed methods may be the best approach when both quantitative and qualitative data, together will provide a richer understanding of the phenomenon being studied because one approach is not enough. So a mixed methods research design is a procedure for mixing both methodologies in a single study to obtain evidence needed to provide a deep understanding of the research problem. This plan includes decisions around sampling and method of data collection.

We know that the research question drives everything, including the methodology and research design. When conducting research, researchers need to follow a plan for how the study will unfold and the various steps taken from data collection tools through data analysis. The function of the design is to ensure that a blueprint is in place and that the researcher has collected enough data and analyzed the findings so that the initial research question(s) can be addressed. In other words, when designing research one needs to ask: given this research question (or theory), what type of data will I need to collect in order to address the research objective? Researchers need to think through carefully what type of information is required to answer the research question(s). One can argue that the validity and reliability of the research findings are directly tied to the amount of upfront logical planning the researcher invested in the design process at the beginning.

The way in which researchers develop research designs is fundamentally affected by whether the research question is descriptive, explanatory, or exploratory as this affects what information is collected. Social researchers ask the following types of research questions:

1. What is going on (Descriptive Research)?
2. Why is it going on (Explanatory Research)?
3. How is it that it is going on (Exploratory Research)?

Descriptive Research

The purpose of this research design is to observe, describe, and document aspects of events as they naturally unfold (Polit & Hungler, 1999). Many scientific disciplines, especially social science and psychology, use descriptive research to obtain a general overview of the subject or characteristics of an organization or community. Thus descriptive research is playing an important role in providing data by providing a snap shot of what is going on at a specific point in time surrounding the research topic under study, focusing on the "what" is happening. As a results, descriptive research is also an effective approach to making predictions of certain outcomes, for example, how many will vote for certain political candidate or how many will purchase a particular product or attend a certain university? Leedy & Ormrod (2010) posit that a descriptive approach is used in order to "see explanations and predictions that will generalize to other persons and places" (p. 95).

Descriptive studies can yield rich data that lead to important recommendations through the data collection tools of surveys, observation, interviews, and portfolio (AECT, para.20). A survey is a research study in which data are collected from the members of a sample, for the purpose of estimating one or more population parameters (Jaeger, 1997, p. 450). For certain studies, subjects can only be observed in a way as to not affect behavior. This type of research design involves observing and describing the behavior of a subject without influencing. Descriptive research has been dismissed as `mere description' (New Your University, n.d., para. 3); however, detailed description is fundamental to the research enterprise and it has added immeasurably to the knowledge of the shape and nature of society. Good description of 'what' is going on provokes the `why' questions of explanatory research.

Explanatory Research

Explanatory research focuses on "why" questions certain events unfold and the nature of relationships. This is a logical way to inquire after things become known or has been described. "For example, it is one thing to describe the crime rate in a country, to examine trends over time or to compare the rates in different countries. It is quite a different thing to develop explanations about why the crime rate is as high as it is, why some types of crime are increasing or why the rate is higher in some countries than in others," namely looking for the reason why a chain of events occur (New York University, n.d., p. 2). Answering the `why' questions involves developing causal explanations. Causal explanations argue that phenomenon Y (for example, level of education) is affected by factor X (for example, parental socioeconomic status).

With explanatory research a researcher conducts causal research or theory-testing. This type of research design is also known as "deductive," "hypothesis-testing," and "predictive." Deductive reasoning starts with a general theory pertaining to the research topic which is then converted into a testable hypothesis(es). This is followed by observations addressing the hypothesis(es). Trochim (2006) notes that "this ultimately leads us to be able to test the hypotheses with specific data -- a *confirmation* (or not)" of the original theory (para. 1). The idea is to test a theory by working out some of the specific implications of the theory (hypothesis) and then collecting data to see if the hypothesis is supported or not. Or, theory-testing often take place after exploratory work has already been done leading to new theory. All studies (except descriptive) are interested in causality as all theories are causal. While causality is complex, the main purpose behind theory-testing studies are basically about testing whether it's true that 'A'causes 'B' and why.

Exploratory Research

Exploratory research seeks to determine "how" is it that things are happening. This type of research design is also known as "inductive" or "theory-building" focusing on "specific observations to broader generalizations and theories" (Troachim, 2006, para. 2). In this kind of study, one moves from specific observations to broader generalizations and theories. The researcher does not start with a theory. Instead, data are collected and analyzed and the data are used to develop a theory. For example, a researcher may want to explore how it is that bullying does not exist in a setting while in other settings it flourishes. A study may then be designed to test the theory (explanatory). So the purpose of most exploratory studies is to develop a causal theory, which can then be tested at a later time.

Figures 1 and 2 outline the steps to designing a research study based on a deductive or inductive approach.

Strategies Associated with Quantitative Design

Quantitative research is a mode of inquiry used often for deductive research, following the scientific method to test theories or hypotheses, gather descriptive information about variables, or examine relationships among variables, or "to determine cause and effect interactions between variables" (Burns & Grove, 2005, p. 23). Furthermore the authors define quantitative research as, "a formal, objective, systematic process in which numerical data are used to obtain information about the world" (p. 23). These variables are measured and yield numeric data that can be analyzed statistically. Social surveys and experiments are frequently viewed as prime examples of quantitative research and are evaluated against the strengths and weaknesses of statistical, quantitative research methods and analysis. Quantitative (mainly deductive) methods are ideal for measuring pervasiveness of "known" phenomena and central patterns of association, including inferences of causality Pasick et al., (2009). "Quantitative data have the potential to provide measurable evidence, to help to establish (probable) cause and effect, to yield efficient data collection procedures, to create the possibility of replication and generalization to a population, to facilitate the comparison of groups, and to provide insight into a breadth of experiences"

Figure 1. Deductive approach
Source: Trochim, 2006

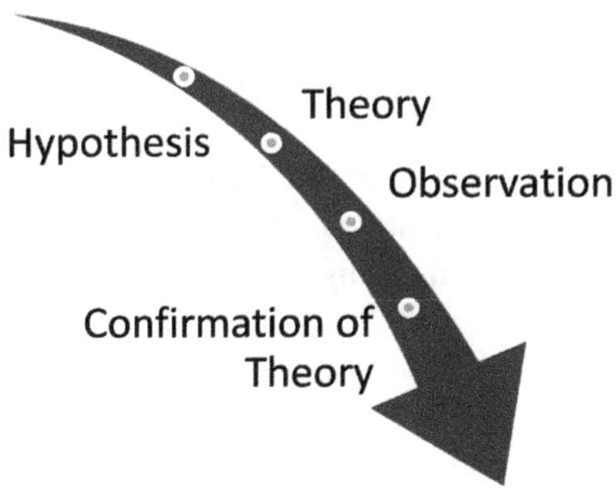

Figure 2. Inductive approach
Source: Trochim, 2006

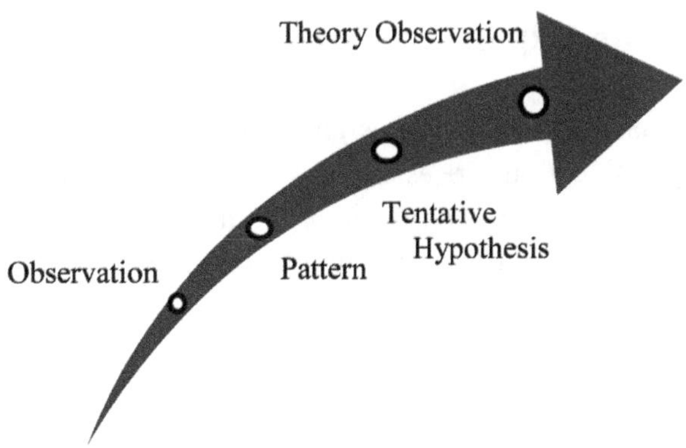

(U.S. Department of Health & Human Services, 2015, para. 6). Creswell (2012) noted the major characteristics for quantitative research:

1. Describing a research problem through a description of trends or a need for an explanation of the relationship among variables.
2. Providing a major role for the literature through suggesting the research questions to be asked and justifying the research problem and creating a need for the direction (purpose statement and research questions or hypotheses) of the study.
3. Creating purpose statements, research questions, and the hypotheses that are specific, narrow, measurable, and observable.
4. Collecting numeric data from a large number of people using instruments with preset questions and responses.
5. Analyzing trends, comparing groups, or relating variables using statistical analysis, and interpreting results by comparing them with prior predictions and past research.
6. Writing the research report using standard, fixed structures and evaluation criteria, and taking an objective, unbiased approach (p. 13).

Typical quantitative approaches used in the social sciences are *true experimental design* and *quasi experimental design.*

True Experimental Design

True experimental design is regarded as the most accurate form of experimental research, in that it tries to find support or lack thereof for a hypothesis mathematically with statistical analysis. In this design, groups are randomly selected and treated in the exact same way controlling the variable. For an experiment to be considered as a true experimental design, it must fit all of the following criteria.

1. The sample groups must be assigned randomly.
2. There must be a viable control group.
3. Only one variable can be manipulated and tested. It is possible to test more than one, but such experiments and their statistical analysis tend to be cumbersome and difficult.
4. The tested subjects must be randomly assigned to either control or experimental groups (Shuttleworth, n.d., para.2).

Quasi Experimental Design

Quasi Experimental design involves selecting groups, upon which a variable is tested, without any random pre-selection processes. The experimental group and the control group are chosen due to convenience and results are compared without the randomization of group assignments used in True Experimental Design. The data collection tool used is usually survey.

Strategies Associated with Qualitative Design

A salient strength of qualitative research is its focus on the contexts and meaning of human lives and experiences for the purpose of inductive or theory-development driven research. It is a systematic and rigorous form of inquiry that uses methods of data collection such as in-depth interviews, ethnographic observation, and review of documents. Qualitative (mainly inductive) methods allow for identification of previously unknown processes, explanations of why and how phenomena occur, and the range of their effects (Pasick et al., 2009). Creswell (1998) notes that: "Qualitative research is an inquiry process of understanding based on distinct methodological traditions of inquiry that explore a social or human problem. The researcher builds a complex, holistic picture, analyzes words, reports detailed views of informants, and conducts the study in a natural setting" (p. 15). Qualitative data help researchers understand processes, especially those that emerge over time, provide detailed information about setting or context, and emphasize the voices of participants through quotes. Qualitative methods facilitate the collection of data when measures do not exist and provide a depth of understanding of concepts. Typical qualitative approaches used in social science are narrative research, phenomenological research, ethnographies, grounded research, and case study.

Narrative Research

Narrative research is a form of inquiry in with the researcher studies the lives of individuals and asks one or more individuals to provide stories about their lives. This information is then retold or restoried by the researcher into narrative chronology. In the end, the narrative combines views from the participant's life with those of the researcher's interpretations in a collaborative narrative.

Phenomenological Research

Phenomenological research is characterized by the researcher identifying the "essence" of human experiences concerning a phenomenon, as described by participants in a study. The researcher describes and interprets the experiences of participants to understand their perspectives based on the belief that there are multiple ways of interpreting the same experience and the meaning of that experience is what constitutes reality. Understanding the "lived experiences" marks phenomenology as a philosophy as well as a method, and the procedure involves studying a small number of subjects through extensive and prolonged engagement to develop patterns and relationships of meaning. In this process, the researcher "brackets" his or her own experiences in order to understand those of the participants in the study. This research design is focused on what is essential for the meaning of the event, episode, or interaction.

Ethnography

Ethnography is an approach in which the researcher studies an intact cultural group in a natural setting over a prolonged period of time by collecting, primarily, observational data culminating in an in-depth description and interpretation of cultural patterns and meanings within a culture or social group. A rich description of the culture of group includes shared patterns of beliefs, normative expectations, behaviors, and meanings. The research process is flexible and typically evolves contextually in response to the lived realities encountered in the field setting.

Grounded Theory

Grounded theory offers the researcher an opportunity to derive a general, abstract theory of a process, action, or interaction grounded in the views of participants in a study. This process involves using multiple stages of data collection and the refinement and interrelationship of categories of information. Two primary characteristics of this design are the constant comparison of data with emerging categories and theoretical sampling of different groups to maximize the similarities and the differences of information.

Case Study

Case study is a design in which the researcher explores a program, an event, an activity, a process, or one or more individuals in-depth. The case(s) are bounded by time and activity, and researchers collect detailed information using a variety of data collection procedures over a sustained period of time culminating in an in-depth analysis of one or more events, settings, programs, groups, or other "bounded systems." Creswell (1998) defines case study as "an exploration of a "bounded system or a case (or multiple cases) over time through detailed, in-depth data collection involving multiple sources of information rich in context" (p. 61). Yin (2003) describes the case study describes the case study: "As a research strategy, the case study is used in many situations to contribute to our knowledge of individual, group, organizational, social, political, and related phenomena" (p.1). He further urges: "In brief, the case study method allows investigators to retain the holistic and meaningful characteristics of real-life events—such as individual life cycles, organizational and managerial processes, neighborhood change, international relations, and the maturation of industries" (p. 2).

Case studies are often seen as prime examples of qualitative research which adopts an interpretive approach to data, studies `things' within their context and considers the subjective meanings that people bring to their situation.

Types of case studies:

1. **Historical Organizational:** Focus on the development of an organization over time.
2. **Observational:** Study of a single entity using participant observation.
3. **Life History (i.e., Oral History):** A first-person narrative completed with one person.
4. **Situation Analysis:** A study of a specific event from multiple perspective.
5. **Multi-Case:** A study of several different independent entities.
6. **Multi-Site:** A study of many sites and participants with the main purpose of which is to develop theory.

RATIONAL FOR MIXED METHODS AND SEQUENCING OF THE DATA COLLECTION

Creswell and Plano Clark (2007) noted: "Mixed methods research is "practical" in the sense that the researcher is free to use all methods possible to address a research problem. It is also "practical" because individuals tend to solve problems using both numbers and words, they combine inductive and deductive thinking, and they (e.g., therapists) employ skills in observing people as well as recording behavior" (p. 10). The rationale for using mixed methods needs to fit the research purpose. Creswell (2009) indicates a mixed methods rationale leads to a better understanding of a "research problem by converging or triangulating broad numeric trends from quantitative research and the detail of qualitative research" (p. 121). A mixed methods approach provides the researcher with an opportunity to utilize both methods and provides a multiple of choices. Creswell and Plano Clark (2011) state the components needed for mixed method research: "In mixed methods research, the data collection procedure consists of several key components: sampling, gaining permissions, collecting data, recording the data, and administering the data collection" (p. 171).

The researcher needs to make decisions about the type of data to be collected and when in the research process it should be collected. For example, starting with quantitative data, a researcher will have the opportunity to test findings during a follow up qualitative data collection phase gaining insights that can help explain initial first phase results. Similarly, collecting qualitative initially, especially when there may be limited understanding on a topic, will provide an opportunity to collect pertinent information about a phenomenon that will aid the researcher in designing a quantitative instrument. This is the implementation stage of the study which pertains to the order in which data are collected. Creswell (2009) asserts a strategic approach to planning for mixed methods giving consideration to four aspects: timing, weighting, mixing, and theorizing (p. 206). Within the context of a mixed methods approach, timing can be sequentially or concurrently; weight or priority may emphasize one type of study over another – qualitative over quantitative or vice versa; mixing the data, integrating, or embedding data relative to the research study and the research question(s); and, theorizing where the existence of a theory is present early in a mixed methods research study impacting the design of the study (Creswell, 2009, pp. 206-208).

Additionally, a mixed methods approach requires multitude of decisions as to how to design the study and how much importance and weight each approach will play in the design. Each methodology can carry equal weight or either approach can carry more weight than the other. Priority of data is "whether greater *priority* or weight is given to the quantitative data and analysis. The priority might be equal, or it might be skewed toward either qualitative or quantitative data." (Creswell, 2009, p. 206). Priority occurs in a mixed methods study through such strategies as whether quantitative or qualitative information is emphasized first in the study, the extent of treatment of one type of data or the other, and use of a theory as an inductive or deductive framework for the study.

Sequencing decisions will also need to be determined. Options include:

1. Collect both quantitative and qualitative data at the same time.
2. Collect quantitative data first followed by qualitative data.
3. Collect qualitative data followed by quantitative data.

For example, a researcher wants to determine the impact of a reading program on student achievement and engagement. She collects pre reading scores (quantitative data) and observes students in their classroom using the traditional reading curriculum (qualitative data). Throughout the implementation period she continues to collect weekly reading scores and observations. In addition, she conducts in-depth interviews with the teachers and parents after the program conclusion. Numerous decisions pertaining to research design had to be made. However, the decision to as to the design is driven by the purpose of the study.

Finally, the researcher connects the data. The "point of interface" (Morse & Niehaus, 2009), or the point where mixing occurs, differs depending on the mixed methods design. This "point" may occur during data collection (e.g., when both quantitative items and qualitative open-ended questions are collected on the same survey), during data analysis (e.g., when qualitative data are converted or transformed into quantitative scores or constructs to be compared with a quantitative dataset), and/or during data interpretation (e.g., when results of quantitative analyses are compared with themes that emerge from the qualitative data).

Creswell and Plano Clark (2011) define the key components that the researcher needs in designing and conducting a mixed methods study:

1. Collects and analyzes persuasively and rigorously both qualitative and quantitative data (based on research questions);
2. Mixes (or integrates or links) the two forms of data concurrently by combining them (or margining them), sequentially by having one build on the other, or embedding one within the other;
3. Gives priority to one or to both forms of data (in terms of what the research emphasizes);
4. Uses these procedures in a single study or in a multiple phases of a program of study;
5. Frames these procedures within philosophical worldviews and theoretical lenses; and
6. Combines the procedures into specific research designs that direct the plan for conduct the study (p. 5).

The following table outlines the strategies associated with quantitative, qualitative, and mixed methods approaches (Figure 3). Three basic strategies are typically associated with mixed methods: The Parallel Convergent Design, Sequential (Explanatory or Exploratory), and Embedded Design.

MAJOR MIXED METHODS DATA COLLECTION AND ANALYSIS APPROACHES

The Parallel Convergent Design

The Parallel Convergent Design is when the researcher converges quantitative and qualitative data in order to provide a comprehensive analysis of the research problem. In this design, the investigator collects both forms of data at the same time during the study and then integrates the information in the interpretation of the overall results. For example, a researcher may want to investigate employees' overall satisfaction with a health care program. Quantitative data are collected through the use of a closed ended instrument measuring overall satisfaction with the program. In addition, at the same time qualitative data are collected through focus groups and in-depth interviews with a random sample of participants in order

Figure 3. Strategies associated with quantitative, qualitative, and mixed methods approaches

Quantitative Research	Qualitative Research	Mixed Methods
• TrueExperimental • Quazi Experimental (Non-Experimental)	• Narrative • Phenomenological • Ethnography • Grounded Theory • Case Study	• Parallell Convergent • Sequential • Embedded

to determine what additional services and programs may be needed or desired. The findings are then converged through statistical analysis of the quantitative data as well as interview findings revealing certain themes of information to best determine the overall satisfaction levels and health care needs and preferences as deemed by the employees. In other words, the data analysis consists of merging data and comparing the two sets of data and results as the objective is to obtain a more complete understanding of two data bases. Figure 4 outlines the steps in designing a Parallel Convergent Design study.

Sequential Procedures (Explanatory and Exploratory) Design

The Sequential Design is when the researcher seeks to elaborate on or expand the findings of one method with another method. The idea is to have one dataset build on the results from the other. These are known as sequential designs, and they may begin by a qualitative exploration followed by a quantitative follow up or by a quantitative analysis explained through a qualitative follow up. This may involve beginning with a qualitative method for exploratory purposes and following up with a quantitative method with a large sample so that the researcher can generalize results to a population. For example, a market researcher can use focus group interview data findings from a sample of consumers to discover scales used in a quantitative instrument administered collecting data from a large sample of the population to determine whether the scales can be generalized.

Alternatively, the study may begin with a quantitative method in which theories or concepts are tested, to be followed by a qualitative method involving detailed exploration with a few cases or individuals. A popular approach in the social sciences is the latter in which qualitative data help to explain in more depth the mechanisms underlying the quantitative results. A researcher exploring caretakers' of cancer patients quality of life may first start with a large sample quantitative instrument and follow up with in-depth interviews with a small sample to investigate further some of the quantitative findings to better understand the responses on the scale(s). "The straightforward nature of this design is one of its main strengths" (Creswell, 2009, p. 211). Figure 5 outlines the steps in designing a Sequential study.

The Embedded Design

A popular design in the social sciences is to use quantitative and qualitative approaches in tandem and to embed one in the other to provide new insights or more refined thinking. These designs are called embedded or nested designs. They may be a variation of a convergent or sequential design. In this form of integration, a dataset of secondary priority is embedded within a larger, primary design. An example is the collection of supplemental qualitative data about how participants are experiencing an interven-

Figure 4. Steps in designing a parallel convergent design study

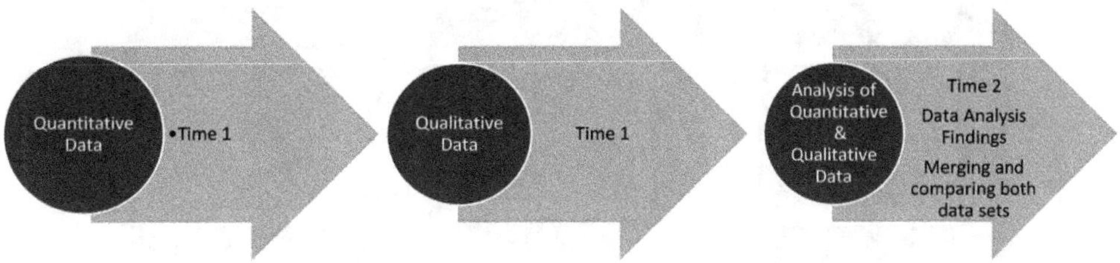

Figure 5. Steps in designing a sequential study

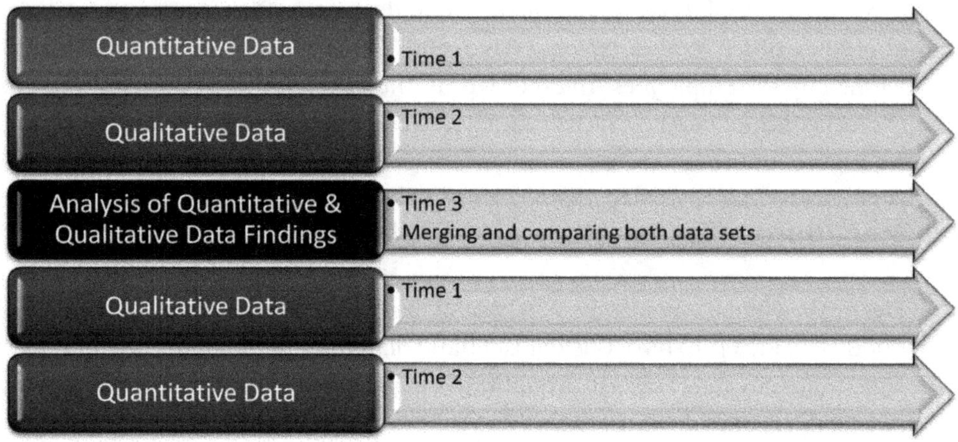

tion during an experimental trial. For example, mentioned previously is the study of student reading achievement and engagement after implementing a reading program. The researcher could decide to include weekly interviews with teachers during the experimental phase to discover the process being experienced by the teachers.

Alternatively, a qualitative data collection may precede an experimental trial to inform development of procedures or follow an experimental trial to help explain the results of the trial. A prototype would be to conduct an intervention study and to embed qualitative data within the intervention procedures to understand how experimental participants experience the treatment. Qualitative data may be used prior to the intervention to inform strategies to best recruit individuals or to develop the intervention, during the experiment to examine the process being experienced by participants, or after the experiment to follow up and better understand the quantitative outcomes. For example, an experimental study of outcomes from an alcohol prevention program might be followed by individual interviews with participants from the experimental group to help determine why the program worked. Figure 6 below outlines the steps in designing an Embedded design study.

Figure 6. Steps in designing an embedded design study

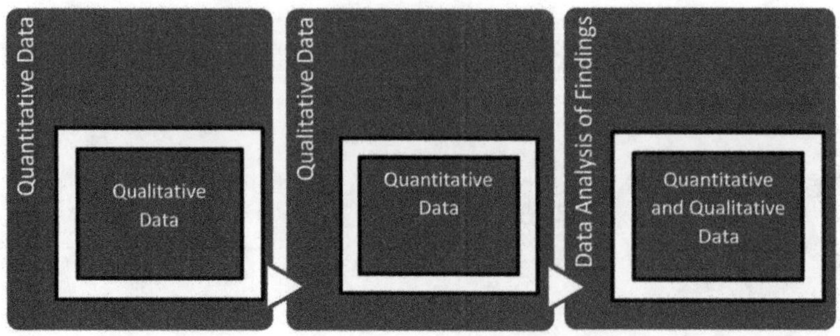

CONCLUSION

This chapter has described the difference between research method and design and provide a rationale for research design. Strategies associated with quantitative and qualitative design have been explained. In addition, a rational for mixed methods research design and sequencing of the data collection have been discussed. Finally, the three basic mixed methods research designs are described: Parallel Convergent design, Sequential (explanatory or exploratory), or Embedded design. In addition, the steps including implementation, weight of findings, and data integration are outlined. Ultimately, the research design is driven by research purpose and not data collection tools.

REFERENCES

AECT. (n.d.). *The Handbook of Research for Educational Communications and Technology*. Retrieved from http://www.aect.org/edtech/ed1/41/41-01.html

Burns, N., & Grove, S. K. (2005). *The practice of nursing research: conduct, critique, and utilization* (5th ed.). St. Louis, MO: Elsevier Saunders.

Creswell, J. W. (1998). *Qualitative inquiry and research design: Choosing among five traditions*. London: Sage.

Creswell, J. W. (2007). *Qualitative inquiry and research design: Choosing among five traditions* (2nd ed.). Thousand Oaks, CA: Sage.

Creswell, J. W. (2009). *Research design, qualitative, quantitative, and mixed methods / approaches* (3rd ed.). London: Sage Publications, Inc.

Creswell, J. W., & Plano Clark, V. (2007). *Designing and conducting mixed methods research*. Thousand Oaks, CA: Sage.

Creswell, J. W., & Plano Clark, V. (2011). *Designing and conducting mixed methods research* (2nd ed.). Thousand Oaks, CA: Sage.

Hartley & Muhit. (2003). Using qualitative research methods for disability research in majority world countries. *Asia Pacific Disability Rehabilitation Journal, 103*(14).

Jaeger, R. M. (1997). *Complementary research methods for research in education* (2nd ed.). Washington, DC: American Educational Research Association.

Leedy, P., & Ormrod, J. (2010). *Practical research: Planning and design* (9th ed.). Boston, MA: Pearson.

Malhotra, N. K. (2004). *Marketing research: an applied orientation* (4th ed.). London: Prentice-Hall International.

Mertens, D. M. (2005). *Research and evaluation in education and psychology: Integrating diversity with quantitative, qualitative, and mixed methods* (2nd ed.). Thousand Oaks, CA: Sage.

New York University. (n.d.). *What is research design? The context of design*. Retrieved from http://www.nyu.edu/classes/bkg/methods/005847ch1.pdf

Pasick, R. J., Burke, N. J., Barker, J. C., Galen, J., Bird, J. A., & Otero-Sabogal, R. (2009). Behavioral theory in a diverse society: Like a compass on Mars. *Health Education & Behavior*, *36*(5), 11S–35S. doi:10.1177/1090198109338917 PMID:19805789

Polit, D. F., & Hungler, B. P. (1999). *Nursing research: Principles and Methods* (6th ed.). Philadelphia: Lippincott.

Shuttleworth, M. (n.d.). True Experimental Design. Retrieved from: https://explorable.com/true-experimental-design

Trochim, W. M. K. (2006). *Research methods knowledge base. Deduction and induction*. Retrieved from: http://www.socialresearchmethods.net/kb/dedind.php

U.S. Department of Health & Human Services. (2015). *The nature and design of mixed methods research*. Retrieved from ttps://obssr.od.nih.gov/scientific_areas/methodology/mixed_methods_research/section2.aspx

Yin, R. K. (2003). *Case study research: Design and methods* (3rd ed.). Thousand Oaks, CA: Sage.

Chapter 5
Designs of Mixed Method Research

Amir Manzoor
Bahria University, Pakistan

ABSTRACT

Mixed methods research is becoming an increasingly popular approach in the discipline fields of sociology, psychology, education and health sciences. Calls for the integration of quantitative and qualitative research methods have been advanced in these fields. A key feature of mixed methods research is its methodological pluralism, which frequently results in research which provides broader perspectives than those offered by mono-method designs. The central premise of mixed methods is that the use of quantitative and qualitative approaches in combination provides a better understanding of research problems and complex phenomena than either approach alone. The purpose of this chapter is to review designs of mixed methods research. The study surveys the common designs of mixed methods research and examine the main characteristics of each in terms of purposes, strengths, and issues, and posits suggestions on the application of these designs.

INTRODUCTION

Mixed method research is a combination technique where statistical information obtained from quantitative measurements is supported and enriched by qualitative information obtained from the explanations provided by the research participants. Mixed method research is one of the recent developments of research methodologies that has gained attention from many educations and social science researchers (Creswell, 2003, Brannen, 2005; Andrew & Halcomb, 2011;Creswell et al., 2011;Creswell, 2014;Denzin, 2012;Kakai et al., 2014;Klassen et al., 2012;Mertens, 2014;Teddlie & Tashakkori, 2011;Venkatesh et al., 2013). Quantitative techniques have also been used to extend qualitative results and research objectives. According to Creswell (2003), "A mixed methods study involves the collection or analysis of both quantitative and/or qualitative data in a single study." In such a study, the quantitative and qualitative data collection can be done either concurrently or sequentially. Each type of data is given a priority level and integrated at one or more stages in the research process. Mixed Methods Research supports the use

DOI: 10.4018/978-1-5225-0007-0.ch005

of multiple research techniques to obtain answers to the research questions and encourage researchers to have a diverse approach towards research method selections (Johnson and Onwuegbuzie, 2004; Klassen, Creswell, Clark, Smith, & Meissner, 2012; Creswell, Klassen, Plano Clark, & Smith, 2011; Kakai et al., 2014).

Quantitative data can be useful to get generalized answers to research questions related to a large group of people. However, quantitative data often is not suitable to get specific answers or explanations related to research questions. On the other hand, qualitative data can provide meaning and context regarding the people and environments of study but unable to provide generalized findings because the number and range of participants is low. Both methods, used together, can complement each other. The rise of mixed method research technique started in 1980s and the article of Campbell and Fiske (1959) can be called as the first formal use of mixed method research technique in the social science methodological literature. In their article, Campbell and Fiske (1959) used the concept of triangulation (i.e. use of more than one techniques to validate the research results and to make sure the variance explained is the result of the underlying phenomenon and not the research technique). According to Bouchard (1976), the triangulation provides convergence of findings and "enhances our beliefs that the results are valid and not a methodological artifact" (p. 268). Webb, Campbell, Schwartz, and Sechrest (1966) extended the idea of (Campbell and Fiske,1959) and defined triangulation as use of multiple measures that "are hypothesized to share in the theoretically relevant components but have different patterns of irrelevant components" (p. 3). Denzin (1978) first outlined how the triangulation works by defining triangulation as "the combination of methodologies in the study of the same phenomenon" (p. 291). Jick (1979) observed that triangulation may not be suitable for all research purposes but it can provide researchers multiple advantages. Some these advantages include more confidence in research results, more creative data collection, richer data, and ability to synthesize or integrate theories, and compare competing theories. According to Morse (1991), methodical triangulations can be of two types: simultaneous or sequential. In simultaneous triangulation, qualitative and quantitative methods are used simultaneously. During the data collection stage, there is limited interaction between the qualitative and quantitative data sources but, during data interpretation stage, the findings complement one another. In sequential triangulation, the results of first approach employed are pre-requisite to plan the next research method used.

According to Rossman and Wilson (1985), there could be multiple reasons for combining quantitative and qualitative research. First, method combination can provide confirmation of the results of each individual method. Second, combination of methods can provide a richer data set. Third, method combinations can result in methods of thinking by analyzing the contradictions that emerge from the two data sources. According to Greene, Caracelli, and Graham Greene (1993), researchers can use mixed method research for various major purposes. Some of these purposes include triangulation (i.e. validation and testing of consistency of findings obtained through different methods), complementarity (i.e. clarification of results from one method with the use of the other method), development (i.e. using results of one method to shape the choice of method to be used in the next step of research), initiation (developing new research questions or challenging results obtained through use of one method), and expansion (i.e. enriching the study and exploring specific features of each research method used). According to Collins, Onwuegbuzie, and Sutton (2006), there could be four rationales for use of mixed method research. These rationales include enrichment of mix of research participants, enhancing validity of research instruments, assessment of the validity of research intervention, and enhancing the significance of research study.

A systematic framework was suggested by Creswell (2003) for conducting mixed methods research. In this framework, mixed method research designs are classified according to two major dimensions:

1. Time order (Time ordering of the qualitative and quantitative phases is another important dimension, and the phases can be carried out sequentially or concurrently).
2. Paradigm emphasis (i.e., deciding whether to give the quantitative and qualitative components of a mixed study equal status or to give one paradigm the dominant status).

Two other dimensions for viewing mixed methods research are the degree of mixing and where mixing should occur (e.g., in the objective[s], methods of data collection, research methods, during data analysis, or data interpretation). Yet another important dimension is whether one wants to take a critical theory/transformative-emancipatory approach or a less explicitly ideological approach to a study. Research studies can involve mixing of qualitative and quantitative approach in a number of ways. This is because there could so many potential dimensions of classification. This key characteristic of mixed method research provides an unlimited potential for future research (Johnson & Onwuegbuzie, 2004).

Each design varies according to the implementation, priority and integration of the data collected (Cameron, 2009). The purpose of this chapter is to review designs of mixed methods research. The study surveys the common designs of mixed methods research and examine the main characteristics of each in terms of purposes, strengths, and issues, and posits suggestions on the application of these designs by taking examples from the literature.

COMMON MIXED METHODS DESIGN

There are four designs of mixed methods research (Creswell & Plano Clark. 2007, p. 59: Kundalaputra. 2007, p. 57-60): Triangulation Mixed Methods Design, Embedded Mixed Methods Design, Explanatory Mixed Methods Design, and Exploratory Mixed Methods Design.

Triangulation Mixed Methods Design

This is the research in which quantitative and qualitative methods are simultaneously implemented in order to analyze and combine the findings of the same question into an overall interpretation. Quantitative and qualitative data analyses are equally applied. After data interpretation, the results are compared or combined.

Triangulation mixed method design is the most well-known and most common mixed method research design that has been used as early as in 1970s (e.g., Jick, 1979). This design is used across disciplines and has been called by many names such as simultaneous triangulation (Morse, 1991), parallel study (Tashakkori & Teddlie, 1998), convergence model (Creswell, 1999), and concurrent triangulation (Creswell, Plano Clark, Gutmann, & Hanson, 2003).

Purpose

According to Morse (1991), the purpose of the triangulation mixed method design is "to obtain different but complementary data on the same topic" (Morse, 1991, p. 122) in order gain better understanding of the research problem. Quantitative and qualitative research techniques posses many strengths and have many non-overlapping weaknesses as well. The triangulation mixed method design brings together these strengths and weaknesses to directly compare and contrast quantitative findings with qualitative findings for corroboration and validation purposes (Patton, 1990). This design can also be used by researchers to prove quantitative findings with qualitative findings and to combine quantitative and qualitative results in order develop a more comprehensive understanding of the research problem.

Philosophical Foundation

In triangulation mixed method design, the researcher collects, analyze, and merge both types of data and results at the same stage. This approach can create issues regarding the philosophical assumptions on which the research is based upon. One approach to resolve this problem is to use triangulation mixed method design as an umbrella model for the research study.

Implementation Procedures

In the implementation process of triangulation mixed method design, some important steps are involved. These steps are shown in Figure 1.

In the first step, both qualitative and quantitative data about the research topic is collected in separate but concurrent stages. Same level of importance is attached to both types of data. In the second step, a separate and independent analysis of both types of data is performed using typical analytical procedures related to each data type. In the third, step the results of both types of data analysis are validated and related. In this step, the researcher can either directly compare the results of both types of data analysis or transform these two results so that the two results could be related during the additional analysis performed by the researcher. In the last and final step, the researcher interprets the results to find out whether the two results converge, diverge from each other, or relate to each other. Researchers also looks at the possibility if a better understanding about the research topic could be gained by combining the two results (Creswell & Plano Clark, 2007).

Figure 1. Triangulation mixed method design

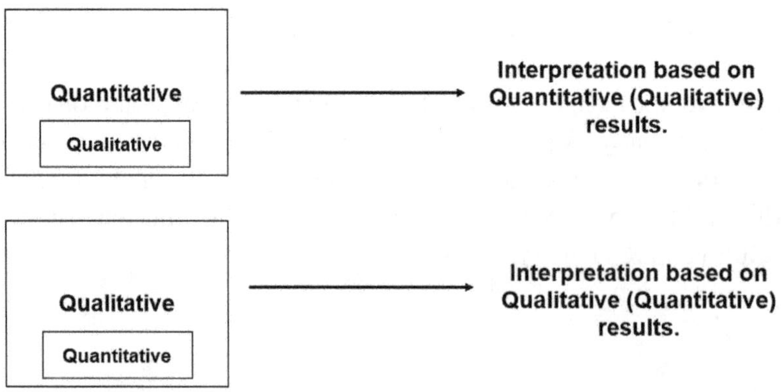

Embedded Mixed Methods Design

The embedded mixed method design combines the collection and analysis of both types of data (quantitative and qualitative) within a traditional research design. This traditional research design could either be qualitative design (such as a case study) or quantitative (such as an experiment) (Caracelli & Greene, 1997; Greene, 2007). The traditional research design is called the primary design or primary research component and data collected is called the primary data. Either quantitative or qualitative component can be secondary (or supplemental) component. The collection and analysis of the secondary data set may occur before, during, and/or after the implementation of the data collection and analysis of primary research data. Both quantitative and qualitative components are implemented simultaneously. The secondary research component is embedded in the larger design of the study in order answer different research questions. Embedded mixed method research designs are suitable in cases where a single research component may not be sufficient to answer all the research questions (Creswell, Fetters, Plano Clark, & Morales, 2009). Figure 2 below shows the embedded mixed method design.

One example of embedded mixed methods design is a research study where researcher is looking to develop peer interventions to help adolescents develop strategies for resisting pressure to drink. In the first step, the researcher may start with a qualitative design (such as focus groups) to understand when adolescents feel pressure to drink and how they resist. Using results of this qualitative data analysis, the researcher can develop an intervention and test it with a quantitative design (such as experiment) by using a students' sample.

Purpose

There are three different premises of the embedded mixed method design. First is that a single data set is not sufficient to answer all the research questions. Second, there may be different research questions to be answered. Third, each research question may need different research component to be answered. A research study may be predominantly quantitative but may include some qualitative data that is needed to answer some secondary research questions. Researchers may embed qualitative data into experiment designs for a variety of reasons such as to improve research results (e.g., Donovan, Mills, Smith, Brindle, Jacoby, & Peters, 2002), examine the process of an intervention (e.g., Victor, Ross, & Axford, 2004),

Figure 2. Embedded mixed methods design

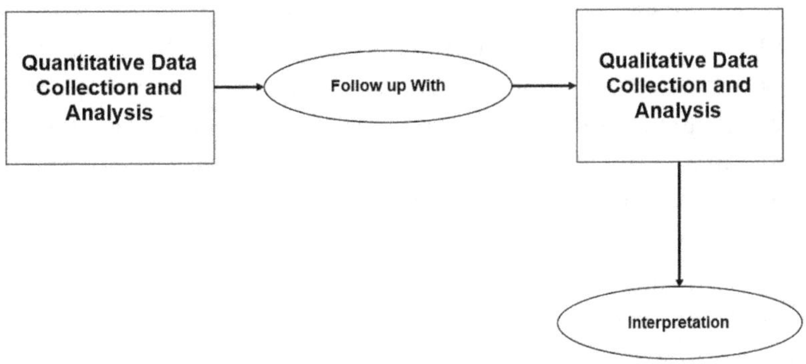

or to explain reactions to participation in an experiment (e.g., Evans & Hardy, 2002a, 2002b). Though purpose of embedding qualitative data is tied to the primary purpose of the quantitative design, the qualitative data can be used to answer some questions other than the research questions sought after by the quantitative design. This distinguishes embedded mixed method design from triangulation mixed method design. In triangulation mixed method design, the researcher uses both quantitative and qualitative methods to address a single research question.

Philosophical Foundation

The purpose of embedded mixed method design is to enhance the application of a traditional research design (quantitative or qualitative). As such, the assumptions of embedded design are guided by the primary research design. The secondary or supplemental method is subservient within the primary research methodology. For example, if the primary research design were quantitative, the assumptions of embedded mixed method design would be guided by the postpositivist approach. According to this approach, a) knowledge can best be gained through a search for regularities and causal relationships among components of the social world. b) regularities and causal relationships can best be discovered if there is a complete separation between the investigator and the subject of investigation, and c) this separation can be guaranteed through the use of the scientific method (Murphy, 2008). Similarly, if the primary research design is qualitative, the assumptions of embedded mixed method design would be guided by the constructivist approach. According to this approach, a) all knowledge is constructed and all learning is a process of that construction, b) individuals construct knowledge as part of a community but each has his/her own invisible world view that he/she believes is the same as everyone else's, and c) knowledge is content dependent so it is important to situate learning in an authentic, relevant and realistic contexts. In either case, the supplemental method is subservient within the primary research methodology.

Implementation Procedures

When thinking about the procedure of embedded mixed method design, one need to decide when supplemental data should be collected and analyzed and what could be the possible reasons to include this supplemental data in the research. The notion of supplemental component (or data) was first introduced by Sandelowski (1996). According to Sandelowski (1996), this supplemental component can occur before, during, or after the primary component of the study. This decision is guided by the purpose of the supplemental component in overall larger design of the study (Creswell, 2009). As such, the embedded mixed method design can be a one-phase or two-phase design.

In embedded mixed method design, a researcher can use either quantitative data or qualitative data as supplemental data. The most common embedded design is the one in which qualitative data is used as supplemental data. The implementation of this type of embedded mixed method design involves some important steps Guest, Namey, and Mitchell, 2012. In the first step, the researcher designs the overall quantitative design and decides the reasons for including the supplemental data i.e. qualitative data. In second step, the researcher collects and analyzes qualitative data to enhance the quantitative design. In third step, researcher analyzes the quantitative data. In the fourth and final step, the researcher interprets how the results of qualitative data analysis can help enhance the quantitative design and or understanding of the results of the quantitative design (Creswell, 2013).

Explanatory Mixed Methods Design

This is the most straightforward mixed method design that is also called sequential model (Tashakkori & Teddlie, 1998), sequential triangulation (Morse, 1991), iteration design (Greene, 2007), and qualitative follow-up approach (Morgan, 1998). It is a two-phase interactive research design that researchers used to identify cause-and-effect-relationships. The first phase is quantitative phase. In the second qualitative phase, researcher analyzes specific results of quantitative to get their in-depth explanation. It is due to this focus on explaining results that we call this method explanatory mixed method design. The two phases are distinct and involves data collection and analysis. Figure 3 provides an overview of the procedural steps used to implement a typical two-phase explanatory mixed method design.

The quantitative data collection and analysis gets priority when addressing the research questions. Once the second qualitative phase is completed, the researcher interprets the qualitative analysis results to see if they help explain the initial quantitative results. For example, a research study may seek to identify significant predictors of adolescent drug use. The researcher would first conduct a quantitative phase in which quantitative data will be collected and analyzed to identify significant predictors of adolescent drug use. This analysis may reveal some surprising findings such as some unusual association between different research variables. To gain additional insight into these results, the research would conduct a qualitative phase (such as detailed interviews with the adolescents) in order explain the unexpected results.

Purpose

Since the overall objective of explanatory mixed method design is to use qualitative phase to explain some unexpected quantitative results (Creswell et al., 2003), this method works well in situations where the researcher need qualitative data to gain additional insight into quantitative results (whether significant or insignificant) that are surprising or outliers (Morse, 2009). This research design can also be used when group results are to be compared or when the quantitative results are to be used for purposeful sampling in qualitative phase (Creswell, Plano Clark, et al., 2003; Morgan, 1998; Tashakkori & Teddlie, 1998).

Figure 3. Explanatory mixed methods design

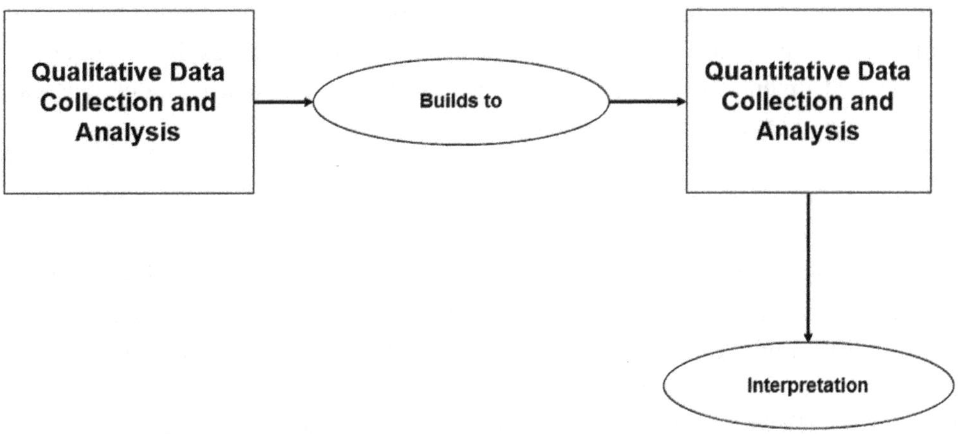

Philosophical Foundation

In this design, the philosophical assumptions of the researcher change from one phase to another. This research design places great emphasis on quantitative aspects that may encourage researchers to use a postpositivist approach in their research. However, researchers use different orientations in each phase of the explanatory mixed method design. A researcher may begin with a postpositivist approach in the quantitative phase to collect data and analyze statistical results. Since the qualitative phase requires considering multiple perspectives and in-depth analysis, a researcher may adopt constructivist approach in the qualitative phase.

Implementation Procedures

In the first step of explanatory mixed method design, the researcher collects and analyzes quantitative data. In the second step, the researcher identifies specific results of quantitative analysis that warrant additional explanation. The researcher uses these results as a basis to develop or refine the qualitative research questions and decide data collection procedures. In the third step, the researcher collects and analyzes the qualitative data. In the fourth and final step, the researcher interprets the qualitative analysis results to understand how these results could provide additional insight into the results of quantitative analysis (Ivankova, Creswell, & Stick, 2006).

Exploratory Mixed Methods Design

This research design is a two-phase sequential research design that starts with a qualitative phase. In this phase, a researcher explores a topic. The researcher use analysis of the results of qualitative phase to build the second quantitative phase to test or generalize the initial exploratory results of the qualitative phase. In many instances, the researcher uses results of qualitative phase to develop a research instrument that is used for data collection in quantitative phase. For this reason, exploratory mixed method design is also called instrument development design (Creswell, Fetters, & Ivankova, 2004) and the quantitative follow-up design (Morgan, 1998). In contrast to the explanatory mixed method design, the exploratory mixed method research design begins with and lay great emphasis on qualitative phase as shown in Figure 4. Once the quantitative phase is completed, the researcher interprets the results to see if they generalize or provide additional insight into the exploratory findings of the qualitative phase. For example, a researcher may conduct a qualitative phase to identify the possible consequences for adolescents who quit drugs. The results of the qualitative phase could provide different research variables that researcher could use to develop a research instrument. The research could then use this research instrument to assess the overall frequency of these variables in a large sample of adolescents that quit drugs.

Purpose

The qualitative phase of the exploratory mixed method design involves few individuals while the quantitative phase involves a large sample of individuals. The purpose of exploratory mixed method design is to generalize the initial findings from qualitative phase. Similar to the explanatory mixed method design, the intention of exploratory mixed method design is to help develop or inform the second, quantitative method (Greene et al., 1989). The premise of exploratory mixed method design is that such exploration

Figure 4. Exploratory mixed methods design

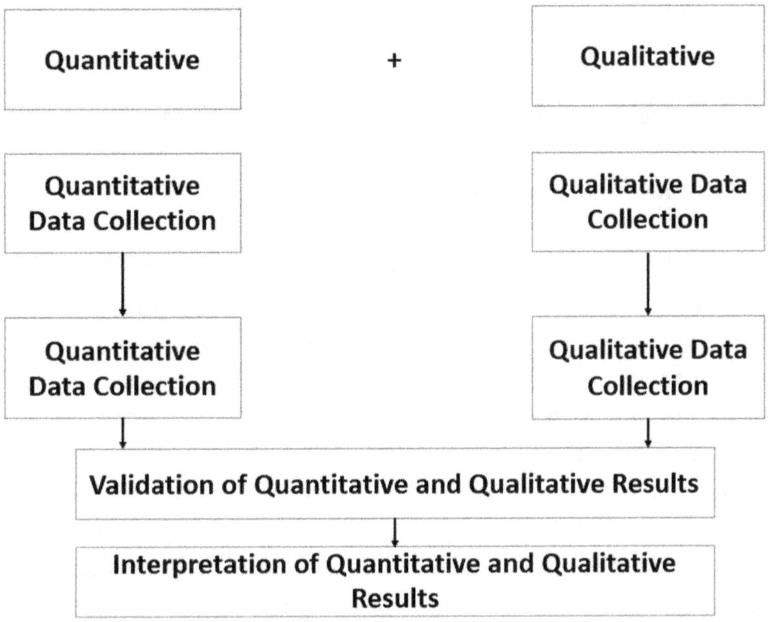

may be needed due to unavailability of research instruments, unknown research variables, and lack of guiding research frameworks. Since this design places great emphasis on qualitative phase, it is well suited for studies that explore a phenomenon (Creswell et al., 2003) or when researcher needs to develop and test a new research instrument (Creswell, 1999; Creswell et al., 2004) or to identify new important but unknown research variables. This design is also appropriate when researcher needs to qualitative results to different groups (Morse, 1991). This design is also suitable when the researcher is seeking testing of an emergent theory or exploring in depth a phenomenon and its associated dimensions (Morgan, 1998).

Philosophical Foundation

Multiple views are used in embedded mixed method research design. Exploratory mixed method design starts with a qualitative component. Qualitative component has more significant and greater priority in formulating the research problem and purpose. The qualitative phase of embedded mixed method design involves analysis of multiple perspectives and gaining deeper understanding of the topic being researched. In this phase, researchers generally adopt constructivist approach. In the quantitative phase, researcher needs to identify and measure variables. In this phase, researchers generally adopt postpositivist approach (Ivankova, Creswell, & Stick, 2006).

Implementation Procedures

There are four major steps of the exploratory mixed method design (Clark & Creswell, 2011). In the first step, the researcher collects and analyze qualitative data. In the second step, the researcher uses the results of qualitative analysis to develop the quantitative component (i.e. developing the quantitative research questions, developing research instrument, and decide data collection procedures. In the fourth and final step, the researcher interprets the results to find out whether the results of quantitative analysis generalize or provide additional insights into the findings of qualitative analysis (Kakai et al., 2014).

STRENGTHS AND CHALLENGES OF MIXED METHODS DESIGN

Triangulation Mixed Method Design

This design has a number of strengths and advantages. This design is intuitive and often the first choice of researchers new to mixed method designs. It was the design first discussed in the literature (Jick, 1979). This design is efficient and provides simultaneous data collection of both types in one phase. Each type of data can be collected and analyzed separately and independently, using their respective techniques. This design is suitable for team research where the team includes individuals having expertise in both quantitative and qualitative research techniques. While most popular, this design is also the most challenging of the major mixed method designs. Use of this design requires much effort and expertise because of the concurrent data collection and the fact that equal priority is generally attached with each data type. To address this challenge, researchers can either form a team of individuals having expertise in quantitative and qualitative research techniques or train a single researcher in both quantitative and qualitative research techniques. Researchers need to consider the consequences of having different samples and different sample sizes when merging the two data sets. Sample sizes may vary because both data types are usually collected for different purposes. To merge two very different data sets and their results can be a daunting challenge. The research study designs should be such that quantitative and qualitative data address the same concepts so that merging of data sets is facilitated. The quantitative and qualitative results may not agree. In this case, Contradictions may provide new insights into the topic. At the same, time these contradictions present additional challenges to the researchers as these differences can be difficult to resolve and may require the collection of additional data. Researchers need to decide as to what type of additional data to collect or to reanalyze. This additional data could be quantitative data, qualitative data, or both. According to Clark & Creswell (2011) researcher would need to either collect additional data or reexamine the existing data to address this challenge.

Embedded Mixed Method Design

Researchers gain many advantages by using embedded mixed method design. Researchers can save time and resources because one research component (i.e. quantitative) is given priority. By using supplemental data, researchers can improve the larger design of the research. Since different research components used in embedded mixed method design address different research questions, a team approach can work here. Each member of a team, depending on his skills and expertise, can focus his work on a particular research question. Such focus also means that the results of both research components (i.e. quantitative and qualitative) can be can be kept and published independently. This design is particularly attractive for donor agencies funding different research projects. Those donor agencies unfamiliar with mixed methods research may find this design useful because of its primary focus on a traditional design that could be either quantitative or qualitative (Teddlie & Tashakkori, 2011).

Embedded mixed method design also has many associated challenges. Besides expertise in mixed methods research, a researcher also needs in one of the two traditional designs (i.e. quantitative and qualitative). It is necessary for the researcher to specify the purposes of collecting supplemental data as part of the large design of the research study. These purposes can be classified as primary and secondary purposes. For example, a researcher may include supplemental data to shape the intervention or to follow up on results of the experiment. Researcher must also decide timing of supplemental data collection (i.e., before, during, after, or some combination). This decision should be guided by the purposes of collecting supplemental data that researcher specified earlier. Since the two research components may be used to answer different research questions, integration of results can be difficult. Contrary to triangulation mixed method design, embedded mixed method design does not merge the two different data sets to answer the same research questions (Teddlie & Tashakkori, 2011; Johnson & Onwuegbuzie, 2004).

Explanatory Mixed Method Design

Explanatory mixed method design provides a two-phase structure that is easy to implement because both phases are independent. Only one type of data is collected at a time. Due to its strong emphasis on quantitative phase, this design is often preferred choice of the researchers. The research results can be compiled and published independently. This design is one of the emergent research design in which the second phase of the research is built on the learning gained in the first phase of the research.

Despite its simplicity, the explanatory mixed method design has many associated challenges. This design is time consuming and implementing qualitative phase can take considerably more time than implementing quantitative phase. Though qualitative phase involves few participants, the collection and analysis of qualitative data can be a lengthy process. To conduct a qualitative phase, normally an approval of institutional review board is needed. This approval can be difficult to obtain. This is because the design of the second phase is dependent on the findings of the first quantitative phase. Therefore, researcher cannot specify how participants will be selected for the second phase. One strategy to deal with this issue is that the researcher can provide a tentative design of the qualitative phase to the institutional review board and inform participants of the possibility of their participation in the second phase of the research. The researcher must decide which results of the quantitative phase would be explained further. This is difficult until the quantitative phase is completed but the researcher may consider the significant results and strong predictors while planning the study. The selection of sample population for qualitative phase is another important decision. The researcher needs to come up with some criterion for this decision. For example, researcher may use groups used in comparisons during the quantitative phase. This is important to address this issue of using individuals from the same sample to provide the best explanations.

Exploratory Mixed Method Design

Exploratory mixed method design follows a two-phase structure and only one type of data is collected at a time. As such, this design shares many of the same advantages of the explanatory mixed method design. There are some advantages specific to this design as well. The description, implementation, and reporting of this design is easy because the two phases are separate. This design can be acceptable to quantitative-biased researchers because, despite a strong emphasis on qualitative phase, a quantitative phase is also included. This design is useful when the results of the qualitative phase warrant a second quantitative phase to gain deeper insight of the research topic. One of the potential products of this design could be a new research instrument.

Implementing this straight forward design can be challenging. This design can be time consuming especially if it involves developing a new research instrument. Since quantitative phase is built on the findings of the qualitative phase, the procedures of the quantitative phase can be difficult to specify when seeking approval of the institutional review board for research study. A researcher may either consider providing a tentative plan of the quantitative phase or submit a separate request of approval for both phases. To avoid bias, a researcher should consider using a small sample in qualitative phase and a large sample of different participants in quantitative phase. The researcher may wish to develop a research instrument in between phases. In this case, the researcher needs to select which data of the qualitative phase would be used to develop the instrument and how various quantitative measure would be generated. Researcher may use procedures for using qualitative themes, codes, and quotes to generate aspects of quantitative instruments. Researcher should a rigorous process to develop instrument and scale in order ensure reliability and validity of scores developed on the instrument (Andrew & Halcomb, 2011).

CRITERIA FOR USE OF MIXED METHODS DESIGNS

Researchers need to take into account a variety of factors when choosing a particular mixed method design for their research. There are number of ways the quantitative and qualitative components of the study relate to each other. A strand is a component of a study that encompasses the basic process of conducting quantitative or qualitative research: posing a question, collecting data, analyzing data, and interpreting results based on that data (Teddlie & Tashakkori, 2009). Mixed methods studies include at least one quantitative strand and one qualitative strand.

A systematic framework was suggested by Creswell (2003) for conducting mixed methods research. According to Creswell (2003), researchers need to consider four important decisions while conducting a mixed methods research. The first decision related to the implementation sequence of data collection i.e. the timing of the quantitative and qualitative strands. Implementation of data collection refers to the sequence that the researcher uses to collect both quantitative and qualitative data. Researcher can adopt either a simultaneous or parallel method of data collection or a sequential or two-phase method. In simultaneous or parallel method of data collection, a researcher gathers both the forms of data at the same time. Using this approach, researcher can compare both forms of data to develop congruent findings. In sequential or two-phase method, researcher gathers either the qualitative or the quantitative data first. The sequence of data collection relates to the objectives being sought by the researcher. When qualitative data is collected first, the intent is to first explore the problem under study and then follow up on this exploration with quantitative data. When quantitative data is gathered first, the intent is to test the variables with a large sample and then carry out a more in-depth exploration of a few cases during the qualitative data. The second decision relates to the priority attached to the method of data collection and analysis. A researcher can either assign equal priority to both quantitative and qualitative research (i.e. equivalent status design), or emphasize more on qualitative or quantitative research (i.e. dominant-less dominant or nested design). There could be various rationales for this priority such as constraints on data collection and the need to understand one form of data first before getting the second form of data.

An equivalent status design uses both the quantitative and qualitative approaches equally to understand the phenomenon under study. A nested design uses one approach dominantly with a small component of the overall study drawn from the other approach. The third decision relates to the integration stage of research finding i.e. the level of interaction between the qualitative and quantitative approaches. Greene (2007) argued that this decision is the "most salient and critical" (p. 120) for designing a mixed methods study. Integration of approaches refers to combining quantitative and qualitative approaches within a given stage of inquiry. For example, the integration of both approaches within research questions can be in the form of having both quantitative and qualitative research questions. Putting open-ended questions on a structured research instrument can provide integration of approaches in data collection stage. Integration of approaches within the data analysis phase can be achieved by transforming qualitative themes into quantitative scales. In the interpretation phase, we can achieve integration by examining both the quantitative and qualitative results. The researcher needs a clear understanding of the sequence of research process and approaches typically taken at each stage. During problem specification, data collection, data analysis, and interpretation, researcher can integrate components of both quantitative and qualitative research. According to Morse & Niehaus (2009), the stage of integration is a point within the process of research where the quantitative and qualitative strands are mixed. Typically, the researcher integrates components of both quantitative and qualitative research at the data analysis and interpretation stages after both quantitative and qualitative data have been collected. The researcher can either merge two data sets, connect from the analysis of one set of data to the collection of a second set of data, embed one form of data within a larger design or procedure, or use a theoretical framework to bind the two data sets. The purpose of the research, the ease with which integration can occur, and the researcher understands of the stages of research helps the researcher decides on which stage or stages integration should occur. The fourth decision relates to the choice of theoretical perspective. Researchers bring to their studies certain assumptions. These assumptions theoretically drive the inquiry. Informally, a theoretical perspective is based on researcher's personal history, experience, culture, gender, and class perspectives. This theoretical perspective informs researcher's point of view about the topics being researched. A formal theoretical perspective informs researcher's detailed point of view about the topic being studies. A formal theoretical perspective takes into account researcher's perspective about gender, culture, lifestyle orientation, critical theory, and class and social status. In a mixed method study, a researcher may use an explicit or implicit theoretical perspective.

A credible and robust mixed methods design addresses the decisions of level of integration, priority, timing, and mixing. The following sub-sections present criteria to be used while selection each of the four mixed methods designs.

Triangulation Mixed Method Design

Triangulation mixed method design requires matching the design to the study's purpose and best suited for group research work or when an individual researcher is able to collect limited quantitative and qualitative data. In addition to that, there exist some additional considerations for use of this design. This design is a good choice in case where there is limited time for data collection or collection of both data types is needed in one visit to the field. In another situation, the researcher feels that collection and analysis of both types of data would be of equal value in understanding the research problem. Triangulation mixed method design is also suitable when researcher posses expertise in both quantitative and qualitative methods of research or can manage extensive data collection and analysis.

Embedded Mixed Method Design

This design is an appropriate choice for researchers investigating different questions requiring different types of data to enhance the application of a quantitative or qualitative design in order address the basic objective of the study. This design is also a good choice when a researcher posses expertise to rigorously implement the planned quantitative or qualitative design. Another case is where the researcher posses little experience in dealing with the supplementary method or does not have adequate resources to attach equal priority to both types of data. Embedded mixed method design would also be suitable in case emergent issues related to the primary research design (quantitative or qualitative) are discovered. In these cases, embedded mixed method design can help gain insight into these issues by using a secondary data set.

Explanatory Mixed Method Design

This design is most useful when the researcher wants to assess trends and relationships with quantitative data but also be able to explain the mechanism or reasons behind the resultant trends. This design can also be useful in cases where the researcher or research question itself is more quantitatively oriented or the researcher knows the important research constructs and have quantitative instruments to measure these constructs. This design could also work well when the researcher can perform a second round of qualitative data collection or can conduct research in multiple phases. This design would also be a good choice for researcher having limited resources and can only afford to collect and analyze or data at a time or the researcher develops new questions based on quantitative results and need qualitative data to answer these questions.

Exploratory Mixed Method Design

The exploratory mixed method design is most useful in studies where the qualitative exploratory results are to be generalized, assessed, or tested for their applicability to a sample and a population. This design is also a good choice when researcher needs to develop an instrument when one is not available (first explore, then develop instrument) or when researcher needs to develop a classification or typology for testing or the researcher wants to identify the most important variables to study quantitatively when these variable are not known. In case the researcher identifies new emergent research questions based on qualitative results that cannot be answered with qualitative data, exploratory mixed method design would be a good choice.

EXAMPLES OF MIXED METHODS DESIGN

The following sub-sections present examples of various studies that use the four designs of mixed methods research.

Triangulation Mixed Methods Design

There exist many examples of triangulation mixed method design such as Pagano, Hirsch, Deutsch, and McAdams (2003), Webb, Sweet, and Pretty (2002), Elliott & Williams (2002), McVea et al., (1996),

Table 1. Published mixed method studies involving triangulation mixed method design

Study Characteristic	Pagano, Hirsch, Deutsch, and McAdams (2003)	Webb, Sweet, and Pretty (2002)	Elliott & Williams (2002)	McVea et al., (1996)	Feldon and Kafai (2008)	Jenkins, J.E. (2001)	Baskerville, Hogg, & Lemelin (2001)
Content area	Parental socialization practices and the values transmitted to school-aged and young adult off-spring, focusing on race and gender issues involved in parental teachings	An evaluation of forensic odontologists with known occupational experience of mass casualty incidents to gain insight into the psychological and emotional impact of conducting work of this nature.	An evaluation of a Fire Brigade counselling service examined the viewpoints of all three stakeholders (clients, counsellors and organization).	An evaluation of the effectiveness of "Put Prevention into Practice" program in family physician private practice settings	To analyze users' avatar-related activities in a virtual world	An examination of high schools students perception of drug resistance difficulties when offered beer, marijuana, and hard drugs.	An evaluation study of intervention practices for implementing prevention guidelines by family physicians.
Rationale for mixing	Multiple data collection strategies were used to ensure comprehensiveness and triangulation of results	The multi-method approach and the depth and width of the evaluation detail produced provides a type of outcome beyond the reach of traditional research methods	To achieve appropriateness-to-context and utility and to allow research to be fused with clinical practice	To ensure comprehensiveness and triangulation of results	To address the methodological and analytical challenges of virtual world research	o ensure comprehensiveness and triangulation of results	To determine intervention quality and to attain a complete understanding
Forms of data collection							
Quantitative	University survey Parenting scales	A Scaled Questionnaire	Encounter questionnaire	Office environment and clinical encounters checklists Chart audits	Surveys	Surveys	Weekly and monthly activity sheets Closed-ended questions during telephone interview Mailed questionnaire
Qualitative	Face-to-face open-ended Interviews with Parents Telephone interviews with Parents	An open-ended questionnaire	Semi-structured interviews	Participant observation Key informant interviews of office staff Semi structured interviews with physicians and office managers	Interviews Online ethnographic observations	Focus Groups	Monthly narrative reports Telephone interviews Interviews at end of intervention

continued on following page

Table 1. Continued

Study Characteristic	Pagano, Hirsch, Deutsch, and McAdams (2003)	Webb, Sweet, and Pretty (2002)	Elliott & Williams (2002)	McVea et al., (1996)	Feldon and Kafai (2008)	Jenkins, J.E. (2001)	Baskerville, Hogg, & Lemelin (2001)
Analytical procedure							
Quantitative	Correlations Logistical regression	Descriptive Statistics	Descriptive statistics	Descriptive statistics	Cluster analysis Descriptive Statistics		Descriptive and content analysis of weekly activity sheets Content analysis of interviews Chi-square analyses
Qualitative	Categories and themes	Content analysis of qualitative data collected from questionnaire	Thematic analysis of Interview transcripts	Categories and themes Descriptive case studies	Content analysis of interviews	Content analysis	Description from monthly narrative reports
Characteristics of design							
Priority	Quantitative	Quantitative	Qualitative	Qualitative		Quantitative	Quantitative
Implementation	Sequential Qualitative → Quantitative	Sequential Quantitative → Qualitative	Sequential Qualitative → Quantitative	Concurrent Qualitative + quantitative	Sequential Quantitative → Qualitative	Sequential Quantitative → Qualitative	Concurrent Quantitative + qualitative
Integration	From data analysis to data collection	From data analysis to data collection	From data analysis to data collection	Analysis and report of results	From data analysis to data collection	From data analysis to data collection	Analysis and report of results
Integrating the quantitative and qualitative data enhanced the interpretation of the interpretation of the results obtained using a single approach (i.e. quantitative or qualitative)	Yes with few exceptions.	Yes	Yes	Yes	Yes with few exceptions.	Yes	Yes

Table 2. Published mixed method studies involving embedded mixed method design

Study Characteristic	Hilton, Budgen, Molzahn, and Attridge (2001)	Skinner, Matthews, and Burton (2005)	Rogers, A., Day, J., Randall, F., & Bentall, R.P. (2003)	Donovan, J., Mills, N., Smith, M., Brindle, L., Jacoby, A., Peters, T., et al. (2002)	Lin, X., & Lehman, J.D. (1999)	Victor, C.R., Ross, F., & Axford, J. (2004)
Content area	To measure the impact of a Canada-based nursing Centre and nurses' work from the perspectives of the clients	"to map the experiences of families meeting the needs of their children with disabilities"	"to illuminate patients' understanding of the nature and purpose and outcomes of a trial designed to improve the management of neuroleptic medication"	To investigate the process of recruitment to randomized trials	To study instruction on control of variables for novel problems	explores the patients' perspective on the meaning and significance of living with arthritis,
Rationale for mixing	"To develop and refine quantitative instruments for assessment of health outcomes" and… "examination of the qualitative and quantitative findings added a further dimensions of rigor"		"to illuminate and add to the quantitative outcomes"	"to help in the interpretation of quantitative results or understanding of trials."	"to illuminate and add to the quantitative outcomes"	"to provide insight into the patients' perspective on the causes and impact of knee OA, individual goals desired from treatment and the quality of care."
Forms of data collection						
Quantitative	Questionnaires	Questionnaires	Questionnaires	Trials	Data collected from student problem-solving activities	Trials
Qualitative	Interviews	Interviews	Interviews	Semi-structured Interviews	Samples of students' responses to prompts in the computer-based environment, Students' experimental designs while using the computer environment Interviews	Baseline Interviews Patient Diaries Group discussion
Analytical procedure						
Quantitative	Descriptive and inferential analysis		Descriptive and inferential analysis	Descriptive and inferential analysis	Analysis of variance	Descriptive and inferential analysis

continued on following page

Table 2. Continued

Study Characteristic	**Hilton, Budgen, Molzahn, and Attridge (2001)**	**Skinner, Matthews, and Burton (2005)**	**Rogers, A., Day, J., Randall, F., & Bentall, R.P. (2003)**	**Donovan, J., Mills, N., Smith, M., Brindle, L., Jacoby, A., Peters, T., et al. (2002)**	**Lin, X., & Lehman, J.D. (1999)**	**Victor, C.R., Ross, F., & Axford, J. (2004)**
Qualitative	Content Analysis		Thematic Analysis	Thematic Analysis	Response Analysis	Thematic Analysis
Characteristics of design						
Priority	Qualitative	Quantitative	Quantitative	Qualitative	Quantitative	Quantitative
Implementation	Sequential Qualitative → Quantitative	Sequential Quantitative → Qualitative	Sequential Quantitative → Qualitative	Sequential Quantitative → Qualitative	Sequential Quantitative → Qualitative	Sequential Quantitative → Qualitative
Integration	From data analysis to data collection	From data analysis to data collection	From data analysis to data collection	From data analysis to data collection	From data analysis to data collection	From data analysis to data collection
Integrating the quantitative and qualitative data enhanced the interpretation of the interpretation of the results obtained using a single approach (i.e. quantitative or qualitative)?	Yes	Yes	Yes with few exceptions.	Yes	Yes	Yes

Table 3. Published mixed method studies involving explanatory mixed method design

Study Characteristic	Igo, Riccomini, Bruning, and Pope (2006)	May and Etkina (2002),	Myers, K.K., & Oetzel, J.G. (2003)	Kutner, J.S., Steiner, J.F., Cotbett, K.K., Jahnigen, D.W., & Barton,P.L. (1999)	Weitzman, P.F., & Levkoff, S.E. (2000)
Content area	"Studying the effect of different modes of note taking on test performance for middle school students with learning disabilities"	"to identify physics students with consistently high and low conceptual learning gains."	to create and validate a measure of organizational assimilation index	A communication study of the information needs of terminally ill patients receiving palliative care from physicians.	A cross-cultural study of family caregiving for demented-affected elders.
Rationale for mixing	"quantitative investigation alone yielded puzzling results"	To enhance the interpretation of the results of quantitative analysis	To enhance the interpretation of the results of qualitative analysis	To identify and describe key themes and to develop a semi structured instrument	To ensure the cultural appropriateness of quantitative measures and to confirm causal relationships to which quantitative data pointed.
Forms of data collection					
Quantitative	Questionnaire	Survey	Survey	Instrument (multiple choice, open-ended) based on qualitative interviews SF-36 Health Survey	Survey
Qualitative	Interviews Documents from the students	Students weekly report	Interviews	Face-to-face open-ended interviews	Interviews
Analytical procedure					
Quantitative	Descriptive and Inferential statistics	Descriptive and inferential analysis	Correlations	Descriptive and inferential analysis	Descriptive and inferential analysis
Qualitative	Phenomenological technique adapted from Groenewald (2004) and Moustakas (1994).	Comparison study of two groups of students' perceptions of learning	Thematic analysis	Template analysis involving coding and common issues across categories of interview data and open-ended instrument data.	Thematic Analysis
Characteristics of design					
Priority	Quantitative	Quantitative	Qualitative	Equal	Quantitative
Implementation	Sequential Quantitative → Qualitative	Sequential Quantitative → Qualitative	Sequential Qualitative → Quantitative	Sequential Qualitative→ Quantitative	Sequential Quantitative → Qualitative
Integration	From data analysis to data collection	From data analysis to data collection	From data analysis to data collection	From data analysis to data collection	From data analysis to data collection
Integrating the quantitative and qualitative data enhanced the interpretation of the interpretation of the results obtained using a single approach (i.e. quantitative or qualitative)?	Yes with few exceptions.	Yes	Yes	Yes	Yes

Feldon and Kafai (2008), Jenkins, J.E. (2001), and Baskerville, Hogg, & Lemelin (2001). Table 1 presents an analysis of these studies from a mixed method perspective.

Embedded Mixed Methods Design

There exist many examples of embedded mixed method design such as Hilton, Budgen, Molzahn, and Attridge (2001), Skinner, Matthews, and Burton (2005), Rogers, A., Day, J., Randall, F., & Bentall, R.P. (2003), Donovan, J., Mills, N., Smith, M., Brindle, L., Jacoby, A., Peters, T., et al. (2002), Lin, X., & Lehman, J.D. (1999), and Victor, C.R., Ross, F., & Axford, J. (2004). Table 2 presents an analysis of these studies from a mixed method perspective.

Explanatory Mixed Methods Design

There exist many examples of explanatory mixed method design such as Igo, Riccomini, Bruning, and Pope (2006), May and Etkina (2002), Myers, K.K., & Oetzel, J.G. (2003), Kutner, J.S., Steiner, J.F., Cotbett, K.K., Jahnigen, D.W., & Barton,P.L. (1999), Mak-L., & Marshall, S.K. (2004), and Weitzman, P.F., & Levkoff, S.E. (2000). Table 3 presents an analysis of these studies from a mixed method perspective.

Exploratory Mixed Methods Design

There exist many examples of exploratory mixed method design such as Goldenberg, Gallimore, and Reese (2005) and Mak and Marshall (2004). Table 4 presents an analysis of these studies from a mixed method perspective.

FUTURE RESEARCH AREAS

The mixed method research is well-established and gaining widespread acceptance. In future, research studies conducted in many areas of research would need to utilize a mixed method design in order improve the proficiency and reliability of research results. This chapter contributes to the understanding of mixed methods research technique by providing a survey of common mixed methods designs, their purpose, philosophical assumptions, procedures, strengths, challenges, criterion of selection, and their use cases. Future research would be needed to analyze the value addition provided by the mixed method research studies. It would also be interesting to analyze how the use of mixed methods designs can improve methodological aspects of research studies (such as construct validity). In addition, future research can also analyze the use and application of mixed methods research in various organizational and management fields. This research would be helpful in expanding upon the contribution made in this chapter.

CONCLUSION

Mixed method research provides several research designs. These designs serve as frameworks for collecting, analyzing, mixing, interpreting, and reporting quantitative and qualitative data. These designs can be very helpful in addressing specific types of research purposes. Researchers willing to use a mixed

Table 4. Published mixed method studies involving exploratory mixed method design

Study Characteristic	Goldenberg, Gallimore, and Reese (2005)	Mak and Marshall (2004)
Content area	To identify new variables and hypotheses about predictors of family literacy practices.	To explore young adults' perceptions about the significance of the self to others in romantic relationships
Rationale for mixing	To test the qualitatively identified variables and relationships.	To test hypotheses based on the theoretical model of the formation and maintenance of perceived mattering
Forms of data collection		
Quantitative	Survey questionnaires Standardized test results Teacher and parent ratings	Open-ended survey
Qualitative	Open-ended interviews Participant and ethnographic observations	Interviews
Analytical procedure		
Quantitative	Path analysis	Correlations
Qualitative	Case study analysis	Content analysis
Characteristics of design		
Priority	Qualitative	Qualitative
Implementation	Sequential Qualitative → Quantitative	Sequential Qualitative → Quantitative
Integration	From data analysis to data collection	From data analysis to data collection
Integrating the quantitative and qualitative data enhanced the interpretation of the interpretation of the results obtained using a single approach (i.e. quantitative or qualitative)?	Yes	Yes with few exceptions.

method design in their research need to make few key decisions in order select the design that best meet their needs. The researcher decides whether the two research components would be independent or interactive, the relative priority of each research component, the timing of research components, and the mixing of research components. These decisions, the logic behind research study, and practical considerations form the foundation on which the researcher selects a particular mixed method design for his study. Researchers can choose among four major mixed methods designs: triangulation, explanatory, exploratory, and embedded. Each design serve different purposes and different philosophical orientations can be involved in different phases of research. Each design includes a specific set of procedures that depicts the logic of the design. Researchers should carefully consider the challenges associated with their design choice and plan strategies for addressing these challenges.

REFERENCES

Andrew, S., & Halcomb, E. (Eds.). (2009). *Mixed methods research for nursing and the health sciences.* Wiley-Blackwell Pub. doi:10.1002/9781444316490

Andrew, S., & Halcomb, E. J. (2011). Mixed Methods Research. *Navigating the Maze of Research: Enhancing Nursing and Midwifery Practice*, 147.

Baskerville, N. B., Hogg, W., & Lemelin, J. (2001). Process evaluation of a tailored multifaceted approach to changing family physician practice patterns improving preventive care. *The Journal of Family Practice*, *50*(3), W242–9. PMID:11252222

Bouchard, T. J. Jr. (1976). Unobtrusive measures: An inventory of uses. *Sociological Methods & Research*, *4*(3), 267–300. doi:10.1177/004912417600400301

Brannen, J. (2005). Mixing methods: The entry of qualitative and quantitative approaches into the research process. *International Journal of Social Research Methodology*, *8*(3), 173–184. doi:10.1080/13645570500154642

Cameron, R. (2009). A sequential mixed model research design: Design, analytical and display issues. *International Journal of Multiple Research Approaches*, *3*(2), 140–152. doi:10.5172/mra.3.2.140

Campbell, D. T., & Fiske, D. W. (1959). Convergent and discriminant validation by the multitrait-multimethod matrix. *Psychological Bulletin*, *56*(2), 81–105. doi:10.1037/h0046016 PMID:13634291

Caracelli, V. J., & Greene, J. C. (1993). Data analysis strategies for mixed-method evaluation designs. *Educational Evaluation and Policy Analysis*, *15*(2), 195–207. doi:10.3102/01623737015002195

Carmen, J. M. (1990). Consumer perceptions of service quality: An assessment of the SERVQUAL dimensions. *Journal of Retailing*, *66*(1), 33–55.

Clark, V. L. P., & Creswell, J. W. (2011). *Designing and conducting mixed methods research*. Thousand Oaks, CA: Sage.

Collins, K., Onwuegbuzie, A., & Jiao, Q. (2007). A mixed methods investigation of mixed methods sampling designs in social and health science research. *Journal of Mixed Methods Research*, *1*(3), 267–294. doi:10.1177/1558689807299526

Collins, K. M. T., Onwuegbuzie, A. J., & Sutton, I. L. (2006). A model incorporating the rationale and purpose for conducting mixed methods research in special education and beyond. *Learning Disabilities (Weston, Mass.)*, *4*, 67–100.

Creswell, J. (2003). *Research design: Qualitative, quantitative and mixed methods approaches* (2nd ed.). Thousand Oaks, CA: SAGE Publications.

Creswell, J. W. (2009). *Research Design: Qualitative, Quantitative, and Mixed Methods Approaches* (3rd ed.). London: Sage Publications.

Creswell, J. W. (2013). Research design: Qualitative, quantitative, and mixed methods approaches. *Sage (Atlanta, Ga.)*.

Creswell, J. W. (2014). *A concise introduction to mixed methods research*. SAGE Publications.

Creswell, J. W., Klassen, A. C., Plano Clark, V. L., & Smith, K. C. (2011). *Best practices for mixed methods research in the health sciences*. Bethesda, MD: National Institutes of Health. doi:10.1037/e566732013-001

Creswell, J. W., & Plano Clark, V. L. (2007). *Designing and Conducting Mixed Methods Research. United State of America*. SAGE.

Creswell, J. W., & Plano Clark, V. L. (2007). Choosing a mixed methods design. *Designing and Conducting Mixed Methods Research*, 58–88.

Creswell, J. W., Plano Clark, V. L., Gutmann, M., & Hanson, W. (2003). Advanced mixed methods research designs. In A. Tashakkori & C. Teddlie (Eds.), *Handbook of mixed methods in social & behavioral research* (pp. 209–240). Thousand Oaks, CA: Sage.

Denzin, N. K. (1978). *The research act: A theoretical introduction to sociological methods*. New York: Praeger.

Denzin, N. K. (2012). Triangulation 2.0. *Journal of Mixed Methods Research*, *6*(2), 80–88. doi:10.1177/1558689812437186

Donovan, J., Mills, N., Smith, M., Brindle, L., Jacoby, A., & Peters, T. et al.. (2002). Improving design and conduct of randomised trials by embedding them in qualitative research: Protetc (prostate testing for cancer and treatment) study. *British Medical Journal*, *325*, 766–769. doi:10.1136/bmj.325.7367.766 PMID:12364308

Donovan, J., Mills, N., Smith, M., Brindle, L., Jacoby, A., Peters, T., & Hamdy, F. et al. (2002). Quality improvement report: Improving design and conduct of randomised trials by embedding them in qualitative research: ProtecT (prostate testing for cancer and treatment) study. *BMJ: British Medical Journal*, *325*(7367), 766–770. doi:10.1136/bmj.325.7367.766 PMID:12364308

Elliott, M. S., & Williams, D. I. (2002). A qualitative evaluation of an employee counselling service from the perspective of client, counsellor and organization. *Counselling Psychology Quarterly*, *15*(2), 201–208. doi:10.1080/09515070210128991

Feldon, D. F., & Kafai, Y. B. (2008). Mixed methods for mixed reality: Understanding users' avatar activities in virtual worlds. *Educational Technology Research and Development*, *56*(5-6), 575–593. doi:10.1007/s11423-007-9081-2

Goldenberg, C., Gallimore, R., & Reese, L. (2005). Using mixed methods to explore Latino children's literacy development. *Discovering Successful Pathways in Children's Development: Mixed Methods in the Study of Childhood and Family Life*, 21–46.

Greene, J. C. (2007). *Mixed methods in social inquiry* (Vol. 9). John Wiley & Sons.

Greene, J. C., & Caracelli, V. J. (1997). Defining and describing the paradigm issue in mixed-method evaluation. *New Directions for Evaluation*, *1997*(74), 5–17. doi:10.1002/ev.1068

Greene, J. C., Caracelli, V. J., & Graham, W. F. (1989). Toward a conceptual framework for mixed method evaluation designs. *Educational Evaluation and Policy Analysis*, *11*(3), 255–274. doi:10.3102/01623737011003255

Groenewald, T. (2004). A phenomenological research design illustrated. *International Journal of Qualitative Methods*, *3*(1). Retrieved, from http://www.ualberta.ca/~iiqm/backissues/3_1/pdf/groenewald.pdf

Guest, G., Namey, E. E., & Mitchell, M. L. (2012). Collecting qualitative data: A field manual for applied research. *Sage (Atlanta, Ga.).*

Hilton, B., Budgen, C., Molzahn, A. E., & Attridge, C. B. (2001). Developing and testing instruments to measure client outcomes at the Comox Valley Nursing Center. *Public Health Nursing (Boston, Mass.), 18*(5), 327–339. doi:10.1046/j.1525-1446.2001.00327.x PMID:11559416

Igo, L. B., Bruning, R. A., & Riccomini, P. J. (2009). Should Middle School Students with Learning Problems Copy and Paste Notes from the Internet? Mixed-Methods Evidence of Study Barriers. *RMLE Online: Research in Middle Level Education, 33*(2), n2. doi:10.1080/19404476.2009.11462065

Ivankova, N. V., Creswell, J. W., & Stick, S. L. (2006). Using mixed-methods sequential explanatory design: From theory to practice. *Field Methods, 18*(1), 3–20. doi:10.1177/1525822X05282260

Jenkins, J. E. (2001). Rural adolescent perceptions of alcohol and other drug resistance. *Child Study Journal, 31*(4), 211–224.

Jick, T. D. (1979). Mixing qualitative and quantitative methods: Triangulation in action. *Administrative Science Quarterly, 24*(4), 602–611. doi:10.2307/2392366

Johnson, R. B., & Onwuegbuzie, A. J. (2004). Mixed methods research: A research paradigm whose time has come. *Educational Researcher, 33*(7), 14–26. doi:10.3102/0013189X033007014

Johnson, R. B., & Onwuegbuzie, A. J. (2004). Mixed methods research: A research paradigm whose time has come. *Educational Researcher, 33*(7), 14–26. doi:10.3102/0013189X033007014

Klassen, A. C., Creswell, J., Clark, V. L. P., Smith, K. C., & Meissner, H. I. (2012). Best practices in mixed methods for quality of life research. *Quality of Life Research: An International Journal of Quality of Life Aspects of Treatment, Care and Rehabilitation, 21*(3), 377–380. doi:10.1007/s11136-012-0122-x PMID:22311251

Kutner, J. S., Steiner, J. F., Corbett, K. K., Jahnigen, D. W., & Barton, P. L. (1999). Information needs in terminal illness. *Social Science & Medicine, 48*(10), 1341–1352. doi:10.1016/S0277-9536(98)00453-5 PMID:10369435

Lin, X., & Lehman, J. D. (1999). Supporting learning of variable control in a computer-based biology environment: Effects of prompting college students to reflect on their own thinking. *Journal of Research in Science Teaching, 36*(7), 837–858. doi:10.1002/(SICI)1098-2736(199909)36:7<837::AID-TEA6>3.0.CO;2-U

Mak, L., & Marshall, S. K. (2004). Perceived mattering in young adults' romantic relationships. *Journal of Social and Personal Relationships, 21*(4), 469–486. doi:10.1177/0265407504044842

May, D. B., & Etkina, E. (2002). College physics students' epistemological self-reflection and its relationship to conceptual learning. *American Journal of Physics, 70*(12), 1249–1258. doi:10.1119/1.1503377

McVea, K., Crabtree, B. F., Medder, J. D., Susman, J. L., Lukas, L., McIlvain, H. E., & Hawver, M. et al. (1996). An ounce of prevention? Evaluation of the'Put Prevention into Practice'program. *The Journal of Family Practice, 43*(4), 361–369. PMID:8874371

Mertens, D. M. (2014). *Research and evaluation in education and psychology: Integrating diversity with quantitative, qualitative, and mixed methods*. Sage Publications.

Morse, J. (1991). Approaches to qualitative-quantitative methodological triangulation. *Nursing Research*, *40*(2), 120–123. doi:10.1097/00006199-199103000-00014 PMID:2003072

Morse, J. M., & Niehaus, L. (2009). *Mixed methods design: Principles and procedures*. Walnut Creek, CA: Left Coast Press.

Moustakas, C. (1994). *Phenomenological research methods*. Thousand Oaks, CA: Sage.

Murphy, R. (2008, September 14). *Post-Positivist Perspectives on Theory Development*. Retrieved from https://davidlobato.wordpress.com/2008/09/14/chapter-3-post-positivist-perspectives-on-theory-development/

Myers, K. K., & Oetzel, J. G. (2003). Exploring the dimensions of organizational assimilation: Creating and validating a measure. *Communication Quarterly*, *51*(4), 438–457. doi:10.1080/01463370309370166

Pagano, M. E., Hirsch, B. J., Deutsch, N. L., & McAdams, D. P. (2003). The transmission of values to school-age and young adult offspring: Race and gender differences in parenting. *Journal of Feminist Family Therapy*, *14*(3-4), 13–36. doi:10.1300/J086v14n03_02 PMID:22323854

Rogers, A., Day, J., Randall, F., & Bentall, R. P. (2003). Patients' understanding and participation in a trial designed to improve the management of anti-psychotic medication. *Social Psychiatry and Psychiatric Epidemiology*, *38*(12), 720–727. doi:10.1007/s00127-003-0693-5 PMID:14689177

Rossman, G. B., & Wilson, B. L. (1985). Numbers and words: Combining quantitative and qualitative methods in a single large-scale evaluation study. *Evaluation Review*, *9*(5), 627–643. doi:10.1177/0193841X8500900505

Skinner, D., Matthews, S., & Burton, L. (2005). Combining ethnography and GIS technology to examine constructions of developmental opportunities in contexts of poverty and disability. *Discovering Successful Pathways in Children's Development: Mixed Methods in the Study of Childhood and Family Life*, 223–239.

Tashakkori, A., & Teddlie, C. (1998). *Mixed Methodology: Combining Qualitative and Quantitative Approaches in Applied Social Research methods Series* (Vol. 46). Thousand Oaks, CA: SAGE.

Teddlie, C., & Tashakkori, A. (2011). Mixed methods research. The Sage Handbook of Qualitative Research, 285.

Venkatesh, V., Brown, S. A., & Bala, H. (2013). Bridging the qualitative-quantitative divide: Guidelines for conducting mixed methods research in information systems. *Management Information Systems Quarterly*, *37*(1), 21–54.

Victor, C. R., Ross, F., & Axford, J. (2004). Capturing lay perspectives in a randomized control trial of a health promotion intervention for people with osteoarthritis of the knee. *Journal of Evaluation in Clinical Practice*, *10*(1), 63–70. doi:10.1111/j.1365-2753.2003.00395.x PMID:14731152

Webb, D. A., Sweet, D., & Pretty, A. (2002). The emotional and psychological impact of mass casualty incidents on forensic odontologists. *Journal of Forensic Sciences*, *47*(3), 539–541. PMID:12051333

Webb, E. J., Campbell, D. T., Schwartz, R. D., & Sechrest, L. (1966). *Unobtrusive measures*. Chicago: Rand McNally.

Weitzman, P. F., & Levkoff, S. E. (2000). Combining qualitative and quantitative methods in health research with minority elders: Lessons from a study of dementia caregiving. *Field Methods*, *12*(3), 195–208. doi:10.1177/1525822X0001200302

ADDITIONAL READING

Bazeley, P. (2009). Editorial: Integrating data analyses in mixed methods research. *Journal of Mixed Methods Research*, *3*(3), 203–207. doi:10.1177/1558689809334443

Bergman, M. M. (Ed.). (2008). *Advances in mixed methods research: Theories and applications*. Sage.

Bryman, A. (1992). Quantitative and qualitative research: further reflections on their integration. *Mixing methods: Qualitative and quantitative research*, 57-78.

Bryman, A. (2006). Integrating quantitative and qualitative research: How is it done? *Qualitative Research*, *6*(1), 97–113. doi:10.1177/1468794106058877

Bryman, A. (2007). Barriers to integrating quantitative and qualitative research. *Journal of Mixed Methods Research*, *1*(1), 8–22. doi:10.1177/2345678906290531

Bryman, A., & Bell, E. (2011). *Business Research Methods 3e*. Oxford university press.

Hall, B., & Howard, K. (2008). A synergistic approach: Conducting mixed methods research with typological and systemic design considerations. *Journal of Mixed Methods Research*, *2*(3), 248–269. doi:10.1177/1558689808314622

Maxwell, J. A., & Loomis, D. M. (2003). Mixed methods design: An alternative approach. In A. Tashakkori & C. Teddlie (Eds.), *Handbook of mixed methods in social & behavioral research* (pp. 241–271). Thousand Oaks, CA: Sage.

Mertens, D. M. (2003). Mixed methods and the politics of human research: The transformative-emancipatory perspective. In A. Tashakkori & C. Teddlie (Eds.), *Handbook of mixed methods in social & behavioral research* (pp. 241–271). Thousand Oaks, CA: Sage.

Tashakkori, A., & Creswell, J. W. (2008). Envisioning the Future Stewards of the social – Behavioral Research Enterprise. *Journal of Mixed Methods Research*, *2*(4), 291–295. doi:10.1177/1558689808322946

Teddlie, C., & Tashakkori, A. (2009). *Foundations of mixed methods research*. Thousand Oaks, CA: Sage.

KEY TERMS AND DEFINITIONS

Embedded Mixed Methods Design: An embedded mixed method design is a design in which one data set provides a supportive, secondary role in a study based primarily on the other data set.

Explanatory Mixed Method Design: It is a mixed methods research design in which data is collected sequentially, with a purposeful sampling of participants for the qualitative component of the study.

Exploratory Mixed Methods Design: It is a mixed methods design in which rich qualitative data is explored so that new research questions emerge that can be answered through quantitative research.

Purposive Sampling: It is a type of non-probability sampling technique in selection of participants is based on some characteristic.

Qualitative Research: Qualitative research is a method of inquiry employed in exploring issues, understanding phenomena, and answering questions by analyzing and making sense of unstructured data.

Quantitative Research: Quantitative research is a more logical and data-led approach of empirical investigation of observable phenomena via statistical, mathematical or computational techniques.

Triangulation Mixed Method Design: It is a mixed method design in which the quantitative and qualitative data are collected at the same time, and emphasis is placed equally on both types of data.

Section 3
Sampling in Research

This chapter explains the role of sampling in research. Sampling plays an important role in any research study and careful consideration needs to be placed on who to include as participants as part of the design process. Researchers need to determine sample sizes for the quantitative and qualitative data and whether or not to include the same participants for both strands of the study. In addition, decisions around random sampling or purposeful sampling must be considered.

Chapter 6
Sampling in Research

Never Mujere
University of Zimbabwe, Zimbabwe

ABSTRACT

Research is aimed at the discovery and interpretation of facts, revision of accepted theories or laws in the light of new facts or practical application of such new or revised theories or laws. A sample provides needed information about the population quickly. However, there is no guarantee that any sample will be precisely representative of the population from which it comes. It is cheaper to observe a part rather than the whole. This chapter is a discussion on sampling in research and it is mainly designed to equip researchers with knowledge of the general issues to consider when sampling. The purpose of sampling in research, dangers of sampling and how to minimize them, types of sampling and guides for deciding the sample size are discussed. For a clear flow of ideas, a few definitions of the terms used are given. They highlight the types and methods of sampling, sampling errors and discusses techniques of sample size determination. Different types of sampling technique, how to carry them out, and their advantages and disadvantages are also introduced.

INTRODUCTION

Scientific research is important to search or investigate exhaustively (Berinstein, 2003). However, populations about which inferences must be made are maybe quite large, costly or inaccessible to reach. This makes it physically impossible to conduct a census. In such cases, selecting a representative sample may be the only way to get the information required about the population. Researchers may choose from a variety of sampling methods (Salant & Dillman, (1994). Choice of the sampling method to be used depends on research goals and whether or not the researcher wants to generalize the findings from the sample. It is vital to be aware of possible errors due to the sample method chosen so that the study is regarded as valid. This is a literature review of what sampling is, how to create a sample, and highlights the advantages and disadvantages or limitations of the sampling techniques.

DOI: 10.4018/978-1-5225-0007-0.ch006

BACKGROUND

To understand population characteristics, it is important to select a sample. Sampling is an important component of any piece of research because of the significant impact that it can have on the quality of research findings. The main reasons for sampling to obtain a sample rather than a complete enumeration (a census) of the population are; economy, timeliness, the large size of many populations, inaccessibility of some of the population and destructiveness of the observation accuracy. To draw conclusions about populations from samples, we use inferential statistics which enables us to determine a population`s characteristics by directly observing only a portion or sample of the population. Taking a sample requires fewer resources than a census.

When the researcher is interested in the entire population (i.e signifies the units that we are interested in studying), it is vital to take a census. It is often impractical and sometimes undesirable to try and study the entire population due to resources constraints such as time (Salant & Dillman, 1994; Frey et al., 2000). Therefore, we choose to study just a sample of the population. A sample consists of only those units (e.g., students) from the population of interest.

SAMPLING

A sample is group of people, objects or items that are taken from a large population for a measurement. The sample should be representative of the population to ensure that we can generalize the findings from the research sample to the population as a whole. (Jopnes, 1955; Salant & Dillman, 1994). Sampling is the act, process, or technique of selecting a suitable sample, or a representative part of a population for the purpose of determining parameters or characteristics of the whole population. In sampling, population units such as people (e.g. students enrolled at a university or studying a particular course), cases (i.e., recruitment agencies, organizations, institutions, countries, etc.) and pieces of data (e.g. customer transactions at a particular supermarket, university applications in a country) are selected. To draw conclusions about populations from samples, one must use inferential statistics, to enable us to determine a population's characteristics by directly observing only a portion (or sample) of the population.

Purpose of Sampling

A sample is needed as it may not be practical and almost never economical to conduct a census of whole population because (Leyman, 1983; Lohr, 1999):

1. The large size of many populations.
2. The time factor- a sample may provide needed information quickly.
3. Inaccessibility (associated with cost or time or just access) of some of the population e.g. prisoners, people with severe mental illness, disaster survivors etc.
4. Destructiveness of the observation e.g. to determine the quality of a fuse and whether it is defective, it must be destroyed. Therefore if you tested all the fuses, all would be destroyed.
5. Accuracy and sampling i.e. a sample may be more accurate than a sloppily conducted census.

SAMPLING ERROR

A sample is expected to mirror the population from which it comes. However, there is no guarantee that any sample will be precisely representative of the population from which it comes. What makes a sample unrepresentative of its population is the sampling error). Sampling error is the degree to which a sample might differ from the population. Sampling error comprises the differences between the sample and the population that are due solely to the particular participants that have been selected. When inferring to the population, results are reported plus or minus the sampling error. There are two basic causes for sampling error (Berinstein, 2003; Salant & Dillman, 1994).

1. **Chance:** The error that occurs just because of bad luck, resulting in untypical choices and unusual units in a population. Chance may dictate that a disproportionate number of untypical observations will be made. Unusual units in a population do exist and there is always a possibility that an abnormally large number of them will be chosen. The main protection against this kind of error is to use a large enough sample.
2. **Sampling Bias:** A tendency to favor the selection of participants that have particular characteristics. Sampling bias occurs when the units that are selected from the population for inclusion in the sample do not reflect the population. Sampling bias is usually the result of a poor sampling plan. The most notable is the bias of non-response when for some reason some participants have no chance of appearing in the sample e.g. no internet access for completion of an online questionnaire. There can be two causes of this type of bias (i) selecting a wrong study population and (ii) poor research design e.g. only one group within the study population agreed to participate in the study. This can lead to the sample being unrepresentative of the population of interest.

Sampling or margin of error, e is estimated as (Smith, 2013):

$$e = z_{\alpha/2}\delta \;/\; \sqrt{n} \tag{1}$$

where: n is the desired sample size, z is the z-score, δ is the standard deviation and e is the margin of error.

SAMPLE SIZE

Sample size depends on the nature of the analysis to be performed, the desired precision of the estimates one wishes to achieve, the kind and number of comparisons that will be made, the number of variables that have to be examined simultaneously and how heterogenous the sampled population is. These factors also influence the sample design and data collection procedures. For example, if the key analysis of a randomized experiment consists of computing averages for experiments and controls in a project and comparing differences, then a sample under 100 might be adequate, assuming that other statistical assumptions hold (Salant & Dillman (1994).

There are several approaches to determining sample size and the most popular of these is the one that studies the power of a test of hypothesis. This will ensure that sample size is large enough to ensure that results are statistically significant but not so big that one could have achieved the same results with a much smaller size (Lohr, 1999). For a pure random sample drawn from an infinite population, the following formula can be taken as the basis for computing the sample size, *n* (Smith, 2013):

$$n = (Z_{\frac{\alpha}{2}})^2 \delta\left(1-\delta\right) / e^2 \qquad (2)$$

where: n is the desired sample size, z is the z-score, δ is the standard deviation and e is the margin of error.

Suppose one wishes to determine the number of respondents in a survey from a 95% confidence level, 0.5 standard deviation, and a margin of error (confidence interval) of +/- 5%. The sample size *n* is estimated as:

$n= ((1.96)^2 * 0.5(0.5)) / (0.05)^2= (3.8416 * 0.25) / 0.0025=.9604 / 0.0025=384.16$

Therefore 385 respondents are needed

Sampling Frame

The sampling frame is very similar to the population being studied, and may be exactly the same. When selecting units from the population to be included in your sample, it is sometimes desirable to get hold of a list of the population from which you select units. This is the case when using certain types of sampling technique (i.e., probability sampling techniques). This list can be referred to as the sampling frame (Babbie, 1990).

SAMPLING METHODS

When one is interested in a population, it is typically vital to study a sample of that population rather than attempt to study the whole population. The purpose of sampling techniques is to help researchers select units to be included in the sample. They differ in the manner in which the elementary units are chosen. Broadly speaking, there are two groups of sampling technique: probability sampling techniques and non-probability sampling techniques. Sampling methods are classified as either *probability* or *non-probability* (Salant & Dillman, 1994). In probability samples, each member of the population has a known non-zero probability of being selected.

Probability Sampling Techniques

Probability sampling techniques use random selection (i.e., probabilistic methods) of units from your sampling frame (i.e., similar or exactly that same as your population) to be included in your sample. These procedures (i.e., probabilistic methods) are very clearly defined, making it easy to follow them. Since the characteristics of the sample researchers are interested in vary, different types of probability sampling technique exist to help you select the appropriate units to be included in your sample. The advantage of probability sampling is that sampling error can be calculated. Probability sampling techniques include simple random, systematic random, stratified random, cluster, multistage and multiphase sampling (Cochran, 1953).

Simple Random Sampling

With the simple random sample, there is an equal chance (probability) of selecting each unit from the population being studied when creating the sample. The aim of the simple random sample is to reduce the potential for human bias in the selection of cases to be included in the sample. As a result, the simple random sample provides us with a sample that is highly representative of the population being studied, assuming that there is limited missing data. A simple random sample can only be carried out if the list of the population is available and complete (Cochran, 1953). To create a simple random sample, there are six steps: (a) defining the population; (b) choosing sample size; (c) listing the population; (d) assigning numbers to the units; (e) finding random numbers; and (f) selecting sample from random number table. Cite?

Advantages of Simple Random Sampling

1. Allows researchers to make generalizations (i.e., statistical inferences) from the sample to the population
2. Generalizations are more likely to have external validity.

Disadvantages of Simple Random Sampling

1. It may be challenging to gain access to that list because the list may be protected by privacy policies or require a lengthy process to attain permissions.
2. There may be no single list detailing the population one is interested in hence, it may be difficult and time consuming to bring together numerous sub-lists to create a final list from which you want to select your sample.
3. Many lists will not be in the public domain and their purchase may be expensive.
4. Some of these populations will be expensive and time consuming to contact, even where a list is available (e.g., postal, telephone, email).

Systematic Random Sampling

In systematic random sampling there is an equal chance (probability) of selecting each unit from within the population when creating the sample. The aim of the systemic random sample is to reduce the potential for human bias in the selection of cases to be included in the sample. As a result, the systemic random sample provides us with a sample that is highly representative of the population being studied, assuming that there is limited missing data (Fink, 1995). To create a systemic random sample, there are seven steps: (a) defining the population; (b) choosing your sample size; (c) listing the population; (d) assigning numbers to cases; (e) calculating the sampling fraction by dividing sample size (s) with total population size (N); (f) selecting the first unit from the random number table; and (g) selecting sample

Advantages of Systematic Random Sampling

1. Sample easy to select.
2. Suitable sampling frame can be identified easily.
3. Sample is evenly spread over entire reference population.
4. Allows us to make statistical conclusions from the data collected that will be considered to be valid.

Relative to simple random sample, selection of units using a systematic procedure is superior because it improves the potential for the units to be more evenly spread over the population.

Disadvantages of Systematic Random Sampling

A systematic random sample can only be carried out if a complete list of the population is available. Attaining a complete list of the population can be difficult for a number of reasons:

1. Even if a list is readily available, it may be challenging to gain access to that list. The list may be protected by privacy policies or require a length process to attain permissions.
2. It may be difficult, expensive and time consuming to bring together numerous sub-lists to create a final list from which you want to select your sample.
3. Many lists will not be in the public domain and their purchase may be expensive.
4. Some populations may be expensive and time consuming to contact by postal, telephone or email.

Stratified Random Sampling

Stratified random sampling is a type of probability sampling technique when one is interested in particular strata (meaning groups) within the population (e.g., males vs. females; houses vs. apartments.). Stratified random sample involves dividing the population into two or more strata (groups). There would an equal chance (probability) that each stratum to be selected from the sample. The stratified random sample also improves the representation of particular strata (groups) within the population, as well as ensuring that these strata are not over-represented. Together, this helps the researcher to compare strata, as well as make more valid inferences from the sample to the population.

To create a stratified random sample: (a) define the population, (b) choose the relevant stratification, (c) list the population, (d) list the population according to the chosen stratification, (e) calculate a proportionate stratification and (f) use a simple random or systematic sample to select sample (Smith, 2013).

Advantages of Stratified Random Sampling

1. Reduce the potential for human bias in the selection of cases to be included in the sample.
2. Provides with a sample that is highly representative of the population being studied, assuming that there is limited missing data.
3. Allows researchers to make statistical conclusions from the data collected that will be considered to be valid.

4. The selection of units using a stratified procedure can be viewed as superior because it improves the potential for the units to be more evenly spread over the population.
5. Where the samples are the same size, a stratified random sample can provide greater precision than a simple random sample.

Because of the greater precision of a stratified random sample compared with a simple random sample, it may be possible to use a smaller sample, which saves time and money.

Disadvantages of Stratified Random Sampling

1. Sampling frame of entire population has to be prepared separately for each stratum.
2. Many lists will not be in the public domain and their purchase may be expensive.
3. It may be challenging to gain access to the data list due to protection by privacy policies or require a length process to attain permissions.
4. There may be no single list detailing the population of interest.
5. It may be difficult and time consuming to bring together numerous sub-lists to create a final list from which to select the sample.
6. When examining multiple criteria, stratifying variables may be related to some, but not to others, further complicating the design, and potentially reducing the utility of the strata.
7. In some cases (such as designs with a large number of strata, or those with a specified minimum sample size per group), stratified sampling can potentially require a larger sample than would other methods.
8. In terms of human populations, some of these populations may be expensive and time consuming to contact by postal, telephone or email even where a list is available.

Cluster Sampling

Cluster sampling, on the surface, is very similar to stratified sampling in that survey population members are divided into unique, non-overlapping groups prior to sampling. These groups are referred to as clusters instead of strata because they are naturally occurring groupings such as schools, households, or geographic units Whereas a stratified sample involves selecting a few members from each group or stratum, cluster sampling involves the selection of a few groups and data are collected from all group members. This sampling method is used when no master list of the population exists but cluster lists are obtainable (Babbie, 1990; Berinstein, 2003). To create a cluster sample, the following steps are taken: (a) sample of areas are chosen, (b) sample of respondents within those areas is selected. (c) population divided into clusters of homogeneous units, usually based on geographical contiguity, (d) Sampling units are groups rather than individuals, (e) sample of clusters is then selected and (f) all units from the selected clusters are studied.

Advantages

1. Cuts down on the cost of preparing a sampling frame.
2. This can reduce travel and other administrative costs.

Disadvantages

1. Sampling error is higher for a simple random sample of same size.

Non-Probability Sampling Techniques

A core characteristic of non-probability sampling techniques is that samples are selected based on non-random manner or the subjective judgment of the researcher. Non-probability sampling represents a group of sampling techniques that help researchers to select units from a population that they are interested in studying. Hence, the degree to which the sample differs from the population remains unknown. Whilst some researchers may view non-probability sampling techniques as inferior to probability sampling techniques, there are strong theoretical and practical reasons for their use. Theoretically, for researchers following a quantitative research design, non-probability sampling techniques can often be viewed as an inferior alternative to probability sampling techniques. In applied social science research it is permissible to consider varieties of non-probability sampling alternatives where it is not practical or theoretically sensible to do random sampling (Trochim, 2006). Practically, non-probability sampling is often used because the procedures used to select units for inclusion in a sample are much easier, quicker and cheaper when compared with probability sampling. Non-probability sampling techniques include: quota sampling, judgment sampling, convenience sampling, purposive sampling, self-selection sampling and snowball sampling. Convenient sampling includes samples selected by a researcher on the basis of ease of accessibility of the sample objects to the researcher.

Convenience Sampling

A convenience sample is one where the units that are selected for inclusion in the sample are the easiest to access. For example, a researcher may simply stand at one of the main campus entrances and invite the first ten students that pass by to take part in the interview. This type of sampling is useful in getting general ideas about the phenomenon of interest. It saves time, money and effort. It is the poorest way of getting samples, has the lowest credibility and yields information-poor cases (Kish, 1965).

Advantages of Convenience Sampling

1. It is easy to carry out with few rules governing how the sample should be collected.
2. The relative cost and time required to carry out a convenience sample are small in comparison to probability sampling techniques.
3. Enables researcher to achieve the sample size in a relatively fast and inexpensive way.
4. Help in gathering useful data and information that would not have been possible using probability sampling techniques, which require more formal access to lists of populations.

Disadvantages of Convenience Sampling

1. Often suffers from biases because a convenience sample can lead to the under-representation or over-representation of particular groups within the sample.

2. Since the sampling frame is not known, the sample is unlikely to be representative of the population being studied thus, undermining the ability to make generalizations from the sample to the population studied.

Quota Sampling

Quota sampling is a type of non-probability sampling technique. With proportional quota sampling, the aim is to end up with a sample where the strata (groups) being studied (e.g., males vs. females students) are proportional to the population being studied. To create a quota sample, there are three steps: (a) choosing the relevant stratification and dividing the population accordingly; (b) calculating a quota for each stratum; and (c) continuing to invite cases until the quota for each stratum is met.

Advantages of Quota Sampling

1. Quota sampling is particularly useful when one is unable to obtain a probability sample, but still trying to create a sample that is as representative as possible of the population being studied. In this respect, it is the non-probability based equivalent of the stratified random sample.
2. Unlike probability sampling techniques, quota sampling is much quicker and easier to carry out because it does not require a sampling frame and use of random sampling techniques. This makes it popular in undergraduate and master's level dissertations where there is a need to divide the population being studied into strata (groups).
3. The quota sample improves the representation of particular strata (groups) within the population, as well as ensuring that these strata are not over-represented.
4. The use of a quota sample, which leads to the stratification of a sample (e.g., male and female students), allows us to more easily compare these groups (strata).

Disadvantages of Quota Sampling

1. Make it impossible to determine the possible sampling error.
2. Ease of access and cost considerations, resulting in sampling bias.
3. It is not possible to make statistical inferences from the sample to the population. This can lead to problems of generalization.
4. Can increase costs and time to carry out the research.

Self-Selection Sampling

Self-selection or volunteer sampling is appropriate when one wants to allow units or cases, whether individuals or organizations, to choose to take part in research on their own accord. The key component is that research subjects volunteer to take part in the research. This allows units to choose to take part in research on their own accord. (http://dissertation.laerd.com/purposive-sampling.php). As a sampling strategy, self-selection sampling can be used with a wide range of research designs and research methods. For example, survey researchers may put a questionnaire online and subsequently invite anyone within a particular organization to take part. It is an effective sampling strategy in experimental research settings. Scientists that conduct experiments using human subjects may advertise the need for volunteers to take part in drug trials or research on physical activity (Kish, 1965). The key component is that research subjects (or organizations) volunteer to take part in the research on their own accord. There may be a wide range of reasons why people (and organizations) volunteer for such studies, including having particularly strong feelings or opinions about the research, a specific interest in the study or its findings, or simply wanting to help out a researcher (Henry, 1990).Creating a self-selection sample involves two simple steps: (a) publicizing your need for units; and (b) checking the relevance of units and either inviting or rejecting them.

Advantages of Self-Selection Sampling

1. This can reduce the amount of time necessary to search for appropriate units; that is, those individuals or organizations that meet the selection criteria needed for your sample.
2. The potential units or cases (individuals or organizations) are likely to be committed to take part in the study, which can help in improving attendance (where necessary), and greater willingness to provide more insight into the phenomenon being studied (e.g., a respondent many be more willing to spend the time filling in qualitative, open-ended questions in an online survey, where others may leave them blank).

Disadvantages of Self-Selection Sampling

1. There is likely to be a degree of self-selection bias. For example, the decision to participate in the study may reflect some inherent bias in the characteristics/traits of the participants (e.g., an employee with a 'chip of his shoulder' wanting to give an opinion).
2. This can either lead to the sample not being representative of the population being studied, or exaggerating some particular finding from the study.

Snowball Sampling

Some populations can be hard-to-reach and/or hidden because they exhibit some kind of social stigma, illicit or illegal behaviors, or other trait that makes them atypical and/or socially marginalized. Such populations include; drug addicts, homeless people, suffers of AIDS/HIV and prostitutes. Individuals that are drug users or prostitutes, for example, are likely to be less willing to identify themselves and take part in a piece of research than many other social groups. Snowball sampling can be used to gain access to such populations (Biernacki & Waldorf, 1981). To create a snowball sample, there are two steps: (a) trying to identify one or more units in the desired population; and (b) using these units to find further units and so on until the sample size is met.

Advantages of Snowball Sampling

1. The unknown and/or secretive nature of some social groups may also make it difficult to identify strata that warrant investigation.
2. In the case of drug users, it may be obvious to identify strata such as gender (i.e. male or female), types of drug user (e.g. causal, addict), and so forth, but others may be unknown to the researcher.

3. May also be helpful in exploring potentially unknown characteristics that are of interest before settling on your sampling criteria.
4. There may be no other way of accessing the sample, making snowball sampling the only viable choice of sampling strategy (Biernacki & Waldorf 1981; Henry, 1990).

Disadvantages of Snowball Sampling

1. It is impossible to determine the possible sampling error and make statistical inferences from the sample to the population.
2. Not representative of the population being studied.
3. It can be difficult to identify units to include in the sample because there is no obvious list of the population interested in e.g. drug users or prostitutes (Biernacki & Waldorf 1981).

Purposive Sampling

Purposive sampling is also known as judgmental, selective or subjective sampling. It reflects a group of sampling techniques that rely on the judgement of the researcher when it comes to selecting the units (e.g., people, cases/organizations, events, pieces of data) that are to be studied (Kish, 1965; Fowler, 1993). The main goal of purposive sampling is to focus on particular characteristics of a population that are of interest, which will best answer the research questions. The sample being studied is not representative of the population, but for researchers pursuing qualitative or mixed methods research designs, this is not considered to be a weakness. Rather, it is a choice, the purpose of which varies depending on the type of purposing sampling technique that is used. For example, in homogeneous sampling, units are selected based on their having similar characteristics because such characteristics are of particular interested to the researcher. By contrast, critical case sampling is frequently used in exploratory, qualitative research in order to assess whether the phenomenon of interest even exists (amongst other reasons).

There are a number of different types of purposive sampling techniques which include maximum variation sampling, homogeneous sampling, typical case sampling, extreme (or deviant) case sampling, total population sampling and expert sampling (Kish, (1965). Patton, 1990). Each of these purposive sampling techniques has a specific goal, focusing on certain types of units, all for different reasons. The different purposive sampling techniques can either be used on their own or in combination with other purposive sampling techniques. During the course of a qualitative or mixed methods research design, more than one type of purposive sampling technique may be used.

Maximum Variation Sampling

Maximum variation sampling, also known as heterogeneous sampling, is a purposive sampling technique used to capture a wide range of perspectives relating to the thing that you are interested in studying; that is, maximum variation sampling is a search for variation in perspectives, ranging from those conditions that are view to be typical through to those that are more extreme in nature. By conditions, we mean the units (i.e., people, cases/organizations, events, pieces of data) that are of interest to the researcher. These units may exhibit a wide range of attributes, behaviors, experiences, incidents, qualities, situations, and so forth. The basic principle behind maximum variation sampling is to gain greater insights into a phenomenon by looking at it from all angles. This can often help the researcher to identify common themes that are evident across the sample.

Homogeneous Sampling

Homogeneous sampling aims to achieve a homogeneous sample; that is, a sample whose units (e.g., people, cases) share the same characteristics or traits (e.g., a group of people that are similar in terms of age, gender, background, occupation). In this respect, homogeneous sampling is the opposite of maximum variation sampling. A homogeneous sample is often chosen when the research question that is being address is specific to the characteristics of the particular group of interest, which is subsequently examined in detail.

Typical Case Sampling

Typical case sampling is a purposive sampling technique used when one is interested in the normality/typicality of the units (e.g., people, cases, events, settings/contexts, places/sites) you are interested, because they are normal/typical. The word typical does not mean that the sample is representative in the sense of probability sampling (i.e., that the sample shares the same/similar characteristics of the population being studied). Rather, the word typical means that the researcher has the ability to compare the findings from a study using typical case sampling with other similar samples (i.e., comparing samples, not generalizing a sample to a population). Therefore, with typical case sampling, you cannot use the sample to make generalizations to a population, but the sample could be illustrative of other similar samples. Whilst typical case sampling can be used exclusively, it may also follow another type of purposive sampling technique, such as maximum variation sampling, which can help to act as an exploratory sampling strategy to identify the typical cases that are subsequently selected.

Extreme (or Deviant) Case Sampling

Extreme (or deviant) case sampling is a type of purposive sampling that is used to focus on cases that are special or unusual, typically in the sense that the cases highlight notable outcomes, failures or successes. These extreme (or deviant) cases are useful because they often provide significant insight into a particular phenomenon, which can act as lessons (or cases of best practice) that guide future research and practice. In some cases, extreme (or deviant) case sampling is thought to reflect the purest form of insight into the phenomenon being studied.

Critical Case Sampling

This is a type of purposive sampling technique that is particularly useful in exploratory qualitative research, research with limited resources, as well as research where a single case or small number of cases can be decisive in explaining the phenomenon of interest. Critical case sampling may be used to investigate whether a phenomenon is worth investigating further, before adopting a maximum variation sampling technique is used to develop a wider picture of the phenomenon. It is this decisive aspect of critical case sampling that is arguably the most important (Patton, 1990).

Total Population Sampling

This is a type of purposive sampling technique where one chooses to examine the entire population that have a particular set of characteristics (e.g., specific experience, knowledge, skills or exposure to an event). In such cases, the entire population is often chosen because the size of the population that has the particular set of characteristics of interest is very small.

Expert Sampling

Expert sampling is a cornerstone of a research design known as expert elicitation. This is used when research needs to glean knowledge from individuals that have particular expertise. This expertise may be required during the exploratory phase of qualitative research, highlighting potential new areas of interest or opening doors to other participants. Alternately, the particular expertise that is being investigated may form the basis of research, requiring a focus only on individuals with specific expertise. Expert sampling is particularly useful where there is a lack of empirical evidence in an area and high levels of uncertainty, as well as situations where it may take a long period of time before the findings from research can be uncovered.

Advantages of Purposive Sampling

1. Provides wide range of sampling techniques that can be used such as homogeneous sampling, critical case sampling and expert sampling. For example, critical case sampling may be used to investigate whether a phenomenon is worth investigating further, before adopting an expert sampling approach to examine specific issues further.
2. Provide researchers with the justification to make theoretical, analytic and/or logical generalizations from the sample that is being studied

Disadvantages of Purposive Sampling

1. Is prone to researcher bias because of its judgmental and subjective component of purpose sampling
2. The subjectivity and non-probability based nature of unit selection (i.e., selecting people, cases/organizations etc.) means that it can be difficult to defend the representativeness of the sample, it can be difficult to convince the reader that the judgment used to select units to study was appropriate and it can also be difficult to convince the reader that research using purposive sampling achieved theoretical/analytic/logical generalization.

CONCLUSION

Research is an organized and systematic way of finding answers to questions. The purpose of this chapter was to describe the use of sampling in research. It explains the sample size determination, different types of sampling techniques, and advantages and limitations of sampling techniques. It has been shown that a sample can be viewed as part or segment of a population which is selected for the study. Using a sample in research saves money and time, if a suitable sampling strategy is used, appropriate sample size selected and necessary precautions taken to reduce on sampling and measurement errors. Then, a sample should yield valid and reliable information. Combination or mixed purposeful sampling combines various sampling strategies to achieve the desired sample. This helps in triangulation, allows for flexibility, and meets multiple interests and needs. When selecting a sampling strategy it is necessary that it fits the purpose of the study, the resources available, the question being asked and the constraints being faced. This holds true for sampling strategy as well as sample size.

REFERENCES

Babbie, E. (1990). *Survey research methods*. Belmont, CA: Academic Press.

Berinstein, P. (2003). *Business statistics on the web: Find them fast – at little or no cost*. Cyber Age Books.

Biernacki, P., & Waldorf, D. (1981). Snowball sampling: Problems and techniques of chain referral sampling. *Sociological Methods & Research*, *10*(2), 141–163.

Cochran, W. G. (1953). *Sampling techniques*. New York: John Wiley & Sons.

Faugier, J., & Sargeant, M. (1997). Sampling hard to reach populations. *Journal of Advanced Nursing*, *26*(4), 790–797. doi:10.1046/j.1365-2648.1997.00371.x PMID:9354993

Fink, A. (1995). *How to sample in surveys* (Vol. 6). London: Sage Publications.

Fowler, F. J. (1993). *Survey research methods* (2nd ed.; Vol. 1). London: Sage Publications.

Frey, L. R., Botan, C. H., & Kreps, G. L. (2000). *Investigating communication: An Introduction to research methods* (2nd ed.). Boston: Allyn and Bacon.

Henry, G. T. (1990). *Practical sampling* (Vol. 21). London: Sage Publications. doi:10.4135/9781412985451

Jones, H. L. (1955). The Application of Sampling Procedures to Business Operations. *Journal of the American Statistical Association*, *50*(271), 763–774.

Kish, L. (1965). *Survey Sampling*. New York: John Wiley & Sons.

Lapin, L. L. (1987). *Statistics for modern business decisions*. Harcourt, Jovanovich, Inc.

Lohr, S. L. (1999). *Sampling: Design and Analysis*. Albany, NY: Duxbury Press.

Lyman, H. R. (1983). *An introduction to statistical methods and data analysis* (4th ed.). Dioxburg Press.

MacNealy, M. S. (1999). *Strategies for empirical research in writing*. New York: Longman.

Patton, M. Q. (1990). *Qualitative evaluation and research methods*. London: New SAGE, Publications.

Salant, P., & Dillman, D. A. (1994). *How to conduct your own survey*. London: John Wiley & Sons, Inc.

Smith, S. (2013). *Determining sample size: How to ensure you get the correct sample size*. LLC, Qualtrics.

KEY TERMS AND DEFINITIONS

Population: A population is a group of individual persons, objects, or items from which samples are taken for measurement for example a population of presidents or professors, books or students.

Research: Scientific research is systematic, controlled, empirical, and critical investigation of natural phenomena guided by theory and hypotheses about the presumed relations among such phenomena.

Sample: A sample is a finite part of a statistical population whose properties are studied to gain information about the whole population. When dealing with people, it can be defined as a set of respondents (people) selected from a larger population for the purpose of a survey.

Sample Frame: Collection of units such as people, objects and events that have a possibility of being selected. It approximates the population as closely as possible and the sample is drawn from the sample frame.

Sample Size: The number of units in the sample.

Sampling: Sampling is the act, process, or technique of selecting a suitable sample, or a representative part of a population for the purpose of determining parameters or characteristics of the whole population.

Strata: Sub-groups within a population.

Section 4
Data Collection Using Innovative Tools

Collecting data is critical to the design of the study and the researcher needs to determine whether the data should be collected concurrently--namely quantitative and qualitative data at the same time, or sequentially--collecting and analyzing one type of data at a time. While the traditional forms of data collection have been stable over the years--collecting archival data, surveys/questionnaires, interviews, and observation--alternative forms of data collection tools are starting to emerge. This allows the researcher to explore an issue at a more holistic level.

Chapter 7
Analyzing Qualitative Data:
Visualizing Lived Experiences through Poems and Photography

Carolyn N. Stevenson
Kaplan University, USA

ABSTRACT

Use of photography and poetry offer a way for participants to express lived experiences through a visual and written means of self-expression. These forms of data collection can provide a rich, thick description of those often overlying on the peripherals of society. Traditional means of qualitative research such as interview and observation can at times create a barrier between the researcher and the participants because of the face-to-face interaction. Participants may be uncomfortable expressing authentic feeling during a formal interview process. By offering participants the opportunity to personally select descriptive photographs and articulate expression in their own voice through poetry, an authentic expression occurs.

INTRODUCTION

This chapter discusses the role of undergraduate photographs and poems in qualitative research to describe and explain social issues in an urban setting. Through use of actual student work, undergraduate student perspectives on critical social issues are presented. The presentation of the data is shown through student photographs and poems. Having an understanding of the visual means of data collection adds to the richness of the lived experiences of the participants. Use of photographs and poems provides a means for participants to express their interpretation of complex social issues in a visual and written format.

Qualitative researchers need to be aware of the visual aspects of data collection and creative use of poetry to allow participants to freely document life in an urban setting and describe formation of self-identity. These methods allow for a more personalized approach to documenting issues related to poverty, violence, and self-understanding. It is through analysis of the photographs and poems a greater understanding of social issues in an urban setting can allow qualitative researchers to gain a broader sense of concerns and ways individuals rise above environmental challenges to reach personal success.

DOI: 10.4018/978-1-5225-0007-0.ch007

This case study involved six undergraduate students from a small, private, urban university in Chicago, Illinois. The participants were instructed to use photography as a vehicle for self-expression on identity and the way they viewed their world and their communities. Participants also created poems to enhance an understanding of the way they view live in an urban setting. This chapter includes example poems, photographs, and detailed description of the social challenge faced in their individual communities and society as a whole.

The exploratory questions that guided the study are:

1. What elements constitute undergraduate student perspectives on social issues in their communities and society?
2. How can photography and poetry be used to illustrate the social issues in their communities and society?
3. What beliefs do these undergraduate students hold which support or negate this perspective?

Findings from this case study will assist qualitative researchers in learning more about using visual data and poetry as a means to present qualitative data. It is through the participants own perception that critical social issues in urban communities can be portrayed through the eyes of outlying, marginalized populations. Through representation of findings in a visual format serious social challenges in urban settings can be called into questions and recommendations for future change can be proposed.

USE OF POETRY AND PHOTOGRAPHY IN QUALITATIVE RESEARCH

The use of photography as a means for presenting data is not new to the field of qualitative research. In fact, "The use of visual research methods has become increasingly widespread throughout the social sciences," Knoblauch, Baer, Laurier, Petschke, Schnettler, p. 1, 2008). Jenkings, Woodward, and Winter (2008) add, "What distinguishes photo elicitation in academia from similar uses of photographs in other areas of media production such as photo journalism is undoubtedly its methodological rigor and, usually, its aim of theorizing from the data," (p.1). While many studies focus one visual means for presenting data, use of poetry and photography provide a more robust means for participants to express views on topics that an interview alone cannot capture. At times, participants may feel uncomfortable expressing their views in a face-to-face interview. Use of poetry and photography provides a way for participants to express their opinions and capture a social phenomenon within their boundaries, allowing their creative expression to serve as the written and visual narrative.

Photography provides a visual representation of the written text. While use of photos can offer a visual image to the reader, the addition of poetry combines both the written and the visual offer a rich view of the issue being presented by the participants.

"Phenomenologically speaking, photography is a visual system of representation, by way of which the visibility of an object not being present is produced. Thus, we may state a relation of similarity between the photography, that is the representing part on the one hand, and the depicted object, that is the represented part on the other hand,"(Christmann, p. 2, 2008). Poetry and photography are a powerful means of presenting a phenomenon in everyday culture as the participants capture images seen at a specific time, in a specific natural setting.

CASE DESCRIPTION

The Organization

The organization selected for this case study was a four-year college in Chicago, IL. The university is a private, not-for-profit and non-denominational institution of higher learning in Chicago's South Loop. The University is accredited by the Higher Learning Commission of the North Central Association of Colleges and Schools (NCA), and provides affordable, quality higher education to students from all ethnic and socio-economic backgrounds.

Eighty percent of the university's students come from disadvantaged minority backgrounds; more than 90% of its students are first-generation college learners. More than 90% of the students come from low-income households eligible for maximum federal and state financial aid. The university's commitment to diversity and multiculturalism is consistently cited as a distinguishing feature by accreditation teams of the NCA. According to the college catalog, the mission of the university, "...fosters equal educational opportunity for all racial, ethnic, and socio-economic groups and offers programs in Liberal Arts and Sciences along with job and career related professional education geared to the service economy of modern times. The university is dedicated to serving humankind with a global, multicultural and future-oriented perspective," (p.3).

The university recruits most of its students from Chicago, which has one of the largest consolidated school districts of the country (District 299), and which exemplifies social and educational problems facing urban youth. All the elementary, middle, and high schools are part of one district—District 299. In the Chicago system:

- The low-income percentage is 84.9%.
- The chronic truancy rate is 31.9%.
- African-American and Hispanic students constitute over 80% of the student population.
- The student readiness for college coursework is 26%.
- The high school graduation rate for the district is 65.4%.
- Majority of the predominantly minority schools have a graduation rate of below 60% and a college enrollment rate of below 50% (National Center for Educational Statistics, 2014).

For this case study, photographs and poems from six participants were selected. Students were from diverse degree programs including Behavioral and Social Science, Business, and Math. The participants' ages ranged between - of 21-38. The course assignment was given in a Group Dynamics undergraduate course offering during summer 2014 and fall 2014. All students were at the sophomore or junior level. With the exception of one student, all the participants attended one the Chicago Public High Schools, all within the same district, District 299. The ethnic backgrounds of the participants included White, non-Hispanic, African American, and Hispanic. Both male and female students agreed to have their photographs and poems presented in this chapter. All of the participating students agreed to discuss their viewpoints of current social issues through a live classroom presentation. Due to space limitations, highlights of the photographs are represented in the vignettes that follow. The poems and photographs were selected based on the relevance to the topic.

Statement of the Problem

Literature and observation reveal that undergraduate students are concerned about social issues facing their communities, especially in an urban setting. In a large metropolitan city such as Chicago, topics such as homelessness, poverty, violence, and lack of quality education are issues challenging daily living for many individuals. For example, Englewood, one of the Southside neighborhoods in Chicago, has seen a 50% increase in the number of violent crimes during the period of September 26, 2014 – October 26, 2014 (City of Chicago, Data Portal, 2014). Englewood ranks 3rd (tied) among Chicago's 77 community areas for Violent Crime reported in this time period (City of Chicago, Data Portal, 2014). This is only one example of the neighborhoods represented in the study but highlights the reality of the magnitude of crime in the neighborhoods represented in this study (tables 1-4).

Quality public education is a concern in Chicago. According to a document published by the Chicago Teacher's Union, "CPS (Chicago Public Schools) schools operate in an environment of intense segregation, a culmination of decades of racist public policies and market forces that have segregated Chicago's communities and neighborhoods" (Chicago Teacher's Union, 2014, p. 22).

The report also states that, "Within CPS, 69% of all African American students (and 42% of Latinos) go to schools that are 'intensely segregated" — schools that have more than 90% of their student body composed of the same ethnicity" (Chicago Teacher's Union, 2014, p.22).

Fair school funding is an issue for Chicago Public Schools and Illinois school in general. According to Baker, Sciarra, and Farrie (2010), Illinois is among the bottom three states when it comes to fairness

Table 1. Demographics of Englewood

	Population (2010)	Area	Population Density	Per-Capita Income	Households below Poverty Level	Unemployed	Living in Crowded Housing	No High School Diploma	Age: Under 18 or Over 64
Englewood	30,654	3.07 sq. mi	9,977 /sq. mi	$11,993	42.2%	21.3%	4.8%	29.4%	43.4%
Chicago	2,695,598	227.6 sq. mi	11,844 / sq. mi	$27,148	18.7%	11.1%	4.7%	20.6%	34%

Source: City of Chicago (2014). Census data: Selected socioeconomic indicators 2008 – 2012.

Table 2. Violent crime, September 26 – October 26, 2014

Crime by Type	30 Days	365 Days
Robbery	31	368
Battery	39	275
Assault	19	172
Homicide	1	15
Sexual assault	2	45

There were 3.0 violent crimes per 1,000 people in Englewood in the past 30 days (up 50% year over year).

Source: City of Chicago, Data Portal, 2014.

Table 3. Property crime, September 26 – October 26, 2014

Crime by Type	30 Days	365 Days
Theft	78	1,135
Burglary	43	380
Motor vehicle theft	18	212
Arson	1	19

There were 4.6 Property crimes per 1,000 people in Englewood in the past 30 days. (up 50% year over year).

Source: City of Chicago, Data Portal, 2014.

Table 4. Quality of life crime, September 26 – October 26, 2014

Crime by Type	30 Days	365 Days
Criminal damage	57	631
Narcotics	57	742
Prostitution	7	63

There were 3.9 Quality of Life crimes per 1,000 people in Englewood in the past 30 days (up 50% year over year).
Source: City of Chicago, Data Portal, 2014.

in funding for education. Illinois school funding — collected largely through local property taxes — is regressive, with communities suffering concentrated poverty receiving less in school funding. Despite having the fifth largest state economy in the U.S., Illinois ranks 33rd in the country in the percent of state Gross Domestic Product (GDP) spent on public education (Baker, Sciarra, & Farrie, 2010).

The disparity in funding creates a greater inequality in the quality of education in Chicago, especially in predominantly minority neighborhoods. "In Illinois, African American and Latino students receive little more than 90% of the per pupil expenditure that white students receive, 93 cents and 91 cents for every dollar, respectively," (Baker, Sciarra, & Farrie, 2010, p. 33).

It is reasonable to suggest there is a need for qualitative researchers to capture the lived experiences of individuals living in communities where crime and poverty are part of the natural landscape. While it is apparent these social issues exit, there is a need for data to be presented through the perspectives of the individuals facing these issues on a daily basis. Through use of visual means data collection concerns can be visually represented through photographs and telling images. Use of poetry allows a format for individuals to creatively express concerns and present social issues through their interpretation.

This case study does not seek to resolve the social problems and concerns of undergraduate students in an urban setting. Rather it seeks to create an awareness of the social issues through the lens of individuals living the experiences in these communities and society as a whole. Additionally, the case study seeks to describe and explain the ways photographs and poetry can be used as a means for data collection and interpretation in qualitative research.

Purpose of the Study

The purpose of this study was to describe and explain ways photography and poetry can be used to illustrate the social issues expressed by undergraduate students. Findings of this study may assist qualitative researchers, instructors of qualitative methods, and college administrators in three ways.

The findings of the study may inform qualitative researchers, instructors of qualitative methods, and college administrators of the value of photographs and poetry in qualitative data collection and analysis. The findings may also inform current and future college administrators and qualitative research faculty members of the ways photographs and poetry can provide an authentic way of presenting data through the lens of participants—-by using visuals such as photographs and in their own voice through poetry. The study may provide the basis for greater understanding of the diversity for data collection methods in qualitative research for researchers and instructors of qualitative methods to encourage participants to express their perspectives through multiple means.

This immersion into the research process actively involves the qualitative researcher in the quest for gaining a deeper understanding of the social phenomena. A case study was selected for this study. Case studies involve an in-depth study of this bounded system and rely on a number of data collection materials.

According to Yin (2003) "the distinctive need for case studies arises out of the desire to understand complex social phenomena" because "the case study method allows investigators to retain the holistic and meaningful characteristics of real-life events," (p.2). The decision to describe and explain student perspectives on social issues facing an urban setting lends itself to the case study approach. In fact, case studies seem to be the preferred strategy when "how or "why" questions are being posed, when the investigator has little control over events, and when the focus is on a contemporary phenomenon within some real-life context (Yin, 1981).

A case study approach was selected for this study to gain an in-depth understanding of undergraduate student perspectives on social issues facing their communities and society as a whole. Willig (2008) asserts, case studies "are not characterized by the methods used to collect and analyze data, but rather its focus on a particular unit of analysis: a case" (p. 74). Stake (2005) argues that the topic of the case can be an individual, but not the means by which the individual engages in a particular practice. He writes, "A doctor may be a case. But his or her doctoring probably lacks the specificity, the boundedness to be a case," (p. 444).

Presentation of the individual cases provides the reader the opportunity to gain understanding of the views, observations, and opinions of the individual undergraduate student participant perspectives on social issues. Actual photographs and poems from the six undergraduate students were used in the case studies as an attempt to portray the participant as an individual entity.

Data Collection

Interviews, observations, researcher field notes, documents, artifacts, transcripts, and researcher reflective journal were the means of data collection. At least two in-depth interviews were conducted with the six participants. In an attempt to gather the rich, descriptive information required for qualitative research, semi-structured interviews with open-ended questions were used. The first step in the data collection process was conducting interviews with the study participants. The information qualitative researchers seek to gain is rich, thick descriptions of the participants in their social settings. Thus, open-ended questions were used to elicit the most complete and thorough responses from the participants.

The nature of qualitative research is flexible, as participants are being studied in their social setting. While variables in the social world cannot be controlled, the researcher can follow a format to help ensure items such as equipment are functioning.

In addition to interviews, participant observation was used to supplement the data collected in the interviews. Janesick (2011) alludes to observation as the immersion into the social setting, which allows the researcher to begin to experience the experiences of the participants. The researcher observed each participant at least one time. Settings for the observations were limited to formal and informal classroom interactions. All of the participants were observed for eleven weeks, the duration of the course. Formal observations were noted behavior and participation in lecture and discussion. Informal observations occurred in small group interaction within the classroom.

Observations also provided a check as to the credibility of the other data collected. Observations do require a series of planned steps. There are limitations as to the amount of information individuals reveal in the interview. Observations served as a means for verifying that the participants' actions match their words.

Document and artifact analysis, researcher field notes, and a researcher reflective journal also served as other sources of data for this case. Field notes consisted of supporting interview and observation notes. Format for the field notes collected during interviews and observations followed suggestions provided by Janesick (2011). A researcher's reflective journal was also kept as another means of data collection. The reflective journal served as means for the researcher to express emotions, ideas, and reactions to the study. During the data analysis stage, the reflective journal provided another resource for identifying emerging themes and sub themes.

The Role of the Researcher

The researcher's role (full participant or observer), issues of entry, reciprocity, personal biography, and ethics (i.e., informed consent, privacy, etc.) must be taken in account during the research process. In this case study, the role of the qualitative researcher was to adhere to the possibility of neutrality (Silva, 2008).

The role of the researcher in this case study follows that of a non-participant observer (Creswell, 1994; Cohen & Manion, 1989). This means that the role as researcher and observer was clearly known by all the participants. The researcher did not interact as a participant in the development, delivery, or activities of this class. Researcher presence was kept as passive as possible, except when necessary to actively pursue additional information from one of the participants. At times this policy of non-interference restricted observations and abilities to seek out more information, which imposed another limitation of the research method, but it also, preserved the natural setting for these observations.

Participant Selection

Six university students attending a small, private college in Chicago, Illinois were selected for this study. The university has approximately 500 full-time students and is mainly a computer school. More than 90% of the university students are first generation college students and receive some form of financial aid. The university is primarily - minority students with a strong social justice mission targeting the under-served. The six participants all were enrolled in a Group Dynamics course in summer 2014 and fall 2014. With informed consent, the participants agreed to share the results of a class projects involving capturing a social issue facing their community and society as a whole. Requirements for the assignment included creating a poem and use of photography to represent the issue and the participants' reflective thoughts and opinions on the issue.

Following the recommendations of Janesick (2011), the researcher relied on collection and analysis of various forms of data. Each pre-university student was interviewed at least two times for a total of eight formal and informal interviews. The university students were audiotaped for the formal interviews. Data were also collected from three observations, eleven documents and artifacts, and sixteen researcher reflective journal entries.

The university students were selected based on their representation of an urban population, their willingness to talk about their experiences, and their ability to provide their poems and photographs, which represented diverse perspectives. Each participant was a full-time student, came from diverse socio-economic backgrounds, and attended the same university in Chicago, Illinois.

There were three males and three females selected for this study. Terrance and Delano were African American males. Jacqueline was a Hispanic female. Tanieka and Sierra were African American females. James was a White, non-Hispanic male. All of the participants were selected because of their willingness to participate in the study, their interest in photography as a means for promoting awareness of social issues, and their connection to the other university students selected for the study. Each participant agreed to use their actual first names in the study. Table 1 shows the basic profile of each participant.

All of the participants were first generation college students from various ethnic backgrounds. With the exception of James, all five participants were born and continue to live in Chicago, Illinois. James was born in South Dakota and moved to Chicago five years ago. He was also the only participant that attended another college in South Dakota where he received his Associate of Arts degree. The other five participants have only attended the University in the study and have no previous college experience.

The six participants live in different neighborhoods in Chicago. The city is very segregated and ethnically diverse. Terrace is from a predominantly African American neighborhood on the city's far west side. Delano is from a predominantly African-American neighborhood on the city's south side. Jacqueline is from a predominantly Latino neighborhood in the city's southwest side. Tanieka and Sierra lives in predominantly African-American neighborhoods on the city's south side. All five participants currently live in the neighborhoods which they lived as a child. James is the only exception, and lives in a nearby Chicago suburb of mixed ethnic backgrounds.

The participants have different college majors, goals, and expectations after graduation (table 5). Terrance is a Behavioral and Social Science major who wishes to pursue graduate school after completing his undergraduate degree. Delano is a Business major who hopes of owning his own not-for-profit business with the intention of giving back to his community. Jacqueline is a Business major who desires to work for a marketing firm upon graduation. Tanieka is a Business major who has her expectations on pursuing a career in the accounting field after completion of her undergraduate degree. Sierra is a Behavioral and Social Sciences major who wishes to continue her graduate education studying psychology. James is a Math major with intentions on continuing onwards to graduate school.

Each participant discussed his or her perspectives on social issues facing their community and the need for change.

Presentation of the Data

All six participants agreed to share their reflective poems and photographs as a means for representing a social issue in their community. The students were assigned a project for a Group Dynamics course and

Table 5. Case ordered matrix, participant characteristics

University Student	Gender	Age
Terrance	Male	Early 20's
Delano	Male	Early 30's
James	Male	Mid 30's
Jacqueline	Female	Early 20's
Sierra	Female	Early 20's
Tanieka	Female	Later 30's

also presented their material to a class of fifteen other students. The researcher served as their instructor for the course. Following are the actual poems and photographs presented by the university students, which capture the social issues faced by students today (see Poems 1-6).

Poem 1. Awoken. By: Terrance.

When sleeping there is Chaos that fills
Of constant war that seems everlasting
An empire soiled with blood and grief.
Stretching beyond the lengths and depths of the earth.
Who will take on the evils of the world?
To clench and grasp by the throat of opportunity.
There is nothing that will stop he who will go and take what is his.

Poem 2. Challenges to our city. By: Delano.

I bet God can get you out
Of This mess,
If anybody believe he
deliver scream "YES"
Street Life
Think back when was running
In the streets
All the things we did up under the
Sun it wasn't sweet

Senseless killing.

continued on following page

Poem 2. Continued

<table>
<tr><td>We were robbers
We were killers
We were thieves</td></tr>
<tr><td>
Gone too soon.</td></tr>
<tr><td>So called Chiefs had us on dark corners,
They programmed our minds had us doing what they wanted.
We were doing everything you can name from A-Z.
Blind to the facts.
Lost in Misery.</td></tr>
<tr><td>
Looking for trouble; nothing to do.</td></tr>
<tr><td>Looking for a way to turn but marked but marked by the beast.
WE were Robots that were Ran by thoughts.
But not aware of the road that we were about to cross.
Then all of a sudden disaster strikes and we end up getting cursed for what we thought was right.
What you get when you live as a misfit
IN and out of trouble like you some type of dip stick
What you get when you live as a misfit
IN and out of trouble like you some type of dip stick
All the blessings God had for us we missed it
Cause we chose to do a bid instead of receiving it
Same people you thought were your back bone
It's the same people that talking about you back home
He ain't this He ain't gone never be that
Prime Example look where he putting himself at
I bet God can get you out
Of This mess,
If anybody believe he
Deliver scream "YES"</td></tr>
</table>

Poem 3. Black Queens. By: Sierra.

Dark Skinned vs. Light skinned when will this battle between us Black Queens ever end. Don't look at me like I am your enemy because I am your sister. Don't want to hurt you, don't want to abuse you, I'm sorry if the light color of my skin has taken some confidence away from you.
But, you have hurt me too! Calling me light bright like as if my skin is white making me seem like I'm not black see that's not even right. You think I'm boogie and mean right off the back yet you don't take the time to get to know me, not realizing that I am still black!
Your mind is still stuck on slavery days, saying how you were the field nigga and I was the house nigga. (Scratching my head) ***AIN'T WE STILL SLAVES!*** *Why does is matter who was outside or who was inside. But if you want to know the truth I would rather be outside, so that I would not be used as a comfort girl for white men to come inside.*
Stripping me of my innocence. Blossoming my flower before its time, constantly going in and out of me not realizing what he's doing to my body is a crime. But baby girl this was you too! You experienced the same pain I do. So you see, you are me, and I am you. Stop letting people divide us, because I am no more than you.
I am sorry for that person that called you dark night, saying how they was glad they child came out light so they won't be dark like you. ***IGNORANCE!*** *Please don't let that ignorance make you start resenting me. I know they told you my skin was fair!? But I was never the fairest in the land because I am still you, you are still me. Fairness was not in my pigmentation, I wasn't fair enough. The Cinderella movie was still sitting in my living room.*
Please stop letting people fool you. When I saw the movie Dark Girls my heart started crying. See my family is different shades of beautiful blackness and we have never discriminated against each other, so please understand that this dark skin and light skin nonsense comes as a surprise to me. Because I will never let someone disrespect you or me.
I am so sorry for that ignorant man and female that called you dirty and ugly and made you feel like you were below the ground. But sweetie, you are a Queen pick up that crown! For you to feel like I am more than you is blasphemy. I am you and you are me. I'm tired of us trying to compete like we are in a race.
What if my DARK BROWN sisters were the eyes, my CARAMEL sisters were the ears, my BROWN skinned sisters were the lips and my RED BONE sisters were the nose.
WHY ARE WE TRYING TO BE LIKE EACH OTHER WHEN WE ARE APART OF THE SAME FACE?! SO LET US EMBRACE EACH OTHER WITH OUR LONG ARMS AS OUR BACKS STAND STRONG AND WE SHOW LOVE LIKE NO ONE HAS EVER SEEN BECAUSE WE ARE SISTERS, AND MOST IMPORTANTLY BEAUTIFUL BLACK QUEENS!

Poem 4. Social Issues: Chicago. By: Jacqueline.

Our Youth
Wanderers of night and day
Walking the streets with no direction.
Confused and misguided
In need of love and affection.
A repeated cycle with no end
Uneducated minds full of rage.
Choosing wrong over right
Ending up in a cage.

Lost on the corner.

continued on following page

Poem 4. Continued

The Forgotten
There is no place to call home
The nights are long and the floor is cold.
Digging through trash
To hopefully find some gold.
A lost soul who has given up
Jobless and full of hunger.
Still living but in the dark
Full of pain and yet to suffer.

The forgotten.

The Beast
Planted in every city and every country
Destroying all businesses in its path.
Conveniently making profits
While workers exploited feeling wrath.
Taking over the world
Remaining in the one percent.
The rest of society fighting for one cent.

The high cost of low prices.

continued on following page

Poem 4. Continued

Bad Economy
Empty rooms full of dust
Wooden windows all around.
No solution it's time to go
Vacant buildings all over town.
The American dream gone bad
Life savings thrown away.
Memories left behind
You can start again they say.

The sign of the times: Vacant and abandoned.

Poem 5. The Hurt of Knowledge Locked Away. By: James.

Excess for capitalistic pleasure.

continued on following page

Poem 5. Continued

In a time when our communities are hurting
Rich, pale men, hiding in their towers
Miserly not in their expenditures, but in their opportunities
Our people struggling
Our communities drying up

Empty lots replace business.

Empty lots where business used to be
Empty schools where children used to learn
Knowledge locked away, like trusts reserved for those with;
Knowledge stolen that could have enriched the victim
Knowledge of our path sadly unavoidable in its absence
Our mistakes unlearned, unchanged, repeated and magnified

Childhood games speak to bleak adult futures.

Even our childhood games tell of what's to come
If only we could see through the conditioning
A child without training, learning, growth
An adult lost and struggling
Those who have, wielding their knowledge as power
Spewing their lies as truth

continued on following page

Poem 5. Continued

Loss of literacy.

Manipulating the uninformed, the uneducated
Texas and its textbooks across the land
The loss of Franklin
The absence of ethnicity's contribution
A dark labor, a dark time, a sad truth

Hope for educating future generations.

Let us tear down their towers
Let us build living, breathing monuments to knowledge
Beautiful and redeeming
Filled with light hearts, laughing voices, and eager minds
Their minds, laughter, and ingenuity will change our future
They will change our nation
They will change our hearts

continued on following page

Poem 5. Continued

The educated will be set free.

Let the misers keep their opportunities,
The educated will make their own
Knowledge free from its cell, given to the young
Hurt no longer felt or feared
A nation, a people; strong enough to stand again

Poem 6. The Future of Our City. By: Tanika.

Beautiful faces of youth.

The fear of rejection in a city of so much class…
The same rejection our ancestors dealt with in the past….

continued on following page

Poem 6. Continued

Loss of resources creates desolate communities.

Having so little resources' provided is an issue in our communities...
Poverty on the rise which to most is no surprise you see...

We can reach our goals together.

Just one change/chance to act fast for our future....

Together we can make a change in our communities.

We'll reach the goals that were set for us to nurture...
As long as we come together and pave a positive way...

continued on following page

Poem 6. Continued

We can make a difference for next generations.

Our children's children will want to stay!

Major Themes

This case study involved a purposeful sampling of the selected student work. Participants were asked to use a camera as a means for capturing social issues of concerns in their neighborhoods. They were provided with instructions on using the camera as the lens for visually representing social issues impacting their communities. Although the participants were from different neighborhoods, age ranges, genders, and ethnicities, the major themes that emerged from the study were woven through similar threads. Education, poverty, violence and homelessness were presented in the photos and poems of all the participants. The theme of hope also ran through their visual voices as a theme of changing the future for the lives of their children and others in the community.

The research questions which guided the study were:

1. What elements constitute undergraduate student perspectives on social issues in their communities and society?
2. How can photography and poetry be used to illustrate the social issues in their communities and society?
3. What beliefs do these undergraduate students hold which support or negate this perspective?

The topics of poverty, violence, and lack of quality education were major themes that consistently appeared in the data. These themes appeared in the transcripts, artifacts (the student work), and the researcher reflective journal. The researcher transcribed the interviews and categories were created and coded. All of the participants expressed a need for safer communities with a strong call to action to stop the violence. This was a reoccurring theme across the data. The participants talked about the lack of quality public education and discussed ways that education can make a positive change in the lives of the youth and the future of the community. Homeless and poverty were major themes that they discussed wanting to do more to help their communities and provide a brighter future for youth.

The findings of the study reveal that although the participants were from different neighborhoods and backgrounds, the social issues and concerns were the same. There was also a strong desire expressed by the participants to create change within society. With the exception of James who lived in a nearby suburb, the other five participants lived in neighborhood where violence and gang activity is a regular experience. All six of the participants viewed quality education as a means for breaking the cycle of violence, poverty, and homelessness.

FUTURE RESEARCH DIRECTIONS

Findings of this study revealed that, regardless of race or age, there is a need to reduce the amount of violence and poverty in the city. Lack of quality education is an area that also needs to be addressed. It is through education and sense of purpose that urban youth can have hope for a brighter future.

Higher education can be an agent for change. Future research needs to be conducted on ways in which higher education can be affordable and accessible to all individuals. Research findings revealed how participants use their knowledge to benefit the community is needed. In addition to completing a course project, actually following up with students to complete a service learning component would benefit to the students and the community. Identification of issues is the first step; creating an action plan for improvement would lead to positive change.

Qualitative researchers should also consider using poems and photography in their research projects when appropriate. The poems allow participants to have their voices heard in a creative manner. Photographs create an outlet for viewing issues through their own lens and capture images participants may not be able to articulate through spoken word. The visual representation allows the reader to see the actual settings as though being a first-hand observer.

EPILOGUE AND LESSONS LEARNED

After conducting this study the researcher learned the there is a strong desire for university students to have their voices heard about social issues facing their communities. While students are willing to talk about these issues, their use of poems and photography provided a more creative approach to sharing their views. All six of the participants enjoyed working on the project and expressed a strong desire to create change in their communities. The reflective poetry allowed students to express their concerns with a heighten level of passion—one that is often times not easy to articulate though conversation alone. Photography also allows the visual representation of the social issues through actually capturing the image of people and scenery depicting challenges in urban communities.

REFERENCES

Baker, B., Sciarra, D., & Farrie, D. (2010). *Is school funding fair? A national report card*. Newark, NJ: Education Law Center.

Chicago Teacher's Union. (2014). *The schools Chicago's students deserve: Research-based proposals to strengthen elementary and secondary education in the Chicago public schools*. Retrieved from: http://www.ctunet.com/blog/text/SCSD_Report-02-16-2012-1.pdf

Christmann, G. B. (2008). The power of photographs of buildings in the Dresden urban discourse. Towards a visual discourse analysis. Forum Qualitative Sozialforschung / Forum: Qualitative. *Social Research*, *9*(3), 11. Retrieved from http://nbn-resolving.de/urn:nbn:de:0114-fqs0803115

City of Chicago (2014). *Census data: Selected socioeconomic indicators 2008 – 2012*. Author.

City of Chicago Data Portal. (2014). *Crimes 2001 – present*. Retrieved from: https://data.cityofchicago.org/Public-Safety/Crimes-2001-to-present/ijzp-q8t2

Cohen, L., & Manion, L. (1989). *Research methods in education* (3rd ed.). London, England: Routledge.

Creswell, J. W. (1994). *Research design: Qualitative and quantitative approaches*. Thousand Oaks, CA: Sage.

Dewey, J. (1934). *Art as experience*. New York: Minton, Malach, and Co.

Janesick, V. J. (2011). *Stretching exercises for qualitative researchers* (3rd ed.). Thousand Oaks: Sage.

Knoblauch, H., Baer, A., Laurier, E., Petschke, S., & Schnettler, B. (2008). Visual analysis. *Qualitative Social Research, 9*(3), Art. 14. Retrieved from http://nbn-resolving.de/urn:nbn:de:0114-fqs0803148

National Center for Educational Statistics. (2014). *College navigator*. Retrieved from: http://nces.ed.gov/collegenavigator/?q=East+West+University&s=all&id=144883

Silva, C. N. (2008). Review: Catherine Marshall & Gretchen B. Rossman (2006). Designing qualitative research. Forum Qualitative Sozialforschung / Forum: Qualitative. *Social Research*, *9*(3), 13. http://nbn-resolving.de/urn:nbn:de:0114-fqs0803137

Stake, R. E. (1995). *The Art of Case Study Research*. Thousand Oaks, CA: Sage.

Willig, C. (2008). *Introducing Qualitative Research in Psychology: Adventures in Theory and Method*. London: Open University Press.

Yin, R. K. (1981). The case study crisis: Some answers. *Administrative Science Quarterly*, *26*(1), 58–65. doi:10.2307/2392599

Yin, R. K. (2003). *Case study research, design and methods* (3rd ed.; Vol. 5). Thousand Oaks, CA: Sage.

ADDITIONAL READING

Anfara, V. A., Brown, K. M., & Mangione, T. L. (2002). Qualitative analysis on stage: Making the research process more public. *Educational Researcher*, *31*(7), 28–38. doi:10.3102/0013189X031007028

Arminio, J. L., & Hultgren, F. H. (2002). Breaking out from the shadow: The question of criteria in qualitative research. *Journal of College Student Development*, *43*(4), 446–460.

Asselin, M. E. (2003). Insider research: Issues to consider when doing qualitative research in your own setting. *Journal for Nurses in Professional Development*, *19*(2), 99–103. doi:10.1097/00124645-200303000-00008 PMID:12679662

Broido, E. M., & Manning, K. (2002). Philosophical foundations and current theoretical perspectives in qualitative research. *Journal of College Student Development*, *43*(4), 434–445.

Bryman, A. (2007). Barriers to integrating quantitative and qualitative research. *Journal of Mixed Methods Research*, *1*(1), 8–22. doi:10.1177/2345678906290531

Cho, J., & Trent, A. (2006). Validity in qualitative research revisited. *Qualitative Research*, *6*(3), 319–340. doi:10.1177/1468794106065006

Coleman, L. J., Guo, A., & Dabbs, C. S. (2007). The state of qualitative research in gifted Education as published in American journals: An analysis and critique. *Gifted Child Quarterly*, *51*(1), 51–63. doi:10.1177/0016986206296656

Côté, L., & Turgeon, J. (2005). Appraising qualitative research articles in medicine and medical education. *Medical Teacher*, *27*(1), 71–75. doi:10.1080/01421590400016308 PMID:16147774

Creswell, J. W. (2012). *Qualitative inquiry and research design: Choosing among five approaches*. Thousand Oaks, CA: Sage.

Creswell, J. W. (2013). *Research design: Qualitative, quantitative, and mixed methods approaches*. Thousand Oaks, CA: Sage.

Creswell, J. W., Hanson, W. E., Plano, V. L. C., & Morales, A. (2007). Qualitative research designs selection and implementation. *The Counseling Psychologist*, *35*(2), 236–264. doi:10.1177/0011000006287390

DeCuir, J. T., & Dixson, A. D. (2004). So when it comes out, they aren't that surprised that it is there: Using critical race theory as a tool of analysis of race and racism in education. *Educational Researcher*, *33*(5), 26–31. doi:10.3102/0013189X033005026

Denyer, D., & Tranfield, D. (2006). Using qualitative research synthesis to build an actionable knowledge base. *Management Decision*, *44*(2), 213–227. doi:10.1108/00251740610650201

Denzin, N. K. (2012). Triangulation 2.0. *Journal of Mixed Methods Research*, *6*(2), 80–88. doi:10.1177/1558689812437186

Denzin, N. K., & Lincoln, Y. S. (Eds.). (2011). *The Sage Handbook of Qualitative Research*. Thousand Oaks, CA: Sage.

Dixon-Woods, M., Shaw, R. L., Agarwal, S., & Smith, J. A. (2004). The problem of appraising qualitative research. *Quality & Safety in Health Care*, *13*(3), 223–225. doi:10.1136/qshc.2003.008714 PMID:15175495

Dixson, A. D., & Rousseau, C. K. (2005). And we are still not saved: Critical race theory in education ten years later. *Race, Ethnicity and Education*, *8*(1), 7–27. doi:10.1080/1361332052000340971

Driessen, E., Van Der Vleuten, C., Schuwirth, L., Van Tartwijk, J., & Vermunt, J. D. H. M. (2005). The use of qualitative research criteria for portfolio assessment as an alternative to reliability evaluation: A case study. *Medical Education*, *39*(2), 214–220. doi:10.1111/j.1365-2929.2004.02059.x PMID:15679689

Easterby-Smith, M., Golden-Biddle, K., & Locke, K. (2008). Working with pluralism determining quality in qualitative research. *Organizational Research Methods*, *11*(3), 419–429. doi:10.1177/1094428108315858

Haggis, T. (2003). Constructing images of ourselves? A critical investigation into 'approaches to learning' research in higher education. *British Educational Research Journal*, *29*(1), 89–104. doi:10.1080/0141192032000057401

Hatch, J. A. (2002). *Doing qualitative research in education settings*. Suny, NY: Suny Press.

Hellawell, D. (2006). Inside–out: Analysis of the insider–outsider concept as a heuristic device to develop reflexivity in students doing qualitative research. *Teaching in Higher Education*, *11*(4), 483–494. doi:10.1080/13562510600874292

Jackson, A. Y., & Mazzei, L. A. (Eds.). (2008). *Voice in qualitative inquiry: Challenging conventional, interpretive, and critical conceptions in qualitative research*. New York, NY: Routledge.

Jenkings, K. Neil; Woodward, Rachel & Winter, Trish (2008). The emergent production of analysis in photo elicitation: Pictures of military identity. Forum Qualitative Sozialforschung / Forum: Qualitative Social Research, 9(3), Art. 30, http://nbn-resolving.de/urn:nbn:de:0114-fqs0803309

Jones, S. R. (2002). (Re) Writing the word: Methodological strategies and issues in qualitative research. *Journal of College Student Development*, *43*(4), 461–473.

Jootun, D., McGhee, G., & Marland, G. R. (2009). Reflexivity: Promoting rigour in qualitative research. *Nursing Standard*, *23*(23), 42–46. doi:10.7748/ns2009.02.23.23.42.c6800 PMID:19263905

Joy, M. (2007). Research methods in education (No. 10). Innovation Way, York Science Park, Heslington, York YO10 5BR: The Higher Education Academy.

Kincheloe, J. L. (2001). Describing the bricolage: Conceptualizing a new rigor in qualitative research. *Qualitative Inquiry*, *7*(6), 679–692. doi:10.1177/107780040100700601

Kitzinger, J. (1995). Qualitative research: Introducing focus groups. *BMJ. British Medical Journal*, *311*(7000), 299–302. doi:10.1136/bmj.311.7000.299 PMID:7633241

Kondracki, N. L., Wellman, N. S., & Amundson, D. R. (2002). Content analysis: Review of methods and their applications in nutrition education. *Journal of Nutrition Education and Behavior*, *34*(4), 224–230. doi:10.1016/S1499-4046(06)60097-3 PMID:12217266

Kuper, A., Reeves, S., & Levinson, W. (2008). Qualitative research: An introduction to reading and appraising qualitative research. *BMJ: British Medical Journal*, 404-407.

Kvale, S., & Brinkmann, S. (2009). *Interviews: Learning the Craft of Qualitative Research Interviewing*. Thousand Oaks, CA: Sage.

Lather, P., & Moss, P. (2005). Introduction: Implications of the Scientific research in education report for qualitative inquiry. *Teachers College Record*, *107*(1), 1–3. doi:10.1111/j.1467-9620.2005.00450.x

Lichtman, M. (2012). *Qualitative Research in Education: A User's Guide: A User's Guide*. Thousand Oaks, CA: Sage.

MacLure, M. (2003). *Discourse in Educational and Social Research*. Open University Press.

Magolda, P., & Weems, L. (2002). Doing harm: An unintended consequence of qualitative inquiry? *Journal of College Student Development*, *43*(4), 490–507.

Marshall, C., & Rossman, G. B. (2010). *Designing Qualitative Research*. Thousand Oaks, CA: Sage.

Mays, N., & Pope, C. (2000). Assessing quality in qualitative research. *BMJ: British Medical Journal*, *320*(7226), 50–52. doi:10.1136/bmj.320.7226.50 PMID:10617534

Myers, M. D., & Avison, D. (2002). An introduction to qualitative research in information systems. *Qualitative Research in Information Systems*, *4*, 3–12.

Noy, C. (2008). Sampling knowledge: The hermeneutics of snowball sampling in qualitative research. *International Journal of Social Research Methodology*, *11*(4), 327–344. doi:10.1080/13645570701401305

Onwuegbuzie, A. J., Johnson, R. B., & Collins, K. M. (2009). Call for mixed analysis: A philosophical framework for combining qualitative and quantitative approaches. *International Journal of Multiple Research Approaches*, *3*(2), 114–139. doi:10.5172/mra.3.2.114

Parker, L., & Lynn, M. (2002). What's race got to do with it? Critical race theory's conflicts with and connections to qualitative research methodology and epistemology. *Qualitative Inquiry*, *8*(1), 7–22.

Payne, G., & Williams, M. (2005). Generalization in qualitative research. *Sociology*, *39*(2), 295–314. doi:10.1177/0038038505050540

Polkinghorne, D. E. (2005). Language and meaning: Data collection in qualitative research. *Journal of Counseling Psychology*, *52*(2), 137–145. doi:10.1037/0022-0167.52.2.137

Pope, C., & Mays, N. (1995). Qualitative research: reaching the parts other methods cannot reach: an introduction to qualitative methods in health and health services research. *BMJ: British Medical Journal*, *311*(6996), 42–45. doi:10.1136/bmj.311.6996.42 PMID:7613329

Pope, C., Ziebland, S., & Mays, N. (2000). Analysing qualitative data. *BMJ: British Medical Journal*, *320*(7227), 114–116. doi:10.1136/bmj.320.7227.114 PMID:10625273

Redmann, D. H., Stitt-Gohdes, W. L., & Lambrecht, J. J. (2000). The critical incident technique: A tool for qualitative research. *Delta Pi Epsilon Journal*, *42*(3), 132–153.

Reeves, S., Kuper, A., & Hodges, B. D. (2008). Qualitative research methodologies: Ethnography. *BMJ*: British Medical Journal *337* (aug07_3), a1020-a1020.

Riege, A. M. (2003). Validity and reliability tests in case study research: A literature review with "hands-on" applications for each research phase. *Qualitative Market Research: An International Journal*, *6*(2), 75–86. doi:10.1108/13522750310470055

Sambell, K., McDowell, L., & Brown, S. (1997). "But is it fair?": An exploratory study of student perceptions of the consequential validity of assessment. *Studies in Educational Evaluation*, *23*(4), 349–371. doi:10.1016/S0191-491X(97)86215-3

Seale, C., Gobo, G., Gubrium, J. F., & Silverman, D. (Eds.). (2004). *Qualitative Research Practice*. Thousand Oaks, CA: Sage.

Seidman, I. (2012). *Interviewing as Qualitative Research: A Guide for Researchers in Education and the Social Sciences*. New York: Teachers College Press.

Shank, G., & Villella, O. (2004). Building on new foundations: Core principles and new directions for qualitative research. *The Journal of Educational Research*, *98*(1), 46–55. doi:10.3200/JOER.98.1.46-55

Solorzano, D. G., & Yosso, T. J. (2002). Critical race methodology: Counter-storytelling as an analytical framework for education research. *Qualitative Inquiry*, *8*(1), 23–44. doi:10.1177/107780040200800103

Willis, J. (Ed.). (2008). Qualitative Research Methods in Education and Educational Technology. Charlotte: IAP: Information Age Publishing.

KEY TERMS AND DEFINITIONS

Action Research: Inquiry or research in the context of focused efforts to improve the quality of practice and is typically designed and conducted by practitioners who analyze the data to improve their own practice.

Case Study: Attempts to shed light on a phenomenon by studying in depth a single case example of the phenomena. The case can be an individual person, an event, a group, or an institution.

Confidentiality: To respect the confidential nature of the information gathered during the research and preserve the anonymity of participants.

Consent: The informed consent of participants and respect the right of respondents to refuse involvement. Must understand the nature of the project, the procedures that will be used and the use to which the results will be put.

Document Analysis: Looks for occurrences of specific words or phrases in a document or interpret text to seek nuances of meaning and to consider context.

Ethnography: Focuses on the sociology of meaning through close field observation of socio-cultural phenomena. Typically, the ethnographer focuses on a community.

Field Observation: The observation of participants in their natural setting. Observation can be direct or indirect, participatory or non-participatory.

Framework Analysis: Where the objectives of the investigation are set in advance and shaped by the information requirements of the funding body; the thematic framework for the content analysis is identified before the research commences (a priori).

Grounded Theory: Aims to generate a theory that is 'grounded in' or formed from the data and is based on inductive reasoning. This contrasts with other approaches that stop at the point of describing the participants' experiences. In terms of data analysis grounded theory refers to coding incidents from the data and identifying analytical categories as they emerge from the data, rather than defining them a priori.

Holistic: Exploration of a research question multi-dimensionally, exhaustively and in its entirety, preserving the complexity of human behavior.

Transferability: The ability to apply the results of research in one context to another similar context. Also, it refers to the extent to which a study invites readers to make connections between elements of the study and their own experiences.

Triangulation: The process by which the area under investigation is looked at from different perspectives. These can include two or more methods, sample groups or investigators. It is used to ensure that the understanding of an area is as complete as possible or to confirm interpretation through the comparison of different data sources.

Validity: Asks whether the interpretation placed on the data accords with common sense and is relatively untainted with personal or cultural perspective.

Section 5
Analyzing Data

Researchers need to give great consideration to the data analysis process, mastering both the deductive and the inductive analytical stages. There are two sources of interpretation and the researcher needs to move swiftly between a pragmatic positivist approach to a more interpretary stance relying on observations and words to better understand the context surrounding the study in order to build theory. Mixed methods researchers operate between statistical enumeration and analysis seeking to confirm hypotheses while coding qualitative data to detect patterns; hence they are required to demonstrate a repertoire for methodologies.

Chapter 8
Analyzing Quantitative Data

Sema A. Kalaian
Eastern Michigan University, USA

Rafa Kasim
Indiana Tech University, USA

ABSTRACT

The main purpose of this chapter is to present a conceptual and practical overview of some of the basic and advanced statistical tools for analyzing quantitative data. Analyzing quantitative data involves two broad analytical methods that serve two main purposes, which are descriptive and inferential statistical methods. The chapter covers both descriptive and inferential quantitative methods. It covers some of the descriptive statistical methods such as mean, median, mode, variance, standard deviation, and graphical methods (e.g., histograms). It also covers inferential statistical methods such as correlation, simple regression, multiple regression, t-test for two independent samples, t-test for two dependent samples, and analysis of variance (ANOVA).

INTRODUCTION

Conducting quality research is dependent on many significant factors and one of these factors is the researcher's ability to analyze the collected data using the most appropriate statistical methods to answer the research questions of the scientific study. Depending on the research question(s) and the nature of the collected data, the researcher might use quantitative analytical methods, qualitative analytical methods, or a combination of both of the data analytical methods. In other words, the researcher must pick the proper quantitative, qualitative, or a combination of both analytical methods (mixed methods) based on the nature and type of the data (quantitative or qualitative) to accurately answer the research questions and draw valid conclusions about the theoretical model put forward for the study.

The focus of this chapter is analyzing quantitative data that are measured using continuous interval and/or ratio scales of measurement. Analyzing qualitative data are covered in another chapter of this book. However, analyzing quantitative data involves two broad analytical methods that serve two purposes; descriptive and inferential statistical methods. This chapter covers both descriptive as well as inferential quantitative methods. Accordingly, the chapter is organized into two sections. Section I covers descriptive statistical methods and Section II covers inferential statistical methods.

DOI: 10.4018/978-1-5225-0007-0.ch008

I. DESCRIPTIVE STATISTICS

Descriptive statistics is a set of statistical methods in the statistical toolbox for describing and providing summary measures of the quantitative data. Any collected original quantitative data is usually overwhelming and uninformative, especially when the amount of the data is large. Therefore, it is necessary for the quantitative data analyst to use descriptive statistics to describe and summarize a large amount of data using simple summary measures.

For example, consider the Grade Point Average (GPA), which is a simple descriptive summary measure used to represent how well each student is performing in an educational institution such as college or high school. This single summary measure, GPA, describes the general performance (achievement) of each student across a wide range of academic courses in the academic institution. So, we need to keep in mind that important details of the data are lost (in this example, the details of each of the courses that had been taken by a particular student) when we describe a large set of data (in this example, achievement scores of each student across all the courses that had been taken) with a single summary measure or indicator (in this example, GPA).

Using descriptive statistical methods for analyzing quantitative data is the first and the most significant step in the quantitative data analysis journey. For example, it helps the researcher to be familiar with the data and discover early in the data analysis process any anomalies in the data set. In the following five subsections, the descriptive statistical methods that can be used for analyzing quantitative data are covered.

1. Describing Data Using Frequency Distributions

The *frequency distribution* for a specific variable in a data set (e.g., the age variable for a group of students) is a tabular representation of the number of instances (occurrences or frequencies) of each numerical value for a specific variable (in this case, the age variable) in the data set.

The frequency distribution is one of the simplest statistical procedures and yet it is one of the most useful tools for describing a quantitative data set.

Creating frequency distributions for each of the variables in the quantitative data set is the first step in analyzing a data set to explore and examine the numerical content and characteristics of each of the variables in the quantitative data. It helps researchers and data analysts to (1) organize and summarize the quantitative data in tabular formats, which helps the data analyst to easily examine the data set; (2) identify and assess the amount of the missing numerical values in a quantitative data set; and (3) detect one or more outliers (unusual extreme values) in the quantitative data (Field, 2009; Kalaian, 2008; Witte & Witte, 2013). For example, a value of a grand point average (GPA) of 0.1 in the frequency distribution of GPA scores, where most of the scores are above 50 on a scale from 1 to100, is considered as being an outlier score. This outlier score requires special treatment from the researcher such as the need to verify its accuracy or to exclude the student with a GPA of 0.1 from the data set.

2. Describing Data Using Graphical Representations

The *graphs (charts)* are important and useful visual representations of a quantitative data set. Generally, graphical presentations of a data set provide a more powerful, informative, and interesting message than the tabular representations. Graphs and charts provide insights and help researchers to visually (1) summarize, describe, and examine the shape of the frequency distribution of a large amount of quantitative data in graphical formats; (2) examine the required assumptions of the statistical methods (e.g., examine the normality assumption); (3) detect one or more outliers (unusual extreme values) in the quantitative data set; and (4) explore and describe the hidden structure or pattern in the data (e.g., relationships between variables). There are many graphical representations for quantitative data such as histograms, frequency polygons, box plots, stem-and-leaf plots, and scatter plots. The histograms and frequency polygons are the two graphical methods that are described and illustrated below:

Histogram

The *histogram* is the most often used graph (chart) of a quantitative data that graphically illustrates the overall pattern and shape of a frequency distribution. A histogram is represented by drawing rectangular bars (columns) of equal width for the numerical categories of a variable in a data set and the height of each rectangular bar represents the frequency of occurrence of each of the numerical categories of a variable in a quantitative data set. In other words, on a histogram, the horizontal axis (x-axis) shows the data values arranged into categories while the vertical axis (y-axis) shows the number of occurrences (frequencies) for each category of a variable. One of the advantages of a histogram is that it provides insights about outliers (extreme values) and the shape of the distribution of the quantitative data set. Examining the shape of a distribution is an important step, for example, to test the normality assumption of most of the inferential statistical methods that are used for analyzing quantitative data (Field, 2009; Witte & Witte, 2013).

A *normal distribution* is characterized of having a bell-shaped symmetrical distribution around the mean (average) value of the numerical values of a variable in a data set. Otherwise, the distribution is characterized to be skewed. A skewness value of zero indicates that the distribution of a given variable in the data set is bell-shaped and symmetrical around the mean of the distribution and therefore it is normal. For example, histogram "A" shown in Figure 1 depicts a numerical data set, which has a normal distribution. A *positive skewness* (shown in histogram "B") indicates a greater number of smaller numerical values in the distribution of the quantitative data set, which are clustered in the left-hand side of the histogram. In contrast, a *negative skewness* (shown in histogram "C") indicates a greater number of larger numerical values in the data set, which are clustered on the right side of the histogram.

Frequency Polygon

The frequency polygon represents a set of continuous straight lines that connect between the midpoints at the top of each of the bars (columns) of a histogram, which represents the frequencies of the numerical values of a variable in a quantitative data set. Frequency Polygons provide insight, especially when

Figure 1. Histograms of frequency distributions displaying different types of skewness

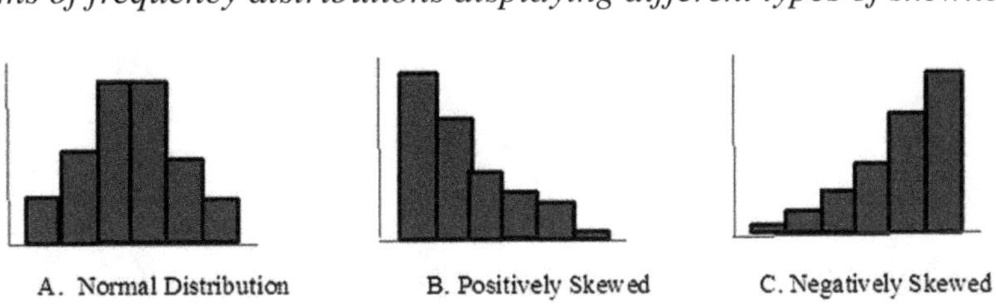

there are large amounts of numerical data. It is also a valuable tool when comparing the distributions of a particular variable for two groups or comparing the distribution of two different variables in a single graph. Graph "A" in Figure 2, shows a frequency polygon for an approximately normally distributed variable. Graphical representation "B" shows a frequency polygon for a positively skewed distribution of a variable in a quantitative data set. Graph "C" shows a frequency polygon of a negatively skewed distribution of a variable.

3. Describing Data Using Measures of Central Tendency

The *central tendency measures* of a distribution are descriptive summary measures that represent the central location of a distribution of a variable in a quantitative data set. Although tables and graphs of the frequency distributions are important basic analytical procedures to describe data, informative summary values such as means (averages) and medians provide additional valuable descriptive and representative measures of the distribution of the data. The most commonly used measures of central tendency are the mode, the median, and the mean. These measures are presented below:

Mode

Mode is the numerical value with the highest frequency in a set of numerical values of a distribution. For example, for the following distribution of numerical values (95, 87, 94, 80, 71, 65, 71, and 68), the mode is 71 because it has the highest occurrence frequency (in this example, it occurred twice).

Although the mode is the easiest and simplest measure of all central tendency measures to find in a set of numerical values, it also has some limitations. One of the key limitations is that the mode focuses only on the most frequently occurring numerical values and ignores all the other numerical values in a distribution. In other words, much information in a data might be lost by using the mode. Therefore, it is considered the least precise measure of central tendency measures. The other limitation is that the mode is not a reliable measure because any change or error in one numerical value can have a dramatic effect on the value of the mode.

It is important to note that it is possible that a set of numerical values of a variable in a data set to have more than one mode. For example, in the data set (95, 87, 68, 97, 80, 71, 65, 71, and 68), the two modes are 68 and 71 because the numerical values 68 and 71 occurred twice. Thus, this data set has a "bimodal" distribution because it has two modes. Keep in mind that if no number appears more than once in a set of numerical values or each of the numerical values of a variable has equal number of occurrences, then that particular variable has no mode.

Figure 2. Frequency Polygons of distributions displaying different types of skewness

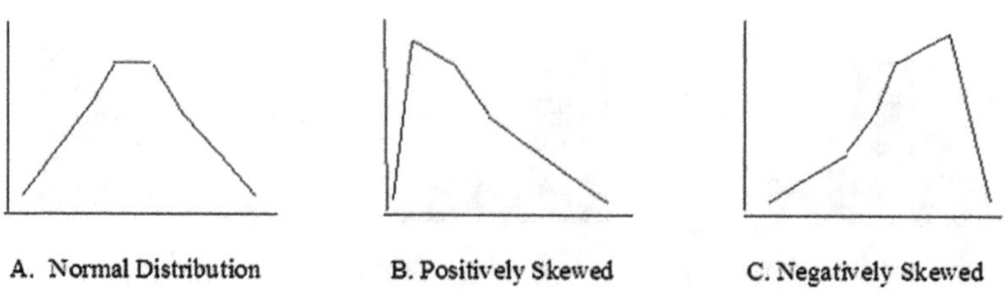

Median

Median is the middle point in a frequency distribution, where fifty percent of the numerical values in a frequency distribution are above the median of a variable in a data set and the other 50% are below the median value (Field, 2009; Witte & Witte, 2013). For example, for the following distribution of odd number of numerical values that are listed in a descending order (95, 87, 80, 71, 71, 68, and 65), the median is 71, which is the middle value of the ordered numerical values (ordered in either descending or ascending order). For a data set of even number of numerical values, the median is the average of the two innermost ordered numerical values. For example, for the following distribution of eight (even number) numerical values, which are listed in a descending order (95, 87, 80, 71, 70, 70, 68, and 65), the median is 70.5 (Field, 2009; Witte & Witte, 2013).

Mean

M*ean* is commonly referred to as average and represents the central value of a quantitative data set. Computationally, it is the total sum of all the numerical values of a variable divided by the number of the numerical values. The formula for the computation of the mean ($\bar{X}$) is as follows:

$$\bar{X} = \sum X / n \tag{1}$$

where, $\sum X$ is the sum of the numerical values of the data; and n is the number of the numerical values in a set of data. For example the mean for the data set (95, 87, 80, 71, 70, 70, 68, and 65) is equal to

$$\bar{X} = (95+87+80+\ldots. +68+65)/8 = 75.75 \tag{2}$$

The mean is the most commonly used central tendency measures for analyzing quantitative data because it is the only one of the three types of the central tendency measures that takes into account all the numerical values in the quantitative data set (Field, 2009; Witte & Witte, 2013). Therefore, it is considered a representative descriptive measure for summarizing a set of numerical values in a data set.

It is important to note that since the mean takes into account all the numerical values in a data set, it is usually influenced by the extreme numerical values (outliers, which have extremely either high or low numerical values). Therefore, in a data set with extreme numerical values (few large or small numerical values), the mean can be deflated (lower) or inflated (higher) depending on the numerical values of the extreme values. In such cases, the mean would misrepresent the central tendency of the distribution of the numerical values in a quantitative data set.

4. Describing Data Using Measures of Variability

Statistical methods for measuring *variability* (spread or dispersion) are very useful tools in descriptive statistics, especially when they are used with measures of central tendency to provide a comprehensive outlook of the distribution of a quantitative data set. As the measures of central tendency, the measures of variability are very powerful descriptive summary measures and yet simple to calculate, understand,

and interpret (Field, 2009; Witte & Witte, 2013). The most commonly used measures of variability are the range, variance, and standard deviation and these measures are presented below:

Range

Range is the difference between the largest and smallest numerical values in a quantitative data set.

Range = Highest numerical Value - Lowest Numerical Value (3)

The range is very easy to calculate and understand because it takes into account only two numerical values (the highest and lowest numerical values) from a quantitative data set. Yet the range provides a quick summary of the variability and overall spread of the distribution of a quantitative data set. One of the key limitations of the range is that its value is based on only two values (the lowest and highest numerical values) of a variable in a data set, which makes vulnerable to the extreme (outliers) values (Field, 2009; Witte & Witte, 2013). Also, it does not provide any information about how all the numerical values of the distribution of a data set is clustered or concentrated around the center of the distribution. We would like to have another measure of variability that takes into consideration every single numerical value in a quantitative data set. The variance and the standard deviation are two such measures of variability.

Variance

Variance refers to the spread (dispersion) of the values of a quantitative variable around their mean. Specifically, it can be defined as the average squared deviations of the numerical values of a quantitative variable from their mean. Similar to the mean of the numerical values of a data set, the variance takes into consideration every single numerical value in a quantitative data set. The formula for calculating the variance of a numerical data set is as follows:

$$s^2 = \sum(X - \bar{X})^2 / (n\text{-}1) \quad (4)$$

where, s^2 is the variance of a set of the numerical values, X; $\bar{X}$ is the mean of a set of the numerical values, X; and n is the number of the numerical values. The value of the variance is always positive because it is based on squared deviations of each of the numerical values of a variable in a data set from the mean of the numerical values.

The variance provides significant information about the spread of the numerical values of the variables in a quantitative data set. The larger the value of the variance the more variability and spread in the numerical values of the data. If the value of the variance is zero, it means that all of the cases have the same value and absolutely no variability exists among the numerical values of a quantitative data set.

Standard Deviation

Standard deviation is the square root of the variance. It is the best measure of the spread of a distribution of a set of numerical values and it is the most commonly used measure of variability (spread, dispersion). As the variance, the value of the standard deviation is always positive because it is simply the square root of the variance, which always has a positive value (Field, 2009; Witte & Witte, 2013).

Similar to the variance, the higher the value of the standard deviation, the more spread are the numerical values of the data. In fact, it is most often used and reported by researchers because it is more useful measure of spread than the variance because it provides significant information about the spread of the numerical values of the variables in a quantitative data set. Unlike the variance, which is stated in measurement units that are squared, the standard deviation is stated in terms of the original units of measurement of each of the variables in a data set. This is the key reason that makes the standard deviation more useful measure of variability to use in reporting and interpreting.

5. Describing Data Using Measures of Relationship

There are several measures for describing the relationships between variables. The choice of any of these measures to describe a relationship depends on the scale of measurements (types of the variables) of the correlated variables. *Pearson product-moment correlation coefficient* is one of these measures of relationships that most commonly used in analyzing quantitative data. It is represented by the lower case, *r*. It measures the degree of the linear relationship between two continuous variables in a quantitative data set. The correlation coefficient as a statistical method for analyzing data named after its developer, Karl Pearson. The product-moment part of the term comes from its calculation method, which focus on summing up the products of the deviations of each of the original numerical values from the mean of the numerical values for each pair of the variables.

The value of the Pearson product-moment correlation coefficient ranges from -1.0 to +1.0 The absolute value of the correlation coefficient represents the magnitude or the strength of the relationship between the pairs of the correlated variables; with zero correlation coefficient indicating no relationship between the correlated variables; and a correlation coefficient of one indicating a perfect relationship. The positive and negative signs describe the direction of the relationship. Positive values of *r* indicate positive association between the two variables. As values of one variable increase, the values of the other variable increase as well.

For example, typically one would expect that students with high GPA in high school also have high college GPA and conversely students with low high school GPA also have low college GPA. Negative values of *r* indicate that as the values of one variable increase the values of the other variable decrease. For example, students with high GPA in high school have low college GPA or conversely students with low high school GPA have high college GPA. Such a relationship with a perfect correlation coefficient of -1.0 is referred to as a high negative correlation between high school GPA and college GPA. It is important to note that the larger the absolute value of the correlation coefficient (e.g., close to absolute 1.00) the greater the relationship between the two variables. A correlation coefficient of zero indicates no relationship between the two variables.

The relationships between two variables can be also presented graphically using *bivariate scatterplots*. A bivariate scatterplot is a graphical representation of the relationship between two quantitative variables. It is created by allocating the x-axis to the numerical values of one variable and the y-axis to the numerical values of the second variable. Individual's (subject's) numerical score on both variables is represented by a dot on the graph. The distance of the dot from the x-axis represents the numerical value of the subject on the variable associated with the y-axis. Similarly, the distance of the dot from the *y*-axis represents the numerical value of the subject on the variable associated with the *x*-axis. In Figure 3, scatterplot "A" represents a strong negative relationship between two variables (for example, students' high school GPA and college GPA); possibly close to -1.0; Scatterplot "B" shows a strong

Figure 3. Scatterplots of bivariate data displaying different types of relationships

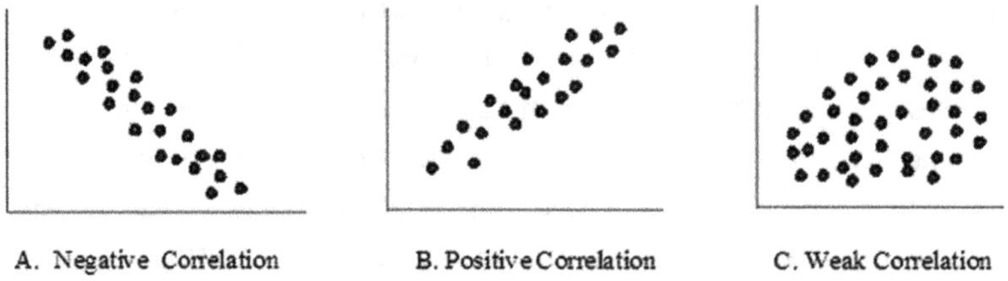

positive relationship between two variables, possibly close to +1.0; Scatterplot "C" shows a weak or no relationship between the two variables, possibly zero or close to zero.

II. INFERENTIAL STATISTICS

Inferential Statistics are statistical procedures used to estimate and test the estimated values of the characteristics (parameters) of a population based on data obtained from a representative sample of individuals randomly drawn from a specified population. Therefore, inferential statistics allow us to make inferences about the characteristics of the population based on a representative sample randomly drawn from the larger population. In other words, it allows us to estimate and test the population parameters (e.g., population mean and variance) based on the sample statistics (e.g., sample mean and variance). Inferential statistical methods include numerous statistical procedures designed to answer different research questions and hypotheses. Some of the basic inferential statistical methods often used for analyzing quantitative data are:

1. Simple Regression Analysis

Simple regression analysis is an inferential statistical method for modeling and analyzing a linear relationship between a single continuous dependent (outcome) variable and a single continuous or categorical independent (predictor) variable. Is there significant effect of each of the multiple independent variables (predictors) on the outcome (dependent) variable?

In a simple linear regression analysis, the linear relationship between a single dependent variable and a single independent (predictor) for *n* individuals (subjects) is represented by the following simple linear regression model:

$$Y_i = \alpha + B\,X_i + e_i \qquad (5)$$

where,

Y_i is the value of the dependent (outcome) variable for the ith individual.

α is a constant numerical value that represents the intercept of the regression line with the y-axis. In other words, it is the value of the dependent variable, Y_i, when the value of the independent variable, X_i, is equal to zero.

B is the regression coefficient (slope) of the regression line. It can be interpreted as the amount of change in the dependent (outcome) variable for one unit change in the independent (predictor) variable.
X_i is the numerical value of the independent (predictor) variable for the ith individual.
e_i is the error of the regression model. It represents the amount by which the observed value of the dependent variable, Y_i, deviates from the predicted value of Y_i for the ith individual.

A scatterplot of the actual data points of a single independent (predictor) variable and a single dependent (outcome) variable is called a bivariate scatterplot (see Figure 3). It is often used to graphically portray the linear relationship between the predictor variable (X) and the dependent variable (Y). A regression line that represents the best fit of the regression model is usually drawn through the scatterplot data points to portray the direction and the extent of the deviations of the actual data values from their predictive values, which are located on regression line (Field, 2009; Kalaian & Kasim, 2015; Tabachnick & Fidell, 2012). In Figure 4, graph "A" represents a linear regression line with a high and negative slope (B) value; and graph "B" represents a regression line with a high and positive slope (B) value.

A t-statistic can be used to test whether or not the associated population parameter of the regression coefficient is equal to zero. In other words, it tests the null hypothesis (H_0: $B = 0$). The t-statistic is computed by dividing the estimated regression coefficient, b, by the standard error (S.E.) of the estimated regression coefficient, b. The formula is represented as follows:

$$t = b / \text{S.E.}(b) \tag{6}$$

As any inferential statistical method, the simple linear regression analysis is based on specific assumptions about the quantitative data. Meeting these assumptions is a necessary condition in order to accurately estimate and test the parameters of the regression model. The assumptions are: (a) The error terms (e_i) are independently and normally distributed with a mean (average) value of zero; (b) The errors of fitting the regression model are not correlated with the single independent (predictor) variable; (c) The relationship between the dependent variable and the independent variable is linear; and (d) The distribution of the errors of the linear regression model for each value of the independent variable (X) is normal and has a constant (equal) variance across all numerical values of the independent (predictor) variable.

Regression analysis aims to (a) predict the values of the dependent (outcome) variable given known values of an independent (predictor) variable, and (b) evaluate the accuracy of the regression model in predicting the unknown values of the dependent variable from known values of the independent variables. That is, evaluating the adequacy of fitting the regression model to the quantitative data (Field,

Figure 4. Regression lines fitted to bivariate data displaying different types of regression lines

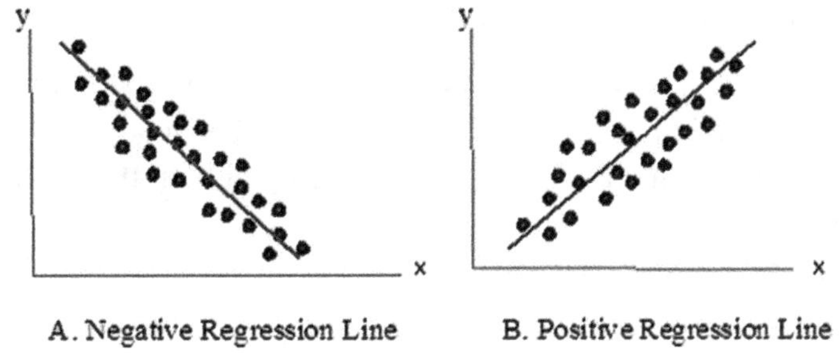

2009; Kalaian & Kasim, 2015). The predicted value (*Yi*) of the dependent variable ($\hat{Y}i$) in simple linear regression is represented as follows:

$$\hat{Y}_i = a + b\,X_i \tag{7}$$

where,

a is the estimated intercept of the population parameter, α.
b is the estimated regression coefficient (slope) of the population's regression coefficient parameter, B.
X_i is the numerical value of the independent (predictor) variable for the i_{th} individual.

The accuracy of the predicted value of a Y_i value can be found by subtracting the predicted value of the dependent variable, $\hat{Y}_i$, for the ith individual from the observed value of the dependent variable, Y_i. The closer the value of the difference to zero the more accurate the prediction is. This difference is called the error of prediction, $\widehat{e_i}$, and it is represented as:

$$e_i = Y_i - \hat{Y}_i \tag{8}$$

2. Multiple Regression Analysis

Multiple linear regression analysis is an extension of the simple linear regression. The main research question addressed in multiple linear regression analysis can be formulated as follows: Are there significant effects of the multiple independent variables (predictors) on the outcome (dependent) variable? However, the linear relationship between a dependent variable and two or more known independent (predictor) variables (X_1, X_2, …., X_k) in the population is represented by the following linear model:

$$Y_i = \alpha + B_1X_{i1} + B_2X_{i2} + \ldots. + B_kX_{ik} + e_i \tag{9}$$

where,

Y_i is the value of the dependent (outcome) variable for individual i.
X_{i1}, $X_{i2,\ldots,}$ X_{ik} are the values of the k independent (predictor) variables for individual i.
α is the intercept of the regression model. In other words, it is the value of the dependent variable, Y_i, when the numerical values of all the independent variables, X_{i1}, $X_{i2,\ldots,}$ X_{ik}, are equal to zero.
B_1, B_2, …, B_k are the regression coefficients for the k independent variables, X_{i1}, $X_{i2,\ldots,}$ X_{ik}, respectively.
e_i represents the amount by which the observed value of the dependent variable, Y_i, deviates from the predicted value, $\widehat{Y_i}$, for the ith individual. Each of the regression coefficients (B_1, B_2, …, or B_k) for a given independent variable is interpreted as the expected change in the value of the dependent variable for one unit change in the independent variable, with all the other independent variables in the specified regression model are held constant.

The multiple linear regression analysis is based on specific assumptions about the quantitative data. Meeting these assumptions is necessary in order to achieve the best linear estimation of the parameters of the regression model. The assumptions of the multiple regression analysis are: (a) The distribution of the dependent (outcome) variable is normal; (b) The relationship between the dependent variable and each of the multiple independent variables (predictor) is linear; (c) The independent variables in the regression model are not correlated; (d) The errors are not correlated with each of the independent (predictor) variables; and (e) The error terms (e_i) are independent and identically normally distributed with an overall mean equal to zero and a constant variance across the values of the independent variables (Kalaian and Kasim, 2015; Mertler & Vannatta, 2013).

An omnibus *F-test* tests the overall model significance. It tests whether or not all the population regression coefficients ($B_1, B_2, \ldots, B_k$) for all of the independent variables in the regression model are simultaneously equal to zero. The null hypothesis associated with the F-test is that none of the independent variables have an effect on the dependent variable. In other words the population regression coefficients are simultaneously equal to zero (H_0: $B_1 = B_2 = \ldots. = B_k = 0$) (Field, 2009; Kalaian & Kasim, 2015; Tabachnick & Fidell, 2012).

A *t-test* tests the unique effect of each independent variable. That is, it tests whether or not the associated population parameter of each of the regression coefficients in the specified regression model is equal to zero. In other words, it tests the null hypothesis (H_0: $B_k = 0$). As in the simple regression analysis, the t-statistic is computed by dividing the estimated regression coefficient, b_k, by the standard error (S.E.) of the estimated regression coefficient, b_k (Field, 2009; Kalaian & Kasim, 2015; Mertler & Vannatta, 2013). The formula is represented as follows:

$$t = b_k \,/\, \text{S.E.}\ (b_k) \tag{10}$$

Generally, the multiple linear regression models are used to (a) predict the values of the dependent variable given known values of the independent (predictor) variables, and (b) evaluate the degree of accuracy in prediction (that is, evaluating the fitting of the regression model to the quantitative data) of the unknown values of a dependent variable from multiple independent (predictor) variables (Field, 2009; Kalaian & Kasim, 2015; Tabachnick & Fidell, 2012). The predicted value ($\hat{Y}$) of the dependent variable (Y) in a multiple linear regression is represented as follows:

$$\hat{Y}_i = a + b_1 X_{i1} + b_2 X_{i2} + \ldots. + b_k X_{ik} \tag{11}$$

where,

a is the estimated intercept of the population parameter, α.

$b_1, b_2, \ldots, b_k$ are the estimated regression coefficients (slopes) of the regression coefficient parameters, $B_1, B_2, \ldots, B_k$ for the independent variables $X_{i1}, X_{i2}, \ldots, X_{ik}$ respectively.

As in simple regression, the accuracy of the predicted values of any Y_i value can be found by subtracting the predicted value of the dependent variable, $\hat{Y}_i$, from the original value of the dependent variable, Y_i, for the ith individual. This difference is called the error of prediction, $\widehat{e_i}$, and it is represented as:

$$\widehat{e_i} = Y_i - \hat{Y}_i \tag{12}$$

Regression analysis is a *least squares* analytic procedure that minimizes the sum of the squares of the residuals (errors), e_i, from the regression line. The residual is the vertical distance from the regression line to each of the original observations of the dependent variable, Y_i. Therefore, regression analysis fits a straight line through the data that yields the smallest sum of the squared residuals (Field, 2009; Kalaian & Kasim, 2015; George & Mallery, 2008).

In addition, as part of the regression model building process, it is necessary to evaluate the goodness of fit of the model (Kalaian & Kasim, 2015). The coefficient of determination, R^2, is a descriptive measure of goodness-of-fit of the regression model. It is the most commonly used statistics to evaluate the fit of the regression model to the quantitative data. R^2 can be interpreted as a measure of the proportion of the total variance in the dependent variable (Y) that is accounted for and explained by the linear combination of the independent variables in the regression model. Simply, it represents the squared correlation between the actual values of the dependent variable (Y_i) and the predicted values ($\hat{Y}_i$) obtained from the regression equation. Its numeric value ranges from 0 to 1 and the closer R^2 statistic is to the value of 1, the better the fit of the estimated regression line to the data (George & Mallery, 2014, Mertler & Vannatta, 2013; Kalaian, & Kasim, 2015; Tabachnick & Fidell, 2012).

3. Two Independent Samples *t*-Test

The *t-test* statistics is used to compare the means (averages) of two independent groups on a single outcome measure (dependent variable) to see if there is a statistically significant difference between the two independent groups. The main research question that is addressed in a research study in order to use two independent groups *t*-test can be formulated as follows: Is there significant difference between the means (averages) of the outcome (dependent) variable for the two independent groups?

Specifically, the t-test assesses the research hypothesis (alternative hypothesis represented as H_A) that the population means of the two independent groups (e.g., a treatment group versus a control group or male versus female groups) are statistically different from each other (Tabachnick & Fidell, 2012; Kalaian, 2008). The research hypothesis is symbolically represented as follows:

$$HA: \mu 1 \neq \mu 2 \tag{13}$$

Here, μ_1 and μ_2 are the population means of the outcome variable for the two groups respectively. The null hypothesis (H_0) associated with this analysis is that there is no difference between the population means of the two groups (H_0: $\mu_{1} = \mu_2$).

The *t*-value will be positive if the first mean is larger than the second and negative if it is smaller. If the null hypothesis that is no difference between the two groups is rejected, the result often described as the average difference between the two groups is statistically significant indicating that it is a true difference in the populations. The formula for the *t*-test can be represented as follows:

$$t = \frac{\bar{X}_1 - \bar{X}_2}{\sqrt{\left[\frac{(n_1 - 1)s_1^2 + (n_2 - 1)s_2^2}{n_1 + n_2 - 2}\right]\left[\frac{n_1 + n_2}{n_1 n_2}\right]}} \tag{14}$$

where the symbols $\bar{X}_1$ and $\bar{X}_2$ represent the means of the dependent variable (outcome) for the two independent groups respectively; s_1^2 and s_2^2 represent the variances of the dependent variable (outcome) for the two groups respectively; n_1 and n_2 are the sample sizes of the two groups respectively. The ratio that is computed is called a t-test and describes the difference between the averages of the dependent variable (e.g., GPA, achievement scores) for the two independent groups relative to the variability of the dependent variable.

As in any quantitative statistical analysis, the *t*-test for difference between two independent samples is based on specific assumptions. The following are the assumptions about the required characteristics of the observations when using *t*-test:

a. The observations (numerical values of the dependent variable) are normally distributed in both populations.
b. The observations (numerical values of the dependent variable) are independent from each other.
c. The variances of the observations (numerical values of the dependent variable) for both populations are equal, which is referred to the equality of the two variances or the homogeneity of variances assumption.

4. Dependent Sample *t*-Test

The *t*-test for two related samples design involves making comparisons between two sets of numerical measures for the same group of individuals. For example, the measures that are taken by the students before administering a treatment or intervention (e.g., pretest scores) and the measures that are taken by the same students after the treatment intervention (e.g., posttest scores). Unlike independent samples t-test, in this research design, the two sets of measurements for a single group of individuals are related, which means that there is a relationship between the two measurements.

The main research question that is addressed in a research study in order to use one-group dependent *t*-test can be formulated as follows: Is there significant change in the means of the outcome (dependent) variable before administering the treatment (*time 1*) and after administering the treatment (*time 2*) to the same sample of individuals? Specifically, the dependent t-test assesses the research hypothesis (alternative hypothesis, H_A) that the population means of two related dependent outcome measures (e.g., a post-test achievement scores and pretest achievement scores, which are measurements over two time points) on average are *statistically* different from each other (H_A: $\mu^{\text{time 1}} \neq \mu^{\text{time 2}}$). Here, μ_1 and μ_2 are the population means of the two related outcome measures (e.g., posttest and pretest) for a single group of individuals (students) respectively. The null hypothesis associated with this analysis is that there is no difference between the population means of the two related outcome measures for a single group of individuals (H_0: $\mu^{\text{time 1}} = \mu^{\text{time 2}}$).

The dependent *t*-value will be positive if the first mean is larger than the second and negative if it is smaller. If the null hypothesis that is no difference between the two outcome measures (post-test and pretest scores) is rejected, the result often described as statistically significant. The formula for the *t*-test can be represented as follows:

$$t = \sum D \,/\, \text{SQRT}\ [(n \sum D^2 - (\sum D)^2) \,/\, (n\text{-}1)] \tag{15}$$

where,

$$D = \overline{X}^{\text{ time 2}} - \overline{X}^{\text{ time 1}} \tag{16}$$

Here, D is the difference between the mean of the measurements taken at two time points (*time 1* and *time 2* measurements); and n is the number of individuals (subjects, cases) with measurements taken at the two time points. As any statistical procedure, the t-test for differences between two dependent samples is based on the assumption that the values of the differences between the measurements taken for *time 1* and *time 2* for a single group of individuals are normally distributed.

5. One-Factor Analysis of Variance (ANOVA)

One-factor analysis of variance (ANOVA) is used to examine the direct main effect of a categorical independent variable (called "factor") on a continuous dependent (outcome) variable. Thus, the research design of one-way ANOVA deals with one categorical independent variable with two or more categories and one continuous dependent variable. It tests whether the categories (e.g., treatment groups) of the independent variable, on average, are different or not on the dependent variable. If the categories seem different, then it is concluded that the independent variable has an effect on the dependent (outcome) variable (Field, 2009; Mertler & Vannatta, 2013).

For example, ANOVA examines the effects of different teaching methods of math as an independent variable on students' math achievement (dependent variable). In other words, it tests the similarities and differences between two or more means of the categories of the independent variable. The main research question that is addressed in a research study in order to use one-factor ANOVA can be formulated as follows: Is there significant differences among the means (averages) of the outcome (dependent) variable for the treatment groups?

An F-test of difference between two or more treatment means is the main statistical test in ANOVA. Specifically, the F-test assesses whether or not the null hypothesis (H_0) that the population means of the dependent variable of two or more categories of the independent variable are *statistically* equal to each other. The null hypothesis for the ANOVA is often stated as there are no significant differences between groups (treatments) on the dependent variable. For example, for three treatment groups, the null hypothesis for ANOVA is represented as follows:

$$H0\text{: } \mu 1 = \mu 2 = \mu 3 \tag{17}$$

where, μ_1, μ_2, and μ_3 are the population means for the three treatments of the independent variable respectively. On the other hand, the research hypothesis (alternative hypothesis, H_A) is represented as follows:

$$H_A\text{: } \textit{At least one of the treatments means is not equal to the other means} \tag{18}$$

Based on the ANOVA F-test results, if the treatment groups are significantly different it is inferred that the independent variable has an effect on the dependent variable. If the treatment group means do not differ significantly then it is concluded that the independent variable did not have an effect on the dependent variable.

As in any quantitative statistical method, ANOVA is based on specific assumptions about the quantitative data. The assumptions of the Analysis of Variance are:

1. The dependent (outcome) variable for each category of the independent variable is normally distributed.

2. The equality of the variances of the dependent variable across all the categories of the independent variable. This is also called the "homogeneity of variance" assumption.
3. The errors (e_i) are independent and identically normally distributed with mean equal to zero and a constant variance.

CONCLUSION

Since the quantitative analysis methods cover a wide-range of analytical and modeling techniques, all the methods are impossible to be covered in a single chapter. Therefore, in this chapter, we chose to cover some of the important descriptive and inferential methods that are most commonly used by researchers across a wide spectrum of disciplines and fields of study. Analyzing quantitative data involves two broad analytical methods; descriptive and inferential statistical methods. Accordingly, descriptive as well as inferential quantitative statistical methods are covered.

Specifically, various descriptive statistical methods for analyzing quantitative data such as frequency distribution, graphical data presentation, measures of central tendency, measures of variability, and measures of relationships were covered. In addition, various univariate inferential statistical methods for analyzing quantitative data such as correlation, simple and multiple linear regression analyses, and analysis of variance (ANOVA) were covered in this chapter. The overview of each of the presented statistical methods included detailed description of the analytic method, research questions, examples, the specific model, the required assumptions, and the computational formulas.

Despite the overview of each of the descriptive and inferential statistical methods for analyzing quantitative data covered in the present chapter, it is important to note that these methods can be expanded to include more information about the statistical methods (e. g., residual analysis for regression analysis, box plots and stem-and-leaf graphical methods for graphical representation of data) and the analytical software that can be used to analyze the data. However, the statistical methods that are covered in this chapter can be found in most of introductory and advanced statistics textbooks. Thus, the reader is recommended to use other statistical textbooks such as Field (2009), George and Mallery (2014), Mertler and Vannatta (2013), Tabachnick and Fidell (2012), Witte and Witte (2013), or any other book for statistical methods to seek additional information such as how to use a specific computer software (e.g., Excel, Minitab, SPSS, SAS) to analyze a quantitative data set and interpret the results of the statistical analysis.

In conclusion, the main purpose of this chapter has been to present a conceptual and practical overview of some of the statistical methods for analyzing quantitative data. These quantitative analytical methods can be used by data analysts and researchers to accurately analyze their quantitative data to answer the research questions of their scientific study and draw reliable and valid conclusions of the research under investigation. The conceptual overview also provides students, researchers, professionals, and analysts in various disciplines and fields of study with the necessary skills to correctly interpret the scientific reports that employed one or more of the quantitative analytic techniques presented in this chapter.

REFERENCES

Field, A. (2009). *Discovering Statistics Using SPSS* (3rd ed.). Thousand Oaks, CA: Sage Publications.

George, D., & Mallery, P. (2014). *IBM SPSS Statistics 21 Step by Step: A Simple Guide and Reference* (13th ed.). Boston, MA: Allyn and Bacon.

Kalaian, S. A. (2008). Frequency Distribution. In P. J. Lavarkas (Ed.), *Encyclopedia of Survey Research Methods* (pp. 292–294). Thousand Oaks, CA: Sage Publications. doi:10.4135/9781412963947.n195

Kalaian, S. A., & Kasim, R. M. (2008). Research Hypothesis. In P. J. Lavarkas (Ed.), *Encyclopedia of Survey Research Methods* (pp. 334–335). Thousand Oaks, CA: Sage Publications. doi:10.4135/9781412963947.n472

Kalaian, S. A., & Kasim, R. M. (2015). Predictive Analytics. In M. Tavana, S. B. Zhou, & S. K. Puranam (Eds.), *Organizational Transformations through Big Data Analytic* (pp. 12–29). Hershey, PA: IGI Global.

Mertler, C. A., & Vannatta, R. A. (2013). *Advanced and Multivariate Statistical Methods* (3rd ed.). Glendale, CA: Pyrczak Publishing.

Tabachnick, B. G., & Fidell, L. S. (2012). *Using Multivariate Statistics* (6th ed.). Pearson Publications.

Witte, R. S., & Witte, J. S. (2013). *Statistics* (10th ed.). John Wiley & Sons, Inc.

Chapter 9
Analyzing Quantitative Data in Mixed Methods Research for Improved Scientific Study

Christopher Boachie
Central University College, Ghana

ABSTRACT

The purpose of the chapter is to review the role of quantitative methods in corporate research and the methods for analysing quantitative data. The study used a secondary data on quantitative research methods and a survey of published articles on schools, businesses and non- profit organizations. The key findings show that exploratory data can be analyzed using graphs and charts and hypothesis testing can be employed to test statements made. Impacts of one variable on another and the relationships between variables can be explained using correlation and regression analysis. The implications are that the value of a quantitative analysis arises when it is possible to identify features that occur frequently across the many participatory discussions aimed at studying a particular research theme.

INTRODUCTION

Data analysis is more than number crunching. The prospect of analysing numerical data sends shivers down the spines of many novice researchers who not only baulk at the thought of statistics but also hold fundamental objections to what they see as 'the mathematisation of nature' (Horkheimer, 1972). Most concepts in mixed research method, some will assert, is a combination of both quantitative and qualitative analysis. Quantitative data analysis has no greater or lesser importance than qualitative analysis. Its use is entirely dependent on fitness for purpose. Arbitrary dismissal of numerical analysis is mere ideology or prejudice.

Quantitative methods are research techniques that are used to gather quantitative data — information dealing with numbers and anything that is measurable e.g. Statistics, tables and graphs, are often used to present the results of these methods. The main aim of this chapter is to establish the methods employed when analysing quantitative data in mixed method research. This chapter is organised as follows: sec-

DOI: 10.4018/978-1-5225-0007-0.ch009

tion one explains the different data and their characteristics followed by exploratory data analysis in section two. Section three discusses hypothesis testing The last section explains the use of regression and correlation for data analysis.

BACKGROUND

Quantitative data analysis is a powerful research form, emanating in part from the positivist tradition. It is often associated with large scale research, but can also serve smaller scale investigations, with case studies, action research, correlational research and experiments.

In this chapter, the author introduces several common statistics used in social research and explains how they can be used to make sense of the "raw" data gathered in research. Such quantitative data analysis, using numbers to discover and describe patterns in data, is the most elementary use of social statistics. Numerical analysis can be performed using software, for example the Statistical Package for Social Sciences (SPSS), Minitab, or Excel. Software packages apply statistical formulae and carry out computations.

With this in mind, extended outlines of statistical formulae are avoided though we do provide details where considered useful. The primary aim is to explain the concepts that underpin statistical analyses and to do this in as user-friendly a way as possible. Lest the approach should raise purist eyebrows, we provide extended treatments in greater detail, signalled where appropriate by web site references. This chapter begins by identifying some key concepts in numerical analysis (scales of data, parametric and non-parametric data, descriptive and inferential statistics, dependent and independent variables). I then address the concept of statistical significance and finally conclude with a brief outline of some simple statistics.

The use of multiple data collection methods dates back to the earliest social science research. It was, however, Campbell and Fiske's (1959) study of the validation of psychological traits that brought multiple data collection methods into the spotlight. In their classic study, the multitrait-multimethod matrix was designed to rule out method effects; that is, to allow one to attribute individual variation in scale scores to the personality trait itself rather than to the method used to measure it.

Over time, mixed methods research has gradually gained momentum as a viable alternative research method. Over the past 15 years, at least 10 mixed methods textbooks have been published (Bamberger, 2000; Brewer & Hunter, 1989; Bryman, 1988; Cook & Reichardt, 1979; Creswell, 2002, 2003; Greene & Caracelli, 1989; Newman & Benz, 1998; Reichardt & Rallis, 1994; Tashakkori & Teddlie, 1998). Recently, the *Handbook of Mixed Methods in Social and Behavioral Research* was published (Tashakkori & Teddlie, 2003). In addition, journals such as *Field Methods* and *Quantity and Quality* are devoted to publishing mixed methods research. International online journals (see *Forum: Qualitative Social Research* at http://qualitative-research.ne) and Web sites (e.g., http://www.fiu.edu/~bridges/people.htm) provide easy access, resources, and hands-on experiences for interested researchers. Despite this growth and development, a number of controversial issues and debates have limited the widespread acceptance of mixed methods research.

Mixed methods researchers have expanded the reasons for conducting a mixed methods investigation (Mertens, 2003; Newman, Ridenour, Newman, & De-Marco, 2003; Punch, 1998). The author agrees with Mertens (2003) and Punch (1998), who suggested that mixed methods investigations may be used to (a) better understand a research problem by converging numeric trends from quantitative data and specific details from qualitative data; (b) identify variables/constructs that may be measured subsequently through the use of existing instruments or the development of new ones; (c) obtain statistical, quantitative data and results from a sample of a population and use them to identify individuals who may expand on the results through qualitative data and results; and (d) convey the needs of individuals or groups of individuals who are marginalized or underrepresented. The method of analysing mixed method research data may be a combination of both qualitative and quantitative approaches. The focus of this chapter, however, is on quantitative data analysis.

DATA AND THEIR CHARACTERISTICS

Quantitative analysis approaches are meaningful only when there is a need for data summary across many repetitions of a participatory process, e.g. focus group discussions leading to seasonal calendars, Venn diagrams, etc. Data summarisation in turn implies that some common features do emerge across such repetitions. Thus the value of a quantitative analysis arises when it is possible to identify features that occur frequently across the many participatory discussions aimed at studying a particular research theme. If there are common strands that can be extracted and subsequently coded into a few major categories, then it becomes easier to study the more interesting qualitative aspects that remain.

For example, suppose it is of interest to learn about people's perceptions of what poverty means for them. It is likely that the narratives that result from discussions across several communities will show some frequently occurring answers like experiencing periods of food shortage, being unable to provide children with a reasonable level of education, not owning a radio, etc. Such information can be extracted from the narratives and coded. Quantitative approaches provide the opportunity to study this coded information first and then to turn to the remaining qualitative components in the data.

These can then be discussed more easily, unhindered by the quantitative components. Quantitative analysis approaches are particularly helpful when the qualitative information has been collected in some structured way, even if the actual information has been elicited through participatory discussions and approaches. The next section reviews the scales of data.

Scales of Data

The values of variable define the scale or level of measurement used to study the variable. Before one can advance very far in the field of data analysis one needs to distinguish the different kinds of numbers with which one is working with. Variables differ in *how well* they can be measured, that is, in how much measurable information their measurement scale can provide. Different measurement scales allow for different levels of exactness, depending upon the characteristics of the variables being measured. There is obviously some measurement error involved in every measurement, which determines the *amount of information* that one can obtain. Another factor that determines the amount of information that can be provided by a variable is its type of *measurement scale*. This takes us to the commonly reported issue of scales or levels of data, and four are identified, each of which, in the order given below, subsumes its predecessor.

Nominal variables allow for only qualitative classification. That is, they can be measured only in terms of whether the individual items belong to some distinctively different categories, but we cannot quantify or even rank order those categories. This scale measures data by name only. For example, all we can say is that 2 individuals are different in terms of a certain variable (for example, they are of different race), but we cannot say which one *has more* of the quality represented by the variable. Typical examples of nominal variables are gender, race, colour, city, marital status, etc. The *nominal* scale simply denotes categories, 1 means such-and-such a category, 2 means another and so on, for example, '1' might denote females, '2' might denote males. The categories are mutually exclusive and have no numerical meaning.

Ordinal variables allow one to rank order the items we measure in terms of which has less and which has more of the quality represented by the variable, but still they do not allow us to say *how much more*. A typical example of an ordinal variable is the socioeconomic status of families. For example, we know that upper-middle is higher than middle but we cannot say that it is, for example, 18% higher. Also, this very distinction between nominal, ordinal, and interval scales itself represents a good example of an ordinal variable. For example, we can say that nominal measurement provides less information than ordinal measurement, but one cannot say *how much less* or how this difference compares to the difference between ordinal and interval scales. The *ordinal* scale not only classifies but also introduces an order into the data. These might be rating scales where, for example, 'strongly agree' is stronger than 'agree', or 'a very great deal' is stronger than 'very little'. It is possible to place items in an order, weakest to strongest, smallest to biggest, lowest to highest, least to most and so on, but there is still an absence of a metric – a measure using calibrated or equal intervals. Therefore one cannot assume that the distance between each point of the scale is equal, i.e. the distance between 'very little' and 'a little' may not be the same as the distance between 'a lot' and 'a very great deal' on a rating scale. Ordinal data include items such as rating scales and Likert scales, and are frequently used in asking for opinions and attitudes.

Interval variables allow us not only to *rank* order the items that are measured, but also to quantify and compare the sizes of differences between them. For example, temperature, as measured in degrees Fahrenheit or Celsius, constitutes an interval scale. One can say that a temperature of 40 degrees is higher than a temperature of 30 degrees, and that an increase from 20 to 40 degrees is twice as much as an increase from 30 to 40 degrees. However, interval scale variables do not have an *absolute zero*. If the temperatures in Singapore and London are 30^0 C and 15^0 C respectively, we cannot say that it is twice as hot in Accra as it is in Berlin. This is simply because it would not be the case if these temperatures were measured in degrees Fahrenheit: 86^0 F and 59^0 F respectively. The *interval* scale introduces a metric – a regular and equal interval between each data point – as well as keeping the features of the previous two scales, classification and order. This lets us know 'precisely how far apart are the individuals, the objects or the events that form the focus of our inquiry'. In fact twice as hot as 50 degrees Fahrenheit is 68 degrees Fahrenheit $((\{50 - 32\}\times 2) + 32)$. Let us give another example. Many IQ tests commence their scoring at point 70, i.e. the lowest score possible is 70. We cannot say that a person with an IQ of 150 has twice the measured intelligence as a person with an IQ of 75 because the starting point is 70; a person with an IQ of 150 has twice the measured intelligence as a person with an IQ of 110, as one has to subtract the initial starting point of 70 $(\{150 - 70\} \div 2)$. In practice, the interval scale is rarely used, and the statistics that one can use with this scale are, to all extents and purposes, the same as for the fourth scale which is the ratio scale.

Ratio variables are very similar to interval variables; in addition to all the properties of interval variables, they feature an identifiable absolute zero point, thus they allow for statements such as x is two times more than *y*. Typical examples of ratio scales are measures of time or space. For example, as the Kelvin temperature scale is a ratio scale, not only can we say that a temperature of 200 degrees is higher than one of 100 degrees; we can correctly state that it is twice as high. Interval scales do not have the ratio property. Most statistical data analysis procedures do not distinguish between the interval and ratio properties of the measurement scales. Height is also a ratio scale variable since, if a person is twice as tall as another, he/she will remain so, irrespective of the units used (centimetres, inches, etc…). The type of quantitative method to employ depends on whether the data is parametric or nonparametric. The next section therefore discusses these types of data.

Parametric and Non-Parametric Data

A parametric test is a statistical test which requires the distribution of the population to be specified. Thus, parametric inferential methods assume that the distribution of the variables being assessed belongs to some form of known probability distribution. Parametric data assume knowledge of the characteristics of the population, in order for inferences to be able to be made securely; they often assume a normal, Gaussian curve of distribution, as in reading scores, for example Wright (2003: 128) suggests that normal distributions are actually rare in psychology. In contrast, a non-parametric test (also known as distribution free test) does not require the distribution to be specified prior to the research, thus this family of tests does not require the assumption on the distribution. Most commonly, non-parametric tests rank the outcome variable from low to high and then analyse the ranks rather than the actual observation.

In practice this distinction means this: nominal and ordinal data are considered to be non-parametric, while interval and ratio data are considered to be parametric data. The distinction, as for the four scales of data, is important, as the consideration of which statistical test to use is dependent on the kinds of data: it is incorrect to apply parametric statistics to non-parametric data, although it is possible to apply non-parametric statistics to parametric data (it is not widely done, however, as the statistics are usually less powerful). Non-parametric data are often derived from questionnaires and surveys (though these can also gain parametric data), while parametric data tend to be derived from experiments and tests (e.g. examination scores).

Choosing the right test will contribute to the validity of the research findings. Improper use of statistical tests will not only cause the validity of the test results to be questioned and do little justification to the research but at times it can be a serious error especially if the results have major implications (for example, if these are used in policy formulation and so on). Parametric tests have greater statistical power compared to their non-parametric equivalent. However, it is not all the time that parametric tests are the best tests to use; parametric test should be used if researcher is sure that the data are sampled from a population that follows a pre-defined distribution (e.g. a normal distribution). On the other hand, a non-parametric test should be used if: (i) the outcome is a rank (e.g. rank order of the test scores); (ii) no assumption is made on the population distribution; and (iii) there are outliers.

Descriptive and Inferential Statistics

Although graphs and tables are important ways of looking at data, we need more precise quantitative measures for analysing data. Sometimes the volume of data is so big that charts may be difficult to draw. Tracking daily share price movements over years and displaying the data as a scatter graphs may indicate trends, but no concrete information.

Let us look at measures of location to identify central values of data and measures of spread to see to see how data deviates from the central values.

- **Mean:** This is the simplest measurement of central tendency and is a widely used measure. Its chief use consists in summarizing the essential features of a series and in enabling data to be compared. It is amenable to algebraic treatment and is used in further statistical calculations.
- **Mode:** It is another form of average and can be defined as the most frequently occurring value in the data. The mode is useful as a measure of location in cases where we desire to know which number comes up most often. The mode is not affected by extreme values in the data and can easily be obtained from an ordered set of data. It can be useful and more representative of the data under certain conditions and is the only measure of central tendency that can be used for qualitative data.
- **Median:** The median is a measure of central tendency and it appears in the centre of an ordered data. It divides the list of ordered values in the data into two equal parts so that half the data will have values less than the median and half will have values greater than the median. If the total numbers of values is odd, then we simply take the middle value as the median
- **Standard Deviation:** This is the most widely used measure of dispersion of a series and is commonly denoted by the symbol σ (pronounced as sigma). Standard deviation is defined as the square root of the average of squares of deviations, when such deviations for the values of individual items in a series are obtained from the arithmetic average.
- **Variance:** It is a measure of how far scores are from the mean, calculated as the average of the squared deviations of individual scores from the mean.
- **Standard Error:** This is the standard deviation of sample means
- **The Skewness:** It is a measure of the asymmetry of the distribution of a real-valued random variable. A measure of dispersion tells us about the variation of the data set. Skewness tells us about the *direction of variation* of the data set. The value of skewness can be positive or negative, or even undefined. Qualitatively, a negative skew indicates that the *tail* on the left side of the probability density function is *longer* than the right side and the bulk of the values (possibly including the median) lie to the right of the mean. A positive skew indicates that the *tail* on the right side is *longer* than the left side and the bulk of the values lie to the left of the mean. A zero value indicates that the values are relatively evenly distributed on both sides of the mean, typically but not necessarily implying a symmetric distribution.
- **Kurtosis:** It is any measure of the peakedness or flatness of the probability distribution of a real valued random variable. It shows how much of the distribution is centered on the distributions mean.

Inferential statistics, by contrast, strive to make inferences and predictions based on the data gathered. These will include, for example, hypothesis testing, correlations, regression and multiple regression, difference testing (e.g. t-tests and analysis of variance, factor analysis, and structural equation modelling). Sometimes simple frequencies and descriptive statistics may speak for themselves, and the careful portrayal of descriptive data may be important. However, often it is the inferential statistics that are more valuable for researchers, and typically these are more powerful (Kothari, 2004).

Dependent and Independent Variables

Research often concerns relationships between variables. A variable can be considered as a construct, operationalized construct or particular property in which the researcher is interested. It is any condition that can vary or change in quantity or quality.

An independent variable is an input variable, that which causes, in part or in total, a particular outcome. It is a stimulus that influences a response, an antecedent or a factor which may be modified (e.g. under experimental or other conditions) to affect an outcome. A dependent variable, on the other hand, is the outcome variable, that which is caused, in total or in part, by the input, antecedent variable. It is the effect, consequence of, or response to, an independent variable. This is a fundamental concept in many statistics. In regression analysis independent variable can be single or multiple parameters. Regression based models continue to be widely used in social sciences (Echambadi & Hess, 2007; Fitzsimons, 2008; Irwin & McClelland, 2001; Judd & Kenny, 2010; Zhao, Lynch, & Chen, 2010). In the simplest bivariate case, a dependent variable is predicted from another independent variable. However, with the increasing complexity of social science phenomena, bivariate models do not adequately capture underlying interrelationships among constructs. In other words, the variance of a dependent variable can be better explained by more than one independent variable. Predicting the dependent variable from more than one independent variable is termed multiple regression.

For example, one may wish to see if economic boom reduces non-performing loans in a nation. One can improve the economic parameters such as exchange rate and rate of inflation and measure the result of non-performing loans (NPL) and we notice, for example, that the non- performing loans reduces. The independent variable has produced a measured outcome. In fact, what may be happening here is that causality may be less in a linear model and more multidirectional and multi related, more like a web than a line. This example indicates a range of issues in the discussion of dependent and independent variables:

- The direction of causality is not always clear: an independent variable may, in turn, become a dependent variable and vice versa.
- The direction of causality may be bidirectional.
- Assumptions of association may not be assumptions of causality.
- There may be a range of other factors that have a bearing on an outcome.
- There may be causes (independent variables) behind the identified causes (independent variables) that have a bearing on the dependent variable.
- The independent variable may cause something else, and it is the something else that causes the outcome (dependent variable).
- Causality may be non-linear rather than linear.
- The direction of the relationship may be negative rather than positive.
- The strength/magnitude of the relationship may be unclear. (Cohen, Manion, & Morrison, 2007).

Reliability

Researchers need to know how reliable their instruments for data collection are. Reliability in quantitative analysis takes two main forms, both of which are measures of internal consistency: the split-half technique and the alpha coefficient. Both calculate a coefficient of reliability that can lie between 0 and 1. The formula for the split-half reliability is given as:

$$r = \frac{2r}{1+r}$$

where $r =$ the actual correlation between the halves of the instrument. This requires the instrument to be able to be divided into two matched halves in terms of content and difficulty.

So, for example, if the correlation coefficient between the two halves is 0.85 then the formula would be worked out thus:

$$r = \frac{2(0.85)}{1+0.85} + \frac{1.70}{1.85} = 0.919$$

Hence the split-half reliability coefficient is 0.919, which is very high. SPSS automatically calculates split-half reliability at the click of a button.

This yields an alpha coefficient of 0.97, which is very high. The alpha coefficients are set out in a table. For the split-half coefficient and the alpha coefficient the following guidelines can be used:

>0.90 very highly reliable
0.80–0.90 highly reliable
0.70–0.79 reliable
0.60–0.69 marginally/minimally reliable
<0.60 unacceptably low reliability (Cohen, Manion, & Morrison, 2007).

Bryman and Cramer (1990: 71) suggest that the reliability level is acceptable at 0.8, although others suggest that it is acceptable if it 0.67 or above 1.

EXPLORATORY DATA ANALYSIS: FREQUENCIES, PERCENTAGES AND CROSS-TABULATIONS

This is a form of analysis which is responsive to the data being presented, and is most closely concerned with seeing what the data themselves suggest, akin to a detective following a line of evidence. Exploratory and confirmatory data analysis can both be viewed as methods for comparing observed data to what would be obtained under an implicit or explicit statistical model. The techniques of exploratory data analysis consist of simple arithmetic and easy-to-draw pictures that can be used to summarize data quickly. The data are usually descriptive. Here much is made of visual techniques of data presentation. Hence frequencies and percentages, and forms of graphical presentation are often used. A host of graphical forms of data presentation are available in software packages, including SPSS and Stata.

Frequencies

A frequency distribution is a tabular summary of data showing the frequency (or number) of items in each of several non-overlapping classes (see table 1). The objective is to provide insights about the data that cannot be quickly obtained by looking only at the original data. An example is guests staying at

Holiday Inn in Ghana were asked to rate the quality of their accommodations as being *excellent, above average, average, below average*, or *poor*. The frequency table (Table 1.) describes the ratings provided by a sample of 200 guests.

A frequency table basically consists of two columns of information. The first column presents the scale of measurement or simply lists the set of categories into which individuals has been assigned. The list of the set of categories from the above table is *excellent, above average, average, below average*, or *poor.* The second column list the frequency, or the number of individuals, located in each category. In this example 26 persons the service quality was poor, 58 thought the service quality was below average, and so on. Other points that can be deduced from the above table are:

- Most respondents thought that the service quality was average (with a response number of 99, i.e. 50 percent); the modal score is that category or score which is given by the highest number of respondents.
- The results were skewed, with only 5 per cent being in the categories 'above average' and 'a 3 per cent excellent'.
- Overall the quality of accommodation appears to have been between average and below average.
- Out of 200 person only 6 thought the quality of service was excellent.

Frequency distribution can be a very effective method for gathering information about a set of scores (Gravetter & Forzano, 2006). The above distribution shows whether the scores are clustered together or spread out across the scale. One can see at a glance if the scores are generally low or generally high. Also, if there are one or two extreme scores, these values are easy to see in the table. However, a frequency distribution is generally considered to be a preliminary method of statistical analysis. As a result, frequency distributions are rarely shown in published research results (Zikmund, Babin, Carr, & Griffin, 2010).

Cross Tabulation

Let us assume that we wished to explore this piece of information further. We may wish to discover, for example, the voting on this item by males and females. This can be presented in a simple cross-tabulation, following the convention of placing the nominal data (male and female) in rows and the ordinal data (the 5-point scale) in the columns. As long as a question deals with only one categorical variable, tabulation

Table 1. Frequency distribution

Rating	Frequency	Relative Frequency	Percentage Frequency
Poor	26	0.13	13
Below Average	58	0.29	29
Average	99	0.50	50
Above Average	11	0.05	5
Excellent	6	0.03	3
Total	200	1.00	100

is probably the best method (Zikmund, Babin, Carr, & Griffin, 2010). Across-tabulation is simply a presentational device, whereby one variable is presented in relation to another, with the relevant data inserted into each cell. It is a statistical process that summarizes categorical data to create a contingency table. They are heavily used in survey research, business intelligence, engineering and scientific research (Zikmund, Babin, Carr, & Griffin, 2010; Gravetter & Forzano, 2006; Kothari, 2004). Cross-tabulation is the appropriate technique for addressing research questions involving relationships among multiple less than interval variables. The table shows the cross-tabulation of the quest in staying at the Holiday Inn Hotel in Ghana.

Table 2 shows that, of the total sample, nearly three times more females (36.5 per cent) than males (13.0 per cent) thought that the quality of accommodation was 'average' or below average. Deductions can be made in terms of both percentages and absolute numbers on both female and male voting pattern.

However, one also has to observe that the size of the two subsamples was uneven. Around three quarters of the sample were female (69.0 per cent) and around one-quarter (31.0 per cent) was male. There are two ways to overcome the problem of uneven subsample sizes. One is to adjust the sample, in this case by multiplying up the subsample of males by an exact figure in order to make the two subsamples the same size (138/62 = 2.23). Another way is to examine the data by each row rather than by the overall totals, i.e. to examine the proportion of males voting such and such, and, separately, the proportion of females voting for the same categories of the variable.

Cross tabulations allow the inspection and comparison of differences among groups based on nominal or ordinal categories. One key to interpreting a cross tabulation table is comparing the observed table values with hypothetical values that would results from pure chance.

Charts and Graphs

With most of these forms of data display there are various permutations of the ways in which data are displayed within the type of chart or graph chosen. While graphs and charts may look appealing, it is often the case that they tell the reader no more than could be seen in a simple table of figures, which take up less space in a report. Pie charts, bar charts and histograms are particularly prone to this problem, and the data in them could be placed more succinctly into tables. Clearly the issue of fitness for audience is important here: some readers may find charts more accessible and able to be understood than tables of

Table 2. Cross tabulation

		Poor	Below Average	Average	Above Average	Excellent	Total
Male	Count	7	20	26	4	5	62
	% of total	3.5%	10.0%	13.0%	2.0%	2.5%	31.0%
Female	Count	19	38	73	7	1	138
	% of total	9.5%	19.0%	36.5%	3.5%	0.5%	69.0%
Total	Count	26	58	99	11	6	200
	% of total	13.0%	29.0%	49.5%	5.5%	3.0%	100.0%

figures, and this is important. Other charts and graphs can add greater value than tables, for example, line graphs, box plots and scatterplots with regression lines, and we would suggest that these are helpful.

Here is not the place to debate the strengths and weaknesses of each type, although there are some guides here:

- Bar charts are useful for presenting categorical and discrete data, highest and lowest.
- Line graphs are useful for showing trends, particularly in continuous data, for one or more variables at a time.
- Multiple line graphs are useful for showing trends in continuous data on several variables in the same graph.
- Pie charts and bar charts are useful for showing proportions.
- Interdependence can be shown through cross tabulations.
- Box plots are useful for showing the distribution of values for several variables in a single chart, together with their range and medians.
- Stacked bar charts are useful for showing the frequencies of different groups within a specific variable for two or more variables in the same chart.
- Scatterplots are useful for showing the relationship between two variables or several sets of two or more variables on the same chart (Cohen, Manion, & Morrison, 2007).

HYPOTHESIS TESTING

Hypothesis is usually considered as the principal instrument in research. Its main function is to suggest new experiments and observations. In fact, many experiments are carried out with the deliberate object of testing hypotheses. Decision-makers often face situations wherein they are interested in testing hypotheses on basis of available information and often take decisions on the basis of such testing. In social science, where direct knowledge of population parameter(s) is rare, hypothesis testing is the often used strategy for deciding whether a sample data offer such support for a hypothesis that generalisation can be made. Thus hypothesis testing enables us to make probability statements about population parameter(s). The hypothesis may not be proved absolutely, but in practice it is accepted if it has withstood a critical testing. Before the author explains the procedure for hypothesis testing, it will be appropriate to explain clearly the meaning of a hypothesis and the related concepts for better understanding for the hypothesis testing techniques.

What Is Hypothesis?

Ordinary, when one talks about hypothesis, one simply means a mere assumption or some supposition to be proved or disproved. But for a researcher hypothesis is a formation question that he intends to resolve. Thus a hypothesis may be defined as a preposition or a set of preposition set forth as an explanation for the occurrence of some specified group of phenomena either asserted merely as a provisional conjecture to guide some investigation or accepted as highly probable in the light of established facts (Kothari, 2004). Quite often a research hypothesis is predictive statement, capable of being tested by scientific methods, that relates an independent variable to some dependent variable. For example, consider statements like the following ones:

- "Students who receive counselling will show a greater increase in creativity than students not receiving counselling," or
- "The automobile A is performing as well as automobile B."

These are hypotheses capable of being objectively verified and tested. Thus, one may conclude that a hypothesis states what we are looking for and it is a preposition which can be put to a test to determine its validity.

Hypotheses Testing Procedure

Hypotheses are tested by comparing the researcher's educated guess with empirical reality. The process can be described as follows:

1. First, the hypothesis is derived from the research objectives. The hypothesis should be stated as specifically as possible.
2. Next, a sample is obtained and the relevant variable is measured.
3. The measured value obtained in the sample is compared to the value either stated explicitly or implied in the hypothesis. If the value is consistent with the hypotheses, the hypothesis is not supported.

A univariate hypothesis consistent with the chapter vignette would be

H_1: The average satisfaction at the Holiday Inn hotel is greater than 3.5.

If the average accommodation satisfaction is 3.4, the hypothesis is not supported. If the average accommodation satisfaction is 3.9 the hypothesis is supported. Univariate hypothesis are typified by tests comparing some observed sample mean against a benchmark value. The test addresses the question, is the sample mean truly different from the benchmark? But how different? If the observed sample mean is 3.45 and the benchmark is 3.50, would the hypothesis still be supported? Probably not! When the observed mean is so close to the benchmark, one does not have sufficient confidence that a second set of data using a new sample taken from the same population might not produce a finding conflicting with the benchmark. In contrast, when the mean turns out well above 3.5, perhaps 3.9, then we could more easily trust that another sample would not produce a mean equal to or less than 3.5.

One-Tailed and Two-Tailed Tests

In using statistics, researchers are sometimes confronted with the decision whether to use a one-tailed or a two-tailed test. Which to use is a function of the kind of result one might predict.

In a one-tailed test one predicts, for example, that one group will score more highly than the other, whereas in a two-tailed test one makes no such prediction. The one-tailed test is a stronger test than the two-tailed test as it makes assumptions about the population and the direction of the outcome (i.e. that one group will score more highly than another), and hence, if supported, is more powerful than a two-tailed test. A one-tailed test will be used with a directional hypothesis (e.g. 'Students who do homework *without* the TV on produce *better* results than those who do homework with the TV playing'). A two-tailed test will be used with a non-directional hypothesis (e.g. 'there is a *difference* between homework done in noisy or silent conditions'). The directional hypothesis indicates 'more' or 'less', whereas the non-directional hypothesis indicates only difference, and not where the difference may lie. Research in a hypothetico-deductive mode often commences with one or more hypotheses. This is the essence of hypothesis testing in quantitative research. Typically hypotheses fall into two types.

The *null hypothesis*, a major type of hypothesis states that, for example, there is *no* relationship between two variables, or that there has been *no* change in participants between a pretest and a post-test, or that there is *no* difference between three school districts in respect of their examination results, or that there is *no* difference between the voting of males and females on such-and-such a factor. The null hypothesis sits comfortably with the Popperian notion of the hallmark of a science being its essential falsifiability. The point here is that by casting the hypothesis in a null form the burden of proof is placed on the researcher *not* to confirm that null hypothesis. The task is akin to a jury starting with a presumption of innocence and having to prove guilt beyond reasonable doubt. Not only is it often easier simply to support a straightforward positive hypothesis, but also, more seriously, even if that positive hypothesis is supported, there may be insufficient grounds for accepting that hypothesis, as the finding may be consistent with other hypotheses.

A second type of hypothesis is termed the *alternative hypothesis*. Whereas the null hypothesis states that there is *no* such-and-such (e.g. change, relationship, difference), the alternative hypothesis state that there *is* such-and-such, for example: there *is* a change in behaviour of the school students; there *is* a difference between students' scores on mathematics and science; there *is* a difference between the examination results of five school districts; there *is* a difference between the pretest and post-test results of such and-such a class. This weaker form of hypothesis is often supported when the null hypothesis is 'not supported', i.e. if the null hypothesis is not supported then the alternative hypothesis is (Kothari, 2004; Cohen, Manion, & Morrison, 2007).

Tests of Hypotheses

As has been stated above that hypothesis testing determines the validity of the assumption (technically described as null hypothesis) with a view to choose between two conflicting hypotheses about the value of a population parameter. Hypothesis testing helps to decide on the basis of sample data, whether a hypothesis about a population is likely to be true or false. Statisticians have developed several test of hypothesis (also known as the test of significance) for the purpose of testing of hypotheses which can be classified as: (a) Parametric tests or standard test of hypotheses; and (b) Non-parametric tests or distribution –free test of hypotheses.

Parametric tests usually assume certain properties of the parent population from which we draw samples. Assumptions like observations come from a normal population, sample size is large assumptions about the population parameters like mean, variance, etc. must hold good before parametric tests can be used. But there are situations when the researcher cannot or does not want to make such assumptions. In such situations one uses statistical methods for testing hypotheses which are called non-parametric tests because such tests do not depend on any assumption about the parameters of the parent population. Besides, most non-parametric tests assume only nominal or ordinal data, whereas parametric tests require measurement equivalent to at least an interval scale. As a result, on-parametric tests need more observations than parametric tests to achieve the same size of Type I and Type II errors. The next section presents correlation and regression analysis.

CORRELATION AND REGRESSION

Much mixed method research is concerned with establishing interrelationships among variables. A researcher may wish to know, for example, how non-performing loan is related to social class background; whether an association exists between the level of qualification and subsequent annual income; whether there is a link between personality and achievement. What, for example, is the relationship, if any, between membership of a public library and social class status? Is there a relationship between social class background and placement in different strata of the secondary school curriculum? Is there a relationship between gender and success or failure in 'first time' driving test results? There are several simple measures of association readily available to the researcher to help her test these sorts of relationships. Of these, the two most commonly used correlations are the Spearman rank order correlation for ordinal data and the Pearson product-moment correlation for interval and ratio data.

Explaining Correlations

The idea of correlation is one of the most important and basic in the elaboration of bivariate relationships. Unlike chi-square, measures of correlation indicate both the strength and the direction of the relationship between a pair of variables. Two types of measure can be distinguished: measures of linear correlation using interval variables and measures of rank correlation using ordinal variables. While these two types of measure of correlation share some common properties, they also differ in some important respects which will be examined after the elucidation of measures of linear correlation.

A correlation coefficient enables you to quantify the strength of the linear relationship between two ranked or numerical variables. This coefficient (usually represented by the letter r) can take on any value between -1 and +1. A value of +1 represents a perfect positive correlation. This means that the two variables are precisely related and that, as values of one variable increase, values of the other variable will increase. By contrast, a value of -1 represents a perfect negative correlation. Again, this means that the two variables are precisely related; however, as the values of one variable increase those of the other decrease. Correlation coefficients between +0.31 and -0.3 represent weaker positive and negative correlations, a value of 0 meaning the variables are perfectly independent. Within business research it is extremely unusual to obtain perfect correlations.

For data collected from a sample you will need to know the probability of your correlation coefficient having occurred by chance alone. Most analysis software calculates this probability automatically. As outlined earlier, if this probability is very low (usually less than 0.05) then it is considered statistically significant. If the probability is greater than 0.05 then your relationship is not statistically significant (Cohen, Manion, & Morrison, 2007; Curwin & Slater, 2007).

Simple Linear Regression

In simple linear regression the model includes one explanatory variable (the independent variable) and one explained variable (the dependent variable). Regression can only interpret what exists physically i.e. there must be a physical way in which independent variable X can affect dependent variable Y. The basic relationship between X and Y is given by

$$\hat{Y} = a + bX$$

where the symbol $\hat{Y}$ denotes the estimated value of Y for a given value of X. This equation is known as the regression equation of Y on X (also represents the regression line of Y on X when drawn on a graph) which means that each unit change in X produces a change of b in Y, which is positive for direct and negative for inverse relationships (Curwin & Slater, 2007).

To Assess the Strength of a Cause-and-Effect Relationship between Variables

In contrast to the correlation coefficient, the coefficient of determination (sometimes known as the regression coefficient) enables one to assess the strength of relationship between a numerical dependent variable and one or more numerical independent variables. Once again, where these data have been selected from a sample, the sample must have been selected at random. For a dependent variable and one (or perhaps two) independent variables you will have probably already plotted this relationship on a scatter graph. If you have more than two independent variables this is unlikely as it is very difficult to represent four or more scatter graph axes visually!

The coefficient of determination (represented by r^2) can take on any value between 0 and +1. It measures the proportion of the variation in a dependent variable (amount of sales) that can be explained statistically by the independent variable (marketing expenditure) or variables (marketing expenditure, number of sales staff, etc.). This means that if all the variation in amount of sales can be explained by the marketing expenditure and the number of sales staff, the coefficient of determination will be 1. If 50 per cent of the variation can be explained, the coefficient of determination will be 0.5, and if none of the variation can be explained, the coefficient will be 0. The next section multiple correlation and regression.

Multiple Correlation and Regression

When there are two or more than two independent variables, the analysis concerning the relationship is known as multiple correlation and the equation describing such relationship as the multiple regression equation. I explain what multiple correlation and regression are by taking into consideration two independent variables and one dependent variable (convenient computer programs exist for dealing with a great number of variables). In this situation the results are interpreted as shown below:

In multiple regression analysis, the regression coefficients become less reliable as the degree of correlation between the independent variables increases. If there is a high degree of correlation between independent variables, we have a problem of what is commonly described as the problem of multicollinearity. In such a situation we should use only one setoff the independent variable to make our estimate. In fact, adding a second variable say X_2 that is correlated with the first variable X_1 distorts the values of the regression coefficients. Nevertheless, the prediction for the dependent variable can be, and even when multicollinearity is present, but in such a situation enough care should be taken in selecting the independent variables to estimate a dependent variable as to ensure that multicollinearity is reduced to a minimum.

Despite the widespread popularity of multiple regression models, researchers have to grapple with several contentious issues. First, multicollinearity (high correlations among independent variables) results in biased parameter estimates and researchers do not always agree on the most appropriate approach for addressing this issue (Echambadi & Hess, 2007; Friedman & Wall, 2005; Grewal, Cote, & Baumgartner, 2004). Second, detection of indirect variable effects in a multiple regression model is not straightforward. For example, an independent variable may affect the dependent variable through a third variable in the model. Typically, a class of intervening variable methods such as mediation analyses, moderating variable techniques, and hierarchical regressions can be used to uncover indirect effects. However, model misspecification remains a major limitation of these approaches because intervening relationships can be incorporated in the initial model only when the researcher has a priori knowledge of theoretical relationships. In the exploratory stages of research, when theories are not well developed, it is difficult to accurately specify indirect effects in the initial regression model. Hence, it is not surprising that a growing debate is currently raging in marketing about appropriate procedures for using intervening variable methodologies (Baron & Kenny, 1986; Iacobucci, 2008; Iacobucci, Saldhana, & Deng, 2007; Judd & Kenny, 2010; Zhao, Lynch, & Chen, 2010). Third, multiple regression models do not calculate parameter estimates simultaneously. For example, if a dependent variable (*Y*) is related to an independent variable (*X1*) through another (mediating) variable (*X2*), three separate regression models have to be estimated. Such non-simultaneous estimation might result in biased parameter estimates (Iacobucci, Saldhana, & Deng, 2007; Zhao, Lynch, & Chen, 2010).

Fourth, multiple regression models assume that independent variables are measured without error. In other words, trait variance in an independent variable should be large in relation to measurement error variance. However, a number of studies have shown that the general assumption of low levels of measurement error is highly questionable in practice.

Finally, regression models do not calculate an overall goodness of fit statistic that indicates the degree of congruence between hypothesized structural paths and the underlying phenomenon of interest. To test theory, researchers typically formulate hypotheses and obtain data by administering scale items to respondents. Next, summary associational information is captured via the variance-covariance matrix of variables.

To Predict the Value of a Variable from One or More Other Variables

Regression analysis can also be used to predict the values of a dependent variable given the values of one or more independent variables by calculating a regression equation. One may wish to predict the amount of sales for a specified marketing expenditure and number of sales staff. You would represent this as a regression equation where: *AoS* is the Amount of Sales, *ME* is the Marketing Expenditure, *NSS* is the Number of Sales Staff, *a* is the regression constant and are the beta coefficients b1 and b2

This equation can be translated as stating:

$$AoSi = a + b1MEi + b2NSSi$$

Amount of Sales, i = value + (b1 * *Marketing Expenditurei*) + (b2 * *Number of Sales Staffi*)

Using regression analysis one would calculate the values of the constant coefficient *a* and the slope coefficients and from data you had already collected on amount of sales, marketing expenditure and number of sales staff. A specified marketing expenditure and number of sales staff could then be substituted into the regression equation to predict the amount of sales that would be generated. When calculating a regression equation one needs to ensure the following assumptions are met:

- The relationship between dependent and independent variables is linear. Linearity refers to the degree to which the change in the dependent variable is related to the change in the independent variables. Linearity can easily be examined through residual plots (these are usually drawn by the analysis software). Two things may influence the linearity. First, individual cases with extreme values on one or more variables (outliers) may violate the assumption of linearity. It is, therefore, important to identify these outliers and, if appropriate, exclude them from the regression analysis. Second, the values for one or more variables may violate the assumption of linearity. For these variables the data values may need to be transformed. Techniques for this can be found in other, more specialised books on multivariate data analysis, for example Anderson (2003).
- The extent to which the data values for the dependent and independent variables have equal variances, also known as homoscedasticity.

Regression Output

In simple linear regression the model includes one explanatory variable (the independent variable) and one explained variable (the dependent variable). Assume the level of achievement of a student in the university depends on the number of hours of study. Therefore the dependent variable is the level of achievement and hours of study the independent variable. Table 3 displays the output for the level of achievement problem. Most computerised software provides similar output for regression analysis.

Interpreting the simple regression output is a simple two-step process.

1. *Interpret the overall significance of the model.* The output will include model *F* and a significance value. When the model *F* is significant (low p-value), the independent variable explains a significant portion of the variation in the independent variable. The coefficient of determination or R^2 can also be interpreted. As mentioned earlier this is the percentage of total variation in the dependent variable accounted for by the independent variable. Another way to think of this is as the extent to which the variances of the independent and the dependent variable overlap
2. *The individual parameter coefficient is interpreted.* The *t*-value associated with the slope coefficient can be interpreted. In this case the *t* of 9.087 is associated with a very low p-value (0.000 to 3 decimal places). Therefore the slope coefficient is significant. For simple regression, the p-value

Table 3. A summary of the R, R square and adjusted R square in regression analysis

			Model Summary	
Model	R	R Square	Adjusted R Square	SE of estimate
I	0.795	0.632	0.625	9.20

a. Predictors: (Constant), Hours of study

for the model F and for the t-test of the individual regression weight will be the same (Zikmund, Babin, Carr, & Griffin, 2006). A t-test for the intercept (constant) term is also provided. However this is seldom of interest since the explanatory power rest in the slope coefficient.

The R of 0.795 indicates a strong direct relationship between the level of achievement and hour of study. The R^2 figure of 0.632 shows that the independent variable (hours of study) accounts for 63.2 per cent of the variance in the dependent variable (level of achievement). Therefore as the number of hours of study increases the level of achievement also increases. Here the adjusted R square is 0.625, and this, again, shows that, in the regression model that we have constructed, the independent variable accounts for 62.5 per cent of the variance in the dependent variable, which is high, i.e. our regression model is robust. Muijs (2004: 165) suggests that, for a goodness of fit with an adjusted R square (table 4):

<0.1: poor fit.
0.11–0.3: modest fit.
0.31–0.5: moderate fit.
>0.5: strong fit.

Table 4 shows the ANOVA output.

Second, SPSS then calculates the analysis of variance (ANOVA). At this stage we will not go into all of the calculations here (typically SPSS prints out far more than researchers may need. The final column, marked 'Sig.'; this is the significance level, and, because the significance is 0.000, we have a very statistically significant relationship (stronger than 0.001) between the independent variable (hours of study) and the dependent variable (level of achievement).

This tells us that it is useful to proceed with the analysis, as it contains important results. SPSS then gives us a table of coefficients, both unstandardized and standardized. It is advisable to opt for the standardized coefficients, the Beta weightings. The Beta weight is the amount of standard deviation unit of change in the dependent variable for each standard deviation unit of change in the independent variable. In the example below the Beta weighting is 0.795; this tell us that, for every standard deviation unit change in the independent variable (hours of study), the dependent variable (level of achievement) will rise by 0.795 (79.5 per cent) of one standard deviation unit, i.e. for every one unit rise in the independent variable there is just over three-quarters of a unit rise in the dependent variable. This also explains why the slope of the line of best fit is steep but not quite 45 degrees – each unit of one is worth only 79.5 per cent of a unit of the other. Table 5 represents the regression output.

Table 4. Anova

Model		Sum of Squares	Df	Mean Square	F	Sig
I	Regression	6988.208	1	6988.208	82.57	0
	Residual	4062.292	48	84.631		
	Total	11050.5	49			
a. Predictors: (Constant), Hours of study						
b. Dependent variable: Level of achievement						

Table 5. Regression output

		Coefficients				
		Unstandardized Coefficient		Standardized Coefficients		
Model		B	SE	Beta	t	Sig
I	(Constant)	26.322	2.982		8.83	0.000
	Hours of study	9.567	1.053	0.795	9.087	0.000
a. Dependent variable: Level of achievement						

In reporting the example of regression one could use a form of words thus: a scatter graph of the regression of hours of study on levels of achievement indicates a linear positive relationship between the two variables, with an adjusted R square of 0.625. A standardized beta coefficient of .795 is found for the variable 'hours of study', which is statistically significant as the $p < 0.001$ for 99 percent level of confidence. The next section explains significance in statistics.

FUTURE RESEARCH DIRECTION

Many researchers have conducted studies on the use of quantitative analysis techniques in a mixed method research (Ye & Zhang, 2011; Vel de, Vermeir, & Corten, 2005). Almost all quantitative researchers employ one or more quantitative techniques to analyse their research data. Quantitative data analysis has no greater or lesser importance than qualitative analysis. Its use is entirely dependent on fitness for purpose. Arbitrary dismissal of numerical analysis is mere ideology or prejudice. It is a powerful research form, emanating in part from the positivist tradition. It is often associated with large scale research, but can also serve smaller scale investigations, with case studies, action research, correlational research and experiments. This chapter concentrated on analysing quantitative data in a mixed method research, yet further research could be carried out on the analysis qualitative data among researchers. In addition, further study should concentrate on the examination of other factors that contribute to use of quantitative data analysis in a mixed method research.

CONCLUSION

There are four levels of measurement scales, nominal, ordinal, interval and ratio. For all four scales one can calculate frequencies and percentages, and we can consider presenting these in a variety of forms. Researchers can also calculate the mode and present cross-tabulations. One can consider combining categories and collapsing tables into smaller tables, providing that the sensitivity of the original data has not been lost.

The median score, which is particularly useful if the data are spread widely or if there are outliers represent a middle figure. For interval and ratio data we can also calculate the mean and the standard deviation; the mean yields an average and the standard deviation indicates the range of dispersal of scores around that average, i.e. to see whether the data are widely dispersed (e.g. in a platykurtic distribution, or close together with a distinct peak (in a leptokurtic distribution). In examining frequencies and percentages one also has to investigate whether the data are skewed, i.e. over-represented at one end of a scale and underrepresented at the other end. A positive skew has a long tail at the positive end and the majority of the data at the negative end, and a negative skew has a long tail at the negative end and the majority of the data at the positive end. An independent variable is an input variable, that which causes, in part or in total, a particular outcome. A dependent variable, on the other hand, is the outcome variable, that which is caused, in total or in part, by the input, antecedent variable. Reliability in quantitative analysis takes two main forms, both of which are measures of internal consistency. Researchers are sometimes confronted with the decision whether to use a one-tailed or a two-tailed test. Descriptive research and causal research designs often climax with hypothesis tests. Hypotheses are defined as formal statements of explanations stated in a testable form.

A correlation coefficient enables one to quantify the strength of the linear relationship between two ranked or numerical variables. The word 'significant' is qualitative and subjective. Whether something is 'significant' is in the eye of the beholder. When we add the word 'statistics', we add a strong quantitative word to the very qualitative word 'significant'. This lends an appearance of credibility and certainty to any experiment that achieves a p-value below 0.05, simply because this is the widely accepted threshold for achieving 'statistical significance'.

REFERENCES

Baron, R. M., & Kenny, D. A. (1986). Moderator-mediator Variables Distinction in Social psychological Research:Conceptual, Strategic, and Strategic Considerations. *Journal of Personality and Social Psychology*, *51*(6), 1173–1182. doi:10.1037/0022-3514.51.6.1173 PMID:3806354

Brown, J. D. (2004). Yates correction factor. SHIKEN. *The JALT Testing & Evaluation SIG Newsletter*, *8*(1), 22–27.

Bryman, A. (1988). *Quantity and Quality in Social Research*. London: Unwin Hyman. doi:10.4324/9780203410028

Bryman, A., & Cramer, D. (1990). *Quantitative Data Analysis for Social Scientists*. London: Routledge.

Campbell, D. T., & Fiske, D. W. (1959). Convergent and discriminant validation by the multitrait-multimethod matrix. *Psychological Bulletin*, *56*(2), 81–105. doi:10.1037/h0046016 PMID:13634291

Cohen, L., Manion, L., & Morrison, K. (2007). *Research Methods in Education*. London, New York: Routledge.

Creswell, J. W. (2003). *Research Design: Qualitative, Quantitative, and Mixed Methods Approaches* (2nd ed.). Thousand Oaks, CA: Sage.

Curwin, J., & Slater, R. (2007). *Quantitative Methods for Business Decisions* (5th ed.). London: Thomson Learning.

Echambadi, R., & Hess, J. D. (2007). Mean-Centering Does Not Alleviate Collinearity Problems in Multiple Regression Models. *Marketing Science*, *26*(3), 438–445. doi:10.1287/mksc.1060.0263

Fitzsimons, G. J. (2008). Death to Dichotomizing. *The Journal of Consumer Research*, *35*(1), 5–8. doi:10.1086/589561

Friedman, L., & Wall, M. (2005). Graphical Views of Suppression and Multicollinearity in Multiple Linear Regression. *The American Statistician*, *59*(2), 127–136. doi:10.1198/000313005X41337

Gay, L. R. (1976). *Educational research: Competencies for analysis and application*. Ohio: Bell & Howell Company.

Gravetter, F. J., & Forzano, L. B. (2006). *Research methods for the behavioural sciences* (2nd ed.). Mason, OH: Thompson.

Greene, J. C., Caracelli, V. J., & Graham, W. F. (1989). Toward a Conceptual Framework for Mixed-method Evaluation Designs. *Educational Evaluation and Policy Analysis*, *11*(3), 255–274. doi:10.3102/01623737011003255

Grewal, R., Cote, J. A., & Baumgartner, H. (2004). Multicollinearity and Measurement Error in Structural Equation Models: Implications for Theory Testing. *Marketing Science*, *23*(4), 519–529. doi:10.1287/mksc.1040.0070

Horkheimer, M. (1972). Critical Theory: Selected Essays. New York: Herder and Herder.

Iacobucci, D. (2008). *Mediation Analysis*. Thousand Oaks, CA: Sage.

Iacobucci, D., Saldhana, N., & Deng, X. (2007). A Meditation on Mediation: Evidence that Structural Equations Models Perform Better Than Regressions. *Journal of Consumer Psychology*, *17*(2), 139–153. doi:10.1016/S1057-7408(07)70020-7

Irwin, J. R., & McClelland, G. H. (2001). Misleading Heuristics and Moderated Multiple Regression Models. *JMR, Journal of Marketing Research*, *38*(1), 100–109. doi:10.1509/jmkr.38.1.100.18835

Judd, C. M., & Kenny, D. A. (2010). In D. Gilbert, S. T. Fiske, & G. Lindzey (Eds.), *Data Analysis in Social Psychology: Recent and Recurring Issues. In the Handbook of Social Psychology* (5th ed.; pp. 233–265). New York: McGraw-Hill.

Kirk, R. E. (1999). *Statistics: An Introduction*. London: Harcourt Brace.

Kothari, C. R. (2004). *Research methodology, Methods and Techniques* (2nd ed.). New Delhi: New Age International.

Kothari, C. R. (2007). *Quantitative techniques*. New Delhi: UBS Publishers Limited.

Mujis, D. (2004). *Doing Quantitative Research in Education with SPSS*. London: Sage Publications.

Mutai, B. K. (2000). *How to write quality research proposal: A complete and simplified recipe*. New Delhi: Thelley Publications.

Sokal, R. R., & Rohlf, F. J. (1981). *Biometry: The Principles and Practice of Statistics in Biological Research* (3rd ed.). Oxford, UK: W. H. Freeman.

Tashakkori, A., & Teddlie, C. (1998). *Mixed Methodology: Combining Qualitative and Quantitative Approaches*. Thousand Oaks, CA: Sage.

Vel de, V. E., Vermeir, W., & Corten, F. (2005). Corporate Social Responsibility and financial performance. *Journal of Finance and Accounting*, *5*, 129–138.

Wright, D. B. (2003). Making friends with your data: Improving how statistics are conducted and reported. *The British Journal of Educational Psychology*, *73*(1), 123–136. doi:10.1348/000709903762869950 PMID:12639281

Yates, F. (1934). Contingency table involving small numbers and the $\chi 2$ test: Supplement. *Journal of the Royal Statistical Society*, *1*(2), 217–235.

Ye, K., & Zhang, R. (2011). Do Lenders Value CSR? Evidence from China. *Journal of Business Ethics*, *104*, 197–206. doi:10.1007/s10551-011-0898-6

Zhao, X., Lynch, J. G. Jr, & Chen, Q. (2010). Reconsidering Baron and Kenny: Myths and Truths about Mediation Analysis. *The Journal of Consumer Research*, *37*(2), 197–206. doi:10.1086/651257

Zikmund, W. G., Babin, B. J., Carr, J. C., & Griffin, M. (2006). Business Research Methods (8th ed.). Cengage Learning.

KEY TERMS AND DEFINITIONS

Contingency Table: A contingency table is a two-dimensional table of counts, usually showing frequencies of two variables, displayed in rows and columns respectively.

Continuous Data: This data measured at least at interval level. It is as precise as measuring instruments will allow.

Dependent Variable: A variable that receives stimulus and measured for the effect the treatment has had upon it.

Discrete Data: Data measured at least at interval level, but only as whole numbers (integers). For example, household size, or number of siblings.

Quantitative Research: It is an empirical research in which the researcher explores relationships using numeric data. Survey is generally considered a form of quantitative research. Results can often be generalized, though this is not always the case.

Reliability: The extent to which a measure, procedure or instrument yields the same result on repeated trials.

Statistically Significant: If a result is 'statistically significant', it implies a statistical test has been carried-out, and the probability of obtaining the observed data (or more extreme) under the null hypothesis, is small – typically less than 0.05.

Chapter 10
How Marketers Conduct Mixed Methods Research:
Incorporating the Exploratory Sequential Design with the Hierarchy of Effects Model

Roger Baran
DePaul University, USA

ABSTRACT

The complimentary nature of qualitative and quantitative research methods are examined with respect to a study assessing the market's view of a training and development institute in the Middle East. The qualitative portion consisted of focus groups conducted with seven distinct market segments served by the institute. The results proved insightful with respect to uncovering and understanding differences of opinion among the seven groups; however, taken alone, the qualitative research would have been very misleading with respect to the institute's standing in the Middle East.

INTRODUCTION: MARKETERS' FREQUENT USE OF THE EXPLORATORY SEQUENTIAL DESIGN IN MIXED METHODS RESEARCH

Mixed Methods research has been widely accepted, used and reported in fields such as sociology, education, and health, among others, and has been practiced by marketers for many decades (Creswell & Plano Clark, 2007; Tashakkori & Teddlie 1998). Even so, most marketers don't use the term "mixed methods" and, in actuality, few mixed methods marketing cases have been published. Consequently, the rich methodology and alternative research designs available in the area of mixed methods have not been as formally adopted by marketers as one might expect—even though qualitative and quantitative approaches have been used by them since the 1950's. Marketers have long realized the advantages of combining quantitative research with qualitative research most often in the form of large sample surveys and focus groups and most often in a QUAL – QUAN sequence.

While one-on-one in-depth interviews are used by marketers when there is a need to study individuals' decision-making processes in depth and for sensitive and personal issues, focus groups have been

DOI: 10.4018/978-1-5225-0007-0.ch010

the qualitative method of choice among marketers for many years. One potential contribution to mixed methods research made by marketers is their careful consideration of segmentation in determining the composition of focus groups. An example is available in Table 1. While not generally using this nomenclature, marketers most often tend to use and begin the sequence with the Exploratory Design in their mixed methods studies. Most often this is done to identify important variables and issues which they will then study quantitatively through survey methodology. Thus, the Exploratory Sequential Design would be the mixed method design of choice in most marketing research studies. The focus group discussions enable marketers to develop a better instrument for use in the quantitative phase in that the variables, propositions and issues have been raised which need to be quantified.

Do marketers tend to emphasize the qualitative or quantitative component when conducting exploratory sequential mixed methods research? This is a difficult question to answer as the case discussed in this paper will reveal. The complimentary nature of qualitative and quantitative research methods is examined in this study which was conducted to assess the market's view of a training and development institute in the Middle East. The qualitative portion consisted of focus groups conducted with seven distinct market segments served by the institute. The results proved insightful with respect to uncovering and understanding differences of opinion among the seven groups; however, taken alone, the qualitative research would have been very misleading with respect to the institute's standing in the Middle East. Subsequent quantitative research, consisting of surveys of users and non-users of the institute's programs, provided an insight critical to management that could never have been uncovered via qualitative means. Similarly, the qualitative findings proved to be much more robust and insightful in many matters than the quantitative findings.

Thus, it is impossible to conclude whether the qualitative or quantitative results of the study were more important. While the advantages of mixed-methods research are well known, this study offers a cautionary note regarding the qualitative component. It also offers a model for questionnaire development for use in the quantitative phase which proves useful in verifying or not verifying results from prior qualitative research.

Background of Study

The training and development (T&D) institute being studied has been located in one, small Middle Eastern country for approximately 30 years. Many consider it to be a world-class provider of T&D, offering

Table 1. Important segments to consider for focus group participation

Current User of Institute	Former User of Institute	Never User of Institute
Heavy user	Currently using competitor	Currently using competitor
Moderate user	No longer looking	Never user currently looking
Light user		Never user not looking
User migrating upward		
User migrating downward		
User with no migration either way		

Note: Companies migrating upward are those using increasing spending at the institute. Those migrating downward are decreasing spending at the institute. Those companies with no migration are maintaining the status quo.

professional qualifications for the financial and insurance sectors with programs in insurance, Islamic banking, leadership, banking and finance. The institute is a multilingual, multicultural organization with an accomplished faculty, most of whom hold internationally recognized professional and academic qualifications, along with extensive market experience. The institute is considered by many to be the best in the world for training in Islamic finance and one of the best educational institutes in banking and insurance. The institute also provides an extensive range of internationally and locally recognized professional qualifications.

The purpose of the study was to assess its position in the marketplace: the strength of its brand and programs among its major segment users and potential users. Oftentimes this is referred to as a SWOT analysis: a study of an organization's strengths, weaknesses, opportunities and threats. As such, qualitative research is conducted prior to quantitative research in order to highlight variables, issues and situations which can be quantified in the subsequent survey research.

THE QUALITATIVE RESEARCH STUDY

Focus group sessions were conducted with each of the institute's main market segments: commercial banks, non-commercial bank financial institutions (NCBFI's), insurance companies, government ministries and prominent country opinion leaders. The first four segments mentioned can provide an assessment of the institute's strengths and weaknesses while prominent country opinion leaders can also do that but in addition discuss its opportunities and threats. This is what Creswell refers to as "purposeful sampling (Creswell & Plano Clark, 2007). These companies and individuals were selected because they had the awareness, knowledge and experience with the institute and could therefore provide necessary information.

Qualitative research generally involves small samples and each focus group consisted of 8-10 individuals. Each group was homogeneous in terms of business segment. Nonetheless, with a total of seven focus groups being conducted, the sample size was approximately seventy—a fairly robust number. The participants in each industry focus group numbered only 7-10 and projections to each industry are not warranted. However, taken together, the seventy focus group participants offer a fairly good representation of the institute's base. The focus groups uncovered dramatic differences of opinion among the sectors but did not uncover a key underlying issue regarding the institute's image which could only be revealed through a quantitative study. First, let us examine the sectors' views of the institute in terms of image, curriculum, course offerings and customer focus.

Sector Opinions of the Institute

Commercial bankers feel the institute has a name which is well established in the country. They view it as a strong bank training institute with high quality offerings and staff, and one which provides an excellent education in banking fundamentals. Opinion leaders viewed the institute in societal terms stating it has a wider reputation than just offering vocational courses. According to opinion leaders in the country, the fact the institute has expanded into executive education and professional certificate programs shows it is meeting the needs of "our" society. They view the institute as filling a gap in a country without too many strong universities. Likewise the heads of the government ministries view the institute as a model of excellence that should expand its financial sector expertise to other sectors such as retail and construction. They encourage their entire staffs to attend the institute's courses and view the certificates they earn as comparable to university degrees. They view the institute as the best educational facility in the entire country.

Insurance companies use 90% of their training budgets at this institute but feel that while the institute excels in basic training, it lacks courses and skills in advanced and specialized training. NCBFI's include securities and investment houses, financial institutions, brokers and investment managers. They feel that the institute has done a great job at entry-level programs but not at levels 2, 3, or 4. They view the institute as being good, solid, safe and reliable but one that does not offer courses that are relevant for their business or for middle and upper management. One from this sector summed up the feelings of the entire group by saying "You aren't meeting our needs but you enjoy enormous credibility."

Sector Opinions of the Institute's Curriculum and Course Offerings

The conclusion from the previous section is that the institute is solid in the basic-core foundation courses and serves the ministries very well, commercial banks fairly well, the insurance industry to some degree, and NCBFI's not very well. However, all have a positive view of the institute and feel it has a strong reputation and reflects well on their country. When it comes to the institute's curriculum and course offerings however, most feel it falls short. Commercial bankers feel it doesn't offer anything of value for employees who have more than five years of service. The insurance sector said they are happy with the institute but have no more courses to attend since there is no advanced curriculum in their field. They view the institute as a shop selling the same "stuff" for more than ten years. The NCBFS feels the institute offers no series of courses paralleling career paths in their industry— not even a single options course. It became obvious that the sectors feel the institute needs to update their courses, curriculum and catalogue by pruning old courses and introducing new ones in growth areas.

Most sectors, especially the NCBFS, feel that if the institute is going to innovate, they must do it now since they are at the peak of the maturity stage. They see the institute as being vulnerable to the many high-powered 2-4 day business workshops being offered by financial experts in Dubai.

Sectors View the Institute's Current Position and Ideal Position

The issue here refers to how the institute should position itself. Is it *(a, the)* leading institute in the *(country, Gulf, entire Middle East)?* Is it a training, development or training & development institute? Which fields, if any, is it a leader in: banking, finance, insurance or management?

Some felt it is the leading training institute in the entire Middle East—but only in banking and finance. The insurance sector felt it was the leading insurance training institute in the Gulf.

As we will see, the results of the quantitative survey render many of these opinions relatively useless.

Most agree that while development should be what the institute does, it is best characterized today as primarily a training or T&D institute. No one positioned the institute as a development institute saying that since it does not target middle and senior management, there is no development focus. All felt the institute should focus on development. One said, "What is the benefit of training? Career development is more important because it includes the reward of training." The feeling is that if the institute wants to move into employee development, it has to work more closely with human resource departments in designing sequenced programs for employees and positions.

Sector Opinions of the Institute's Customer Relationship Efforts

All sectors viewed the institute as lacking customer-centricity. They complained that they never receive visits, see few marketing efforts, lack knowledge of program offerings, and feel the institute doesn't understand their educational and training and development needs. They feel the institute needs to sequence course offerings to match career growth within each sector. To do that, the institute needs to become a partner with companies in each sector—not merely an acquaintance.

THE QUANTITATIVE SURVEY RESULTS: A REAL SURPRISE

Focus group participants were intimately involved with the institute and its training of their employees; in addition, focus group sessions were held at the institute's meeting room. It is well understood that opinions generated in focus group sessions may not represent the entire marketplace since participants basically "self-select" their attendance at a session and consequently may be different from those who don't attend. In addition, the number of participants is small and they may not reflect the larger whole. Finally, opinions can sometimes be channeled by dominant participants in ways that don't reflect the market. Consequently there is a need to quantify results using a larger, probabilistically-selected sample.

As a result, two larger surveys were conducted in this research study—one with users of the institute (128 companies) and one with non-users (146 companies). Ninety percent of the respondents in the users group were employed at companies located in the institute's small home country. Half were directors/ managers/ department heads, a fifth were general managers or assistant general managers, fifteen percent worked in Human Resources departments and ten percent worked in T&D as managers or assistant managers. These were highly qualified respondents.

When designing a questionnaire such as the one used in these surveys, it is important to cover the domain of issues pertinent to the organization. Consequently it is recommended that the Hierarchy of Effects Model be used to insure that all key issues are addressed. Basically this model consists of six components around which questions can be arrayed. It also mirrors the stages in the buying or consumption process (Lavidge & Steiner, 1961). The first two are cognitive components—awareness and knowledge. Since there are different degrees of awareness, measures of "aided recall" and "recognition" should both be taken. Aided recall refers to the prominence of the company or brand once the category is given. For example, "Once you think of jogging shoes, which brands come to mind?" Recognition refers to the prominence of the brand name or symbol. For example, which of the following brands of jogging shoes have you seen before: Nike, Adidas, etc.?"

It was taken for granted that participants in the focus groups would be well aware of the institute in that they were pre-screened for awareness, knowledge and usage of the institute and the groups were conducted at the institute. In the quantitative survey of the institute's users, it was also assumed awareness of the institute would be extremely high; after all, all respondents were middle and upper-level managers in companies whose employees were being trained at the institute. Consequently it was a shock when these users were asked the following aided awareness question: "Are you aware of any professional training bodies in the Gulf?" Only 26 of the 128 (20%) users responded with the institute's name and this is after they were informed that the institute was the sponsor of the survey! Among non-users, less than four percent mentioned the name of the institute. This result was not even remotely anticipated. The implications for marketing strategy were far reaching—instead of fine-tuning the image of the institute it was the name of the institute that would have to be marketed! Before attitudes toward the institute can be formed, modified or changed there must be awareness and knowledge of the institute—and clearly, on the basis of such low scores, the institute was not prominent even among company users! Further, since it was found that less than 20% of the companies in the Gulf have ever heard of the institute, the current feasible market for the institute should not be all companies in the Gulf but rather selected, targeted segments—for example, insurance or banking.

This finding, while surprising, is perhaps more typical than atypical and highlights the reason why most marketing research studies of this nature need to be characterized by the mixed-methods approach. The qualitative portion focuses on an organization's attributes, strengths and weaknesses and, as such, respondents are selected who are intimately familiar with the organization—very often heavy users or previous heavy users. The quantitative portion, however, is needed, at the very least, to determine if the organization has a "footprint" in the marketplace; i.e., to determine if it is salient. This is best done by measuring awareness and knowledge of the organization—the first two stages in the hierarchy of effect. As this study shows, awareness and knowledge cannot be taken for granted even among heavy users! Only 20% of the institute's users mentioned the name of the institute when asked if they knew of any professional training bodies in the Gulf.

Another aided awareness question was asked which focused on the specialty of the institute: "When you think of banking and finance, what professional training bodies come to mind?" Thirty percent did not name any, which was very surprising since the institute was the known sponsor of the survey. The other 89 companies named a total of 122 different banking and finance training bodies in the Gulf with 58% of respondents mentioning the institute. No other company received a significant number of mentions with second place going to a company recording 7% of responses. The results of the aided awareness question was very instructive for the institute in that the market's awareness was not nearly as strong as anticipated from the focus group results. Only 41% of the institute's users mentioned the name of the institute when asked what banking and finance instates they can think of. Further, the institute had no major competitor; rather, its challenge is to compete against a hundred small training organizations each of which has strong ties with a few companies.

Since participants in focus group sessions are selected based on their familiarity with the company, product or brand, awareness is going to be one-hundred percent. As this study shows, in the marketplace a company's awareness may be extremely low even among users. The challenge then is not to build strategies and tactics around attitudes and image; rather the challenge is to simply build awareness. This finding will never result from qualitative research and points out the need for quantitative research as well—best done in a mixed methods Qual-Quant exploratory sequential study.

Once awareness has been measured in the quantitative study, the next stage in the Hierarchy of Effects is to measure the market's knowledge of the company, product or brand. Where is the company located? What brands and products does it offer? What are the price points? What is its differential advantage? In what country are its products produced? In this survey, respondents were given the acronym of the institute and 93% correctly knew what the institute's acronym stood for. Granted this is only one measure of knowledge regarding the institute and many other questions could be asked to get a more accurate reading of the market's knowledge of the institute. The important thing to keep in mind is that the market's attitude toward a company can be based on correct or incorrect information. Before one can change attitudes based on incorrect information it would be necessary to get the correct information out there. Thus, measuring and correcting the knowledge component precedes any tactics regarding attitudinal change.

The survey instrument thus far has measured the cognitive components (awareness and knowledge). The next step in the instrument is to collect information regarding "affects"; i.e., liking and preference. To this end, respondents were asked "When you think of (name of institute), what is the first thing that comes to your mind? This was deliberately left open-ended and the two prominent responses were "professional training" (32%) and "quality, reputation, credibility" (16%). With nearly half of all respondents associating the institute with broader capabilities than simply finance and banking, the results of the survey indicates the institute could broaden its offerings, with the expansion being viewed as credible. Additional attitudinal questions assessed user satisfaction with the way the institute trained their employees, overall satisfaction with the institute and its courses, deficiencies, quality of staff, quality of course materials, quality of facilities, etc.

Preferences toward types of new programs and courses were assessed as well as preferences toward a positioning statement regarding the sphere of influence for the institute. Forty-four percent of users regarded the institute as being **the** leading institute in the country, Gulf or Middle East. It was decided, therefore, that the institute had the credibility to market outside its country. Further, respondents were asked what the institute was at its core: training, development or T&D. Nearly half saw the institute as T&D and a fourth as merely training.

Results from the focus groups proved more instructive with respect to how to position the institute than results from the quantitative surveys. From the focus groups it was found that "development" really had two components: first, developing students for careers where manpower is scares in the country; and second, developing individual skills to move participants from entry level to senior management positions (Mumford, 2006). It was agreed that the institute was not really doing either but had the capability for the second type of development. Further, if the market views the institute as only being into "training", it would be impossible to upgrade its image. Thus, based on both the qualitative and quantitative findings it was decided to position the institute as a professional training and development institute. The subtleties in the nature of training and development came through in the qualitative research but not in the quantitative research. The feeling that in-depth responses to issues such as these can be achieved in surveys is more hope than reality. Open-ended questions are used to encourage respondents to "open-up" in ways not restricted by alternatives given in closed-end questions; however, the reality is that for full and rich responses, qualitative research is best. The focus group participants in this study discussed the meanings of training and development from many angles and with a depth of thinking that wouldn't be achieved in a survey conducted over the phone.

Numerous improvements were suggested in the areas of courses, programs, communication and customer relationship management. These were viewed as tactics which would strengthen the relationship between the institute and their company clients leading to greater enrollments. These questions comprise the last phase in the Hierarchy of Effects model—the conative or action-tendency phase (intentions and behavior) in this case intentions to enroll company employees and actually enrolling them in the institute's programs. Enrollment would be based on a number of factors including price, time offered, course content, level of content, etc. and preferences toward these factors would be determined through quantitative survey questions geared to the conative constructs.

The Need for Both Qualitative and Quantitative Research

Institutions need to position themselves appropriately for the markets they serve. Oftentimes qualitative research is conducted with users to determine the position of the institution and the feasibility of adopting a new position. The focus is on the attributes the institution needs to adopt in order to move toward the "ideal" position. In this study, focus group participants brought up what subsequently was found to be the key attribute and key issue facing the institute—how can it become a training and development institute instead of just a training institute. If the quantitative study had not been preceded by the qualitative study, this finding would never have emerged and never been explored. The essence of what this organization was striving for would have never been uncovered.

The assumption is made, based on the fact that all focus group participants have been selected because of their familiarity with the institution, that others in the marketplace are as familiar with the institution. Consequently institutions oftentimes mistakenly allocate resources to change their image or position when, in fact, a better use of their funds would be in investing in an awareness campaign in order to increase the salience of the institution in the marketplace.

Fine-Tuning the Qualitative Research Component in Mixed Methods Research through Segmentation

Careful consideration should be given to the selection of participants in each focus group. It is common practice to make each group homogeneous on participant characteristics related to the issue at hand (Patton, 2005). In this study, five main user groups were identified (banking, insurance, government ministries, etc.) and each focus group consisted of managers from only one of each of the main user groups. It will be argued that there are other important characteristics which should be considered as the basis for selecting members of each group. Table 1, indentified these. (Note: "usage" refers to usage of training and development program).

Given that there are eleven possible participant segments based on usage characteristics alone, it is not likely that those companies in the former-user-no-longer-looking category and never-user not-looking category would be of interest. That leaves nine segments of possible interest based on the "Usage" characteristic alone. The differences among these groups and findings based on their behavior may determine the future of the organization. For example, one might feel that the focus on the heavy user segment is of top priority. However, it could be argued that such a focus would be a mistake. After all, the heavy user segment–insurance companies, banks and non-bank financial institutions–said they only use the institute for introductory/fundamental and basic courses. The institute will never grow (if the number of such businesses are at its peak) if it only solicits comments from heavy users in its focus groups. It may garner better competitive insight if it conducts groups with companies that use them then switch to competitors for higher-level courses, or if it conducts groups who are migrating upward and determines why they are doing so.

Similarly, companies migrating downward should be interviewed to determine why their business is declining: is it due to dissatisfaction or simply due to lack of employees needing training. Former users currently using competitors should be interviewed to determine why they left (defection analysis) and what is needed to regain them (winback analysis). Finally, never- users of the institute's programs should be interviewed to determine whether they simply are not aware of the institute or whether they are aware but don't see a match between their needs and the institute's offerings. One could argue that the greatest contribution of marketers in helping organizations grow is their focus on segmentation. It is felt that this contribution also holds in their selection of segments for participation in focus groups.

Mixed Methods Research and the Hierarchy of Effects Model

In business and marketing, many models exist showing the stages that final consumers and businesses progress through before actually purchasing a product or service. The AIDA (awareness, interest, desire and action) model has been used in business-to-business marketing and consumer marketing for many years as a model of consumer response (Strong, 1925) and the Hierarchy of Effects Model has been used in marketing and advertising since the 1960's.

As previously mentioned the Hierarchy of Effect Model consists of six stages:

1. Awareness
2. Knowledge
3. Liking
4. Preference
5. Intention
6. Usage

Depending upon the marketing issue at hand, tables similar to Table 1, which is based on usage, could be developed based on the 5 other variables with varying levels of complexity. For example, since "awareness" is basically a dichotomy, participants would be those either aware or not aware of the institute. Focus groups are seldom if ever conducted with those who are "unaware" of a company, organization, product, brand or issue, because if they are unaware, they would have nothing to say about the 5 other components which follow in the hierarchy of effects model. Groups selected on the basis of the "liking" variable would be those highly favorable, neutral and not favorable toward the institute. Using "intention" as a selection variable, groups could be selected based on those who are highly likely to retain an institute for T & D in the near future, those who are likely, somewhat likely, not too likely, and not likely at all. Marketers often find the "usage" characteristic to be most relevant, however, because of its stronger relationship to purchase volume, revenue or profits.

Considering the hierarchy of effects model in conjunction with the qualitative and quantitative components of mixed-methods research, it is suggested that quantitative research can most often be conducted alone when the issue is measuring "awareness" and "knowledge" in the marketplace. It is also necessary in establishing your standing in the marketplace on the other 4 variables in the hierarchy of effects. However, when one wants to understand what "motivates" individuals or companies to like, prefer, intend to use, or explain differences in usage, then the qualitative approach is needed. For example, an understanding of the distinction between training and development was only uncovered through qualitative research.

On some occasions, quantitative measurement alone will suffice. Are there occasions when qualitative measurement alone will suffice? It is suggested that in the following situations, self-contained focus groups without quantitative research follow-up could be sufficient:

- To see users first hand again,
- To generate ideas,
- To develop and screen concepts,
- To detect deficiencies,
- To test clarity and comprehension of materials,
- To collect expert opinions in order to generate ideas and forecasts.

In business, particularly marketing, the sequence in mixed-methods research is generally qualitative followed by quantitative. Very often, however, quantitative surveys are conducted without any qualitative work preceding them. When results from the survey prove confusing, however, there may be qualitative follow-up which results in the quantitative-qualitative sequence. Such was the case when a publishing company did survey research on its 700 page book for emergency technicians. They said they didn't like the "order" of the book. What did this mean? Did it mean they didn't like the chapter sequence; mix of pictures, charts, tables and text or something else. Through qualitative research they found what the "order" problem was—a finding that wasn't uncovered in the quantitative portion.

SUMMARY AND CONCLUSION

A mixed-methods research study was discussed in which the key attribute and key issue facing a regionally renowned institute in the Middle East was uncovered through qualitative research. However, taken alone without a subsequent quantitative follow-up, the research results would have been extremely misleading with disastrous consequences for the institute. It would have positioned itself on dimensions and attributes which would have been rejected by its constituency, and spent resources in ill-advised marketing efforts. This study offers a cautionary note regarding the singular use of qualitative research. Using a model called the Hierarch of Effects, it is shown that there are certain key dimensions that are better purviews for both qualitative and quantitative research methods. Further, it offers the suggestion that to maximize the strategy implications of qualitative research, the composition of the groups themselves should be based on finely-tuned usage segments as well as other possible segments based on Hierarchy of Effects variables. Further, participants in focus groups are selected based on their familiarity with the company, product, brand or issue at hand. This research found that even among heavy users, awareness may be extremely low.

REFERENCES

Creswell, J. W., & Clark, V. L. P. (2007). *Designing and conducting mixed methods research* (2nd ed.). Thousand Oaks, CA: Sage Publications.

Lavidge, R. J., & Steiner, G. A. (1961, October). A model for predictive measurements of advertising effectiveness. *Journal of Marketing*, *25*(6), 59–62. doi:10.2307/1248516

Mumford, A. (2006). Management Development - Strategies for Action. Development and Learning in Organizations: An International Journal, 20(2).

Patton, M. Q. (2005). *Qualitative research.* John Wiley & Sons, Ltd.

Strong, E. K. (1925). *The Psychology of selling*. New York: McGraw-Hill.

Tashakkori, A., & Teddlie, C. (1998). *Mixed methodology: Combining qualitative and quantitative approaches*. Thousand Oaks, CA: Sage Publications.

Chapter 11
Making Sense of All the Words:
Analyzing Qualitative Data

Janice E. Jones
Cardinal Stritch University, USA

A. J. Metz
University of Utah, USA

ABSTRACT

This chapter provides an introduction to the process of qualitative analysis and to use step by step examples to provide an idea of how the process of qualitative analysis actually works. Crabtree and Miller, 1992, note that there are many different strategies for analysis, in fact, they suggest there are as many strategies as there are qualitative researchers. This chapter is intended to give the researcher a place to begin and to inspire a deeper dive into this rewarding form of data analysis. Stake, (1995) writes that qualitative data analysis is "a matter of giving meaning to first impressions as well as to final compilations. Analysis essentially means taking something apart. We take our impressions, our observations, apart… we need to take the new impression apart, giving meaning to the parts"(p. 71). While qualitative data analysis can be time consuming the rewards that come from immersion in the data far outweigh the time spent doing so.

INTRODUCTION

The goal of this chapter is to provide an introduction to the process of qualitative analysis and to use step by step examples to provide an idea of how the process of qualitative analysis actually works. Crabtree and Miller, 1992, note that there are many different strategies for analysis, in fact, they suggest there are as many strategies as there are qualitative researchers. This chapter is intended to give the researcher a place to begin and to inspire a deeper dive into this rewarding form of data analysis. Stake, (1995, p. 71) writes that qualitative data analysis is "a matter of giving meaning to first impressions as well as to final compilations. Analysis essentially means taking something apart. We take our impressions, our observations, apart… we need to take the new impression apart, giving meaning to the parts". While qualitative data analysis can be time consuming the rewards that come from immersion in the data far outweigh the time spent doing so.

DOI: 10.4018/978-1-5225-0007-0.ch011

SOURCES OF QUALITATIVE DATA

Before beginning a discussion of qualitative data analysis, we must first discuss sources of qualitative data. This section is a brief overview of qualitative data collection, please further recommendations section at the end of the chapter for a more resources to help with this aspect. Most people naturally think of interviews as the most common source of qualitative data, however, focus groups, field observations, comments on surveys, historical records, and secondary data. The common thread that runs through these forms of qualitative data is words. Words of the interviewee or focus group participant who responds to an open ended question that the interviewer poses. Qualitative data is most commonly words. Patton (2002) reports that frequently the sources of qualitative date includes interviews, documents and documents which provide the researcher with a myriad of WORDS that will need to be analyzed. In quantitative analysis we would say that we are crunching the numbers, however in qualitative data analysis, we cannot crunch the words.

Normally the interviewer audio records the session, whether it be a one on one interview or a focus group of 6 to 8 people who are united by a common experience whether it be work, trauma, or illness, to name a few examples. Someone will then transcribe the session and prepare a document that has the questions asked along with the responses of the participant (See sample at end of chapter). This is a very labor intensive step of the process and if the researcher is fortunate to have grant money or other source of funding available to pay a transcriptionist they are lucky. Otherwise it is the researcher's job to listen intently to the recording and type carefully to capture each spoken word accurately. If audio recording is not possible the researcher will write down the responses that the participants have to his or her questions. If handwriting responses, it helps tremendously to prepare ahead of time with a document that is formatted with two columns, one for the actual verbatim responses and two with a place for the researcher to provide memos to him or herself, recording thoughts, feelings and/or observations (see sample at end of the chapter).

It is also necessary to transcribe the field notes that a researcher has taken which would include the researchers thoughts, feelings and observations of occurrences such as body language of participants or other nonverbal cues that can be seen during the interview or focus group but not seen or heard on a transcribed document (Yin, 2016). Many times a researcher will carry a special notebook specifically for field notes. These field notes can be documented by the researcher at the end of the interview, during a quiet period of reflection. Yin (2016) also discusses the importance of field notes and the importance of a researcher developing a kind of "transcribing language" (p. 170) where the researcher uses a common code such as a shorthand as it were to speed up your field note taking process. Consistency will be the key here, for example, one might use a large letter A to indicate anger. Or a large letter L for laughter. These field note images of laughter or anger will help the researcher better remember the context that is surrounding the participant's experience. Yin (2016) also describes the importance of processing the field notes as soon as possible after the interview or focus group experience as memories fade or blend together and as researchers we want the richest and most accurate data possible. Field notes are important but should not be confused with field observations.

Field observations are observations made in the field by the researcher of, for example, a classroom where the researcher is purposefully observing behavior of a student or the teacher as they go about their day. A researcher making field observations will record the observation and perhaps even count the number of times, for example, the teacher says "open your books". Data that might be recorded along with the number of times the teachers says "open your books" might be tone of voice, inflection, facial expression, students body language, moans and groans of students, etc. The field observations would be transcribed in a similar fashion as mentioned previously to prepare for the data analysis.

Additionally, survey comments, or any responses on a survey to an open ended question would be handled the same way. They would be gathered, transcribed and prepared for coding. Open ended questions at the end of a survey are becoming more common as researchers understand the need for respondents to be offered an opportunity to explain themselves and/or their responses.

Historical records and secondary data will be combined here for brevity of discussion but it should be noted that they are two distinct categories of data collection. Historical records are records that have been kept usually archived of events from the past. If possible, the researcher would obtain copies however sometimes that is limited due to age of original documents, no available copier or other issue. Then the researcher will take notes on the historical document. Secondary data would be similar in that it is data that has been gathered by someone else for another purpose other than the current qualitative study. The data from secondary data could come to the researcher in an electronic format or paper depending on the original data collection project. While there may be other methods of data collection, this section was intended to provide the researcher with an overview of data collection techniques before beginning a discussion of data analysis.

DATA ANALYSIS

One of the joys of qualitative data analysis is that the process of data analysis allows the researcher to stay close to the data as he or she literally immerses themselves into the data, analyzing word for word, line for line and experience for experience. By staying close to the data, the researcher becomes sensitive to emerging themes and theories and begins to understand what is unfolding in front of them. Remembering that qualitative data analysis is not a linear process, meaning the researcher begins analyzing data after the first data gathering experience, the researcher has the benefit of beginning to understand the research and also to look for holes in the data. While in quantitative data analysis where the researcher typically waits until all of the data is gathered, authors such as Glaser and Strauss (1967) provide a framework with the constant comparison method of data analysis as a part of the Grounded Theory process of qualitative data analysis. Using the constant comparative method (Glaser and & Strauss, 1967) to analyze the data, codes are generated and utilized throughout the analysis, from the first interview through the final interview. Each incident is compared as it was discovered in the data and themes occur or emerge as they become apparent to the researcher (Glaser & Strauss, 1967; Lincoln & Guba, 1985). The categories are examined for possible relationships among the categories. So now the researcher has pages and pages of transcribed data, what will be done with it? How will the data be analyzed? Given the nature of Mixed Methods Research, data analysis is an extremely important component. As has been presented in other chapters in this book, qualitative and quantitative go hand in hand, each a necessary component of mixed methods research.

To better understand the procedure associated with qualitative data analysis, this chapter will suggest the following plan. First the researcher organizes the data that has been gathered. Second he or she immerses him or herself in the data, meaning a review of the transcripts by a careful, thoughtful reading over and over again. This leads naturally to the third step, that of generating categories or themes. Followed by, fourth, coding the data, following a coding scheme supported by the themes. Next a typical analytic procedures includes an understanding of the memoing that has taken place. The researcher then will ask him or herself what other meanings could be associated or found with this data. Are there other interpretations? Finally a report is written that highlights the data analysis and the findings.

The researcher will be advised to avoid weak results. Caution will be important when considering how to address weak results from either quantitative or qualitative. It is important to remember to try not to support weak results from one methodology by adding the other methodology on top. If there are weak quantitative results, do not try to support those with qualitative results, go back and strengthen the quantitative first. Make sure the results are meaningful and accurate before adding MORE data. A helpful metaphor is building a house. A fancy roof will not be supported by a weak foundation. Both the roof and the foundation of the house needs to be strong so that the house stands safe and sound. In other words, both the quantitative and qualitative data analysis needs to be strong, complete and accurate.

Coding

In order to make sense of the data, the researcher needs to apply some coding scheme to the data in order to prepare categories and themes. By doing so, he or she utilizes some system that makes the most sense for the researcher. Some researchers use abbreviations of words, some develop a list of key words, some use numbers, some may even use colored dots to highlight the words chosen. The process of coding may take several forms, the researcher decides what works best for him or herself. The main organizational point to remember here is to be consistent, that is, take careful note of what code you are using and why. As is with anything else, the process of coding will be interrupted by other work, other commitments and life in general, so notes to oneself help keep the data analysis procedure organized, consistent and scientific in nature. Simple office tools like sticky notes, highlighters, index cards and big sheets of paper to hold the sticky notes or other creative manner to organize your quotes and other data points will make the coding easier. As the researcher begins to code the data, categories, patterns and themes will emerge from the data. This process is time consuming but very necessary. Coding is really another way to think about how the researcher will take the data and create meaning out of a large volume of words. Coding allows the researcher to organize the data. By doing so, patterns of word repetition may appear, significant phrases or ideas held by participants will rise to the surface, which in turn creates themes that the researcher will then take and compare to the next transcript, and so on and so on.

It is common practice for the researcher to develop major codes and for those major codes to have sub codes. For example, if a researcher was conducting a study on students with disabilities (see sample code book at end of chapter), possible codes and subcodes could be as follows:

- Concern for safety of their children (CODE)
  - Potential for violence against people with disabilities (SUB CODES)
  - Three buses to get to school (SUB CODES)
  - Possible bullying (SUB CODES)
  - Behavior problems (SUB CODES)

Eventually the researcher will provide quotes from the participants that support each subcode. For example, "Three buses to get to school", could be supported with a quote from a mother who said "I worry all the time about Andrea, she rides three buses to get to school. She is diagnosed with LD and I worry that she will miss the bus and be stranded somewhere waiting and waiting for the next bus, what if it doesn't come? What if she doesn't understand? How will I know? I am turning into a nervous Nellie, I call the school every day to make sure that she has arrived. I don't what I will do when the day comes and I call and she is not there". By reading this quote, the reader can generate a picture in his or her own head about why the researcher chose the sub code "Three buses to get to school" and why it is falls under the Code of "Concern for safety of their children".

When reviewing the transcripts and immersing oneself in the data as codes become clear, patterns will also present themselves. Based on informed ideas and suppositions, the researcher might identify some patterns. Tentative patterns are acknowledged and continuous review of the data that has been collected will help the researcher decide if the additional data are consistent with those patterns identified. Patterns can be described and defined by examination of categories, perhaps expanding, creating new or combining existing categories to create patterns or themes.

The work done between coding and the first draft should feel a little overwhelming in that the researcher is working to exhaust his or her thinking about the data and its meaning. Themes help to summarize the findings and create categories so that eventually the data can be presented in a sensible fashion to the reader.

Generating Categories and Themes

The researcher begins the qualitative data analysis process by reading and reading and then reading some more. The data for qualitative research are normally pages and pages of transcription. As the data generates meaning for the researcher, the researcher seeks data that has internal convergence and external convergence to form categories (Guba, 1978). Categories need to be internally consistent yet need to be different from each other. This is called inductive analysis. Patton (2002) describes the processes of inductive analysis as "discovering patterns, themes, and categories in one's data, in contrast with deductive analysis where the analytic categories are stipulated beforehand, according to an existing framework."

The Importance of Memos and Memoing

Memos are extremely important tools to help the researcher throughout this entire process. Writing notes to oneself as ideas and insights bubble to the surface will become invaluable tools to help the qualitative researcher, take the work from basic analysis to reflective and insightful analysis of the data. Memos can become an auxiliary albeit nonexistent researcher, as they provide reminders of important data points and provide a rich context to help make sense of the words being analyzed.

Offering Interpretations

As themes and categories are developed through by constant scrutiny and examination of the transcripts the researcher begins to form impressions and develop interpretations of what he or she is reading, in other words, themes emerge and the story presents itself. This allows the researcher to make meaning out of the words, to make sense of the story presenting itself and allows the lived experience of the participant to be shared with first the researcher and then eventually, the reader. At this time it is important for the researcher to ask him or herself, this question. What else could the data be telling me?

Utilizing the quotes from the interviews or focus groups to support the codes helps to answer the question, what else could the data be telling me. Observations, memoing, and field notes provide another layer to support the work. It is imperative that the researcher report evidence that is to the contrary and to provide multiple perspectives. Individual experiences may generate contradictions and tensions within the data.

Demonstrating Credibility

Throughout qualitative data analysis one will hear different terms used to describe what in quantitative analysis would be called reliability and validity. Words such as trustworthiness, credibility, and confirmability are used instead. Trustworthiness is defined by Yin (2016) as an infusion of an attitude throughout the qualitative research process that demonstrates a sense of credibility for the reader. This can be accomplished by having transparent data collection and analysis procedures that are defined for the reader without a hidden or poorly defined steps along the way. Credibility Following is a description of other steps and terms used by qualitative researchers to demonstrate credibility.

Prolonged Engagement

Qualitative research is different from quantitative in many areas but most specifically in the amount of time the researcher spends with the participants. A quantitative survey provides a snapshot in time, a one shot deal as it were. Qualitative data collection is a prolonged experience with the participants, the interviews are longer, over a longer period of time, the researcher traditionally spends more time observing, getting to know the participant and seeking to better understand each participants lived experience of the phenomenon being studied.

Triangulation

Thinking of a triangle, one can imagine data on each point of the triangle forming a strong bond that will ensure a tight presentation of data which will in turn inform the final report. This step in demonstrating credibility allows the reader to see, through the written explanation that the data is working together to inform the reader. Often researchers will use literature for one point, data for another and theory for the third. All three working in concert to present a uniform presentation of the findings. Yin (2006, p. 87) suggests that the researcher keep a "triangulating mind" that is, to "always seek to develop converging lines of inquiry about all of your research actions and assumptions".

Peer Review

A peer review may simply be colleagues or peers who are able to review the data and the findings before the final report is written. Consensus may be sought or independent review is fine as well. The important component is to ensure that this process is clearly explained in the final report. Another step in credibility that provides the reader with more confidence that the data that is informing the final report has been gathered in a scientific manner.

Negative Case Analysis

To support the rigor of the qualitative data analysis, negative case analysis provides the researcher with an opportunity to analyze data points that are different than what the researcher expected to find. These are wonderful, rich nuggets of information that challenge what the other participants are saying simply by being different. The negative cases strengthen the findings as they force the researcher to ask him or herself WHY? Why is this different? What about this participant would have made him or her say that? What is different about the lived experience of this participant? (Brodsky, 2008).

Researcher Bias

What does that mean in qualitative research? In qualitative research it is imperative that the researcher acknowledge any bias he or she holds before beginning the project and again during the data analysis process. By acknowledging bias the researcher is bolstering the trustworthiness of the research (Elo, Kaarlainen, Kanste, Polkki, Utriainen & Kyngas, 2014). Specifically focusing on content analysis, Elo, Kaarlainen, Kanste, Polkki, Utriainen and Kyngas (2014) provide a series of questions that the researcher can ask him or herself such as "did I lead the participant?" or "did I ask too broad of questions?" (p. 4) as the researcher is progressing through the research process. These type of questions can be asked at any time during the research process but seem to especially good after a pilot of the interview questions and again during the data analysis phase. A simple way of identifying researcher bias is for the researcher to identify blind spots, past experiences, past accomplishments, status, and values before beginning the research and reflecting on the same throughout the research process. This might the time to ask for feedback from a peer who can help assist in the identification of blind spots. It is always easier to find those in someone else rather than in our own self.

Another important step in addressing researcher bias is bracketing. Bracketing tasks the researcher with acknowledging ideas, values and beliefs that society takes for granted in the world we live in. The focus of bracketing is to acknowledge ideas that we know exist but that we may not be able to change. By bracketing, the researcher makes very clear first to him or herself as the researcher and then again in the written report of that data analysis what assumptions are underlying the thoughts, values and ideals that society holds.

Member Checking

An integral step in the review of the data is member checking. Member checking has been described as an opportunity for the participants in the study to review what they have said, to provide an opposing view or a clarification if needed (Miles, Huberman & Saldana, 2014). By doing this, the researcher allows for a clearer picture of the data to come forth and this step provides another level of credibility for the data. Traditionally, this step, of member checking is outlined in the procedures section and the consent documentation.

Rich, Thick Description

The researcher uses many quotes from the participants to support the themes suggested by the research. In addition, the researcher writes in vivid detail, providing and shaping a context for the reader to fully understand and visualize what is happening, what the participant is feeling, to better understand the lived experience of the participant. The reader should feel as though he or she can recreate the setting, the mood, the feel of the data gathering experience. The reader should feel as though he or she is part of the story being told. Additionally, throughout the qualitative research arena, one will hear expressions such as "I knew I had enough data when I began hearing the same thing from different participants". There is no magic number, however one will hear that the researcher has reached a point of saturation which refers to the feeling that one is reading similar experiences over at which point the researcher feels as though he or she is saturated by the data.

To summarize this section on how the researcher can demonstrate that he or she has good results, it is important to remember to accurately describe the procedures used through the study. The timing and placement of the procedures should be explained, how the data was gathered, steps in the process are clear and well defined. Additionally, the researcher has done a good job of explaining his or her bias, worldview, assumptions and epistemology and how those will be impacting the analysis of the data. Utilizing some or all of the methods described above, the researcher outlines a clear process with clear descriptions of how he or she arrived at the conclusions and findings presented in the final report.

Computer Software and Data Analysis

This chapter has focused on the more traditional way of coding data, that is, by hand, with the researcher utilizing paper and pen and assorted office supplies to organize, code and create themes out of the data. It is worthwhile to note that there are many different computer software programs available for the researcher to use. A Google search will start the researcher on his or her path to finding a program that is both affordable and viable for the researcher. What is most important to note is that computer software programs do not do the analysis for the researcher, instead, they label and organize data which allows the researcher to search among the data points to create the themes. If there are large amounts of data, or the data collection is ongoing, a computer software program might be a great investment for the researcher to keep track of the data and ease the organization of it.

HOW WILL I PRESENT MY QUALITATIVE DATA?

Part of the joy of being a qualitative researcher is the freedom that comes from presenting your data in a manner that makes sense to you. One way to begin the presentation of your data might be a concept chart (see sample at end of chapter) which shows your reader how the concepts are all connected or how they fit together in your research. Another interesting tool is a flow chart (see sample at end of chapter) which demonstrates for your reader how the concepts work together through time or how they might even be steps in a process. There are a myriad of free programs online that can help the researcher create a concept map or a flow chart. In addition to the use of these tools in the data analysis portion of qualitative research, many researchers use these tools for the research design and/or the literature review. If one is not comfortable with the use of a computer to aid in the creation of these tools, a simple sketch with paper and pen works just as well. Sticky notes on pieces of paper make an excellent starting point as the sticky notes can be moved as the flow as interpreted. Another alternative method for presenting your qualitative data might be through the use of a table (see sample at end of chapter) that you make that has concepts on one axis and participants on the other, with meaningful quotes from your participants to support the identification of the concepts. All of these examples may be used with traditional written format of the presentation of the data. A recommendation that comes from years of practice is to use these graphical representations of the process or data, which ever it might be, in isolation from the rest of the data. Hang the graphical representation on the refrigerator, keep a copy on your desk or even on a mirror. The idea here is that you will look at the flow, the graph or the concept at different times and reflect on what it means, how can it be improved, or even what can be done differently. In order to answer the research question that is presented in the study, data presented will include quotes from the participants that clearly support the themes that the researcher has identified as being the most salient.

In sum, while the work needed to analyze qualitative data may seem to be more than quantitative data, the results are well worth the effort. For social scientists, sharing findings and the lived experiences of participants who might otherwise be marginalized is a noble goal. Combining the qualitative results with quantitative results will provide a richness and completeness to data that has been gathered to allow the researcher to assemble a quality mixed methods research study. Following this chapter is a list of recommended further reading along with some samples of work product using a study that focuses on parental influence on the vocational development of adolescents with a disability in an urban setting.

REFERENCES

Brodsky, A. E. (2008). Negative case analysis. In L. M. Given (Ed.), *The Sage encyclopedia of qualitative research methods*. Thousand Oaks, CA: Sage Publications. doi:10.4135/9781412963909.n283

Creswell, W. (1998). *Qualitative inquiry and research design: Choosing among five traditions*. Thousand Oaks, CA: Sage Publications.

Eisner, E. W. (1991). *The enlightened eye: Qualitative inquiry and the enhancement of educational practice*. New York: Macmillan.

Elo, S., Kaarlainen, M., Kanste, O., Polkki, T., Utriainen, K. & Kyngas, H. (2014). Qualitative content analysis: A focus on trustworthiness. *Sage Open,* 1-10. DOI: 10.1177/2158244014522633

Glaser, B. G., & Strauss, A. L. (1967). *The discovery of grounded theory: Strategies for qualitative research*. Chicago, IL: Aldine Publishing Company.

Lincoln, Y., & Guba, E. (1985). *Naturalistic inquiry*. Beverly Hills, CA: Sage Publications.

Stake, R. E. (1978). The case study method in social inquiry. *Educational Researcher*, *7*(2), 5–8. doi:10.3102/0013189X007002005

Strauss, A., & Corbin, J. (1990). *Basics of qualitative research: Grounded theory procedures and techniques*. London: Sage.

Yin, R. K. (2016). *Qualitative research from start to finish*. New York, NY: Guilford Press.

ADDITIONAL READING

Miles, M. B., Huberman, A. M., & Saldana, J. (2014). *Qualitative data analysis: A methods sourcebook*. Thousand Oaks, CA: Sage Publications.

Wertz, F. J., Charmaz, K., McMullen, L. M., Josselson, R., Anderson, R., & McSpadden, E. (2011). *Five ways of doing qualitative analysis: Phenomenological psychology, grounded theory, discourse analysis, narrative research and intuitive inquiry*. New York, NY: Guilford Press.

Yin, R. K. (2016). *Qualitative research from start to finish*. New York, NY: Guilford Press.

APPENDIX

Figure 1. Basic flow chart

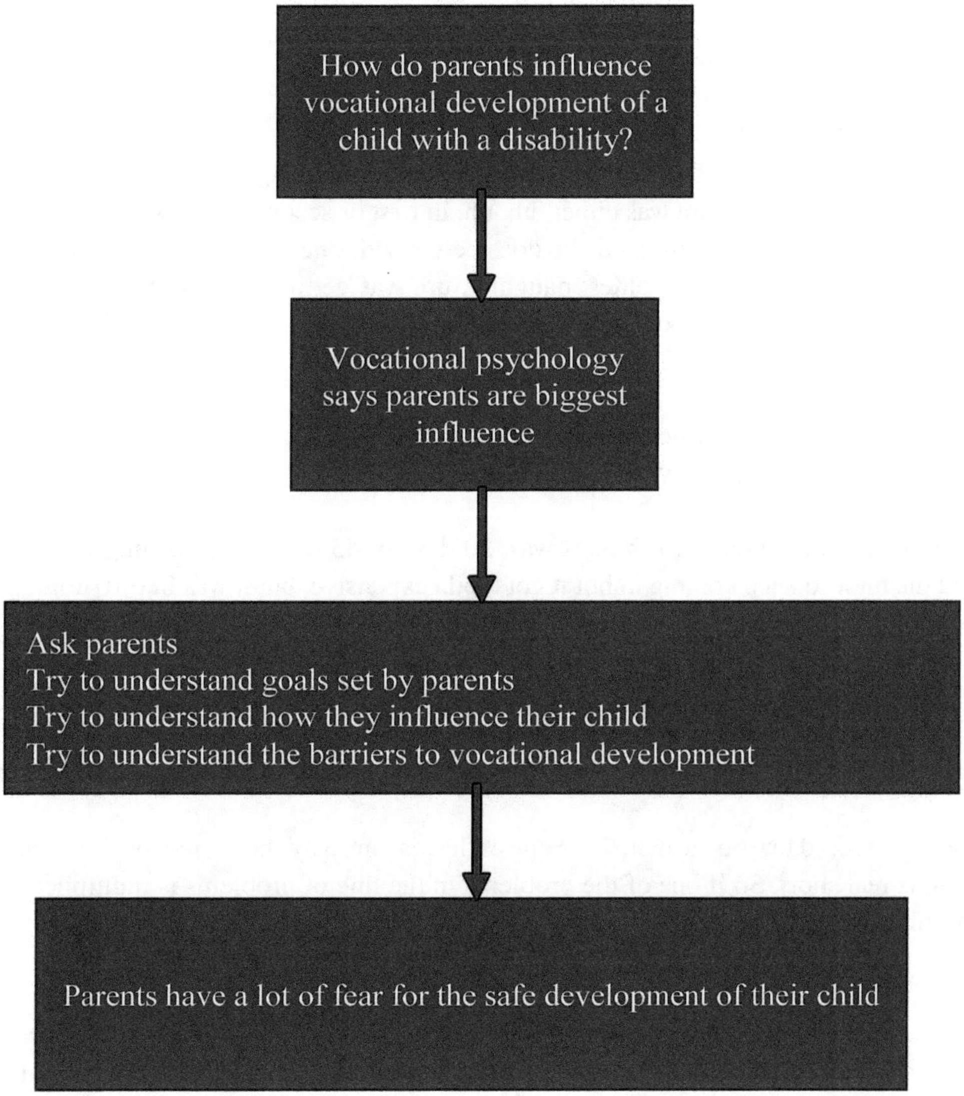

Transcribed Interview: First Question

Interviewer comments in italics.

1. When did you first find out about your son or daughter's disability? (gather information about disability specifics, strengths and needs)

First or second grade.

Did the school tell you or how did you find that out?

It was just from her behavior. The behavior problems she was having, I have two daughters who have a learning disability. I think Mindra was either, uh, uh, in first or second grade when she was tested and all of the tests and all of her M teams said she does better with one on one learning and I haven't been able to get her that. I hired well, my oldest daughter, um was getting into trouble in school, behavior problems, and she went back to school and she told the teacher that I had slapped her and the teacher called Child Protective Services

Oh dear.

And when Child Protective Services came in and saw that my youngest daughter, Mindra, was having problems with the MTeam and everything, they gave her a tutor.

Oh great.

But the tutor was only good for, I think it was, 30 days or 45 days or something like that, but after that we hired the tutor to keep coming in but it got kinda expensive, but it was like 10 dollars, but it was supposed to be like 20 dollars, and Mindra did really, really well, and the part that made me mad with the school was they seen from the time that she had the tutor and they seen she was in 2nd or 3rd grade level but she was in the 7th grade, from the time she got the tutor to the time she got the tutor again she had brought all of grades, her reading, and everything was at level. But the school board did not want to pay for her a tutor, so they would never admit it, the tutor went with me to the MTEAM meetings

Good.

The reason she scored low on the math tests one of the reasons why she scored low is because Mindra's attention span is real short. So if one of the problems in the line of problems is multiplication, Mindra did every problem like it was multiplication, so if it was two additions and two subtractions, you know they mix them up,

Sure.

The first one on the line if it was multiplication she multiplied all of them and of the second line was addition, she added all of them and that was the way she did, she multiplied all of them in a row. But she got a failing grade on the math test, and of the tutor hadn't a been there, to bring that to there attention, they would have had left that grade like that. But the tutor was there and she said, do you see what she did, she wasn't paying attention, during the first problem and she took it to the whole line, every problem, she multiplied all of them. The only thing she did differently and correctly was the division problem because the division problem looks different. So that was the only one she did absolutely right was the division problem. The multiplication, the adding, the subtraction…She has LD and ADHD.

Sample

Code Book

- **Career Aspirations Held by Parents for Child**
    - o Lack of information about options
    - o Parents choosing same job they have for their child
    - o Financial concerns
    - o College
- **Concern for safety of their children**
    - o Potential for violence against people with disabilities
    - o Three buses to get to school
    - o Possible bullying
    - o Behavior problems
- **Learned hopelessness**
    - o Marginalization
    - o Lack of options
    - o Lack of financial resources
- **Lack of support**
    - o Self reliant
    - o Need for self advocacy on part of parents
    - o Expert Model or deification of expert
    - o Rely on one teacher to help
    - o Lack of knowledge about how or where to get resources
    - o Education professionals lack sensitivity about transition needs
    - o Transition plans not well implemented, executed or understood
    - o Lack of information
- **Family system difficulties**
    - o Difficulties between husband and wife
    - o Difficulties between parent and child
- **Get married**
    - o From a disability perspective-someone else to look after daughter with disability
    - o Cultural nuance-protective measure-have daughter out of "harm's way" or risk
- **Ideas for vocational development**
    - o Talk to child
    - o Help them develop goals
    - o Responsibility
    - o Manners
    - o Budgeting
    - o Being on time
    - o Help with applications
- **Parental involvement at schools**
    - o Invited to meetings

- o Phone calls
- o Emails
- o They haven't involved me
- o Need more communication

- **Future Planning**
  - o Hope for independence
  - o Need for continued involvement

Sample

Interview form for notetaking and place for memos (table 1).

Participant #________

When did you first find out about your son or daughter's disability? (gather information about disability specifics, strengths and needs)

Table 1.

Memos	Participant's Response
No eye contact Looking at floor, seems uncomfortable	First or second grade. *Did the school tell you or how did you find that out?* It was just from her behavior. The behavior problems she was having, I have two daughters who have a learning disability. I think Mindra was either, uh, uh, in first or second grade when she was tested and all of the tests and all of her M teams said she does better with one on one learning and I haven't been able to get her that. I hired well, my oldest daughter, um was getting into trouble in school, behavior problems, and she went back to school and she told the teacher that I had slapped her and the teacher called Child Protective Services.

Figure 2. Concept map

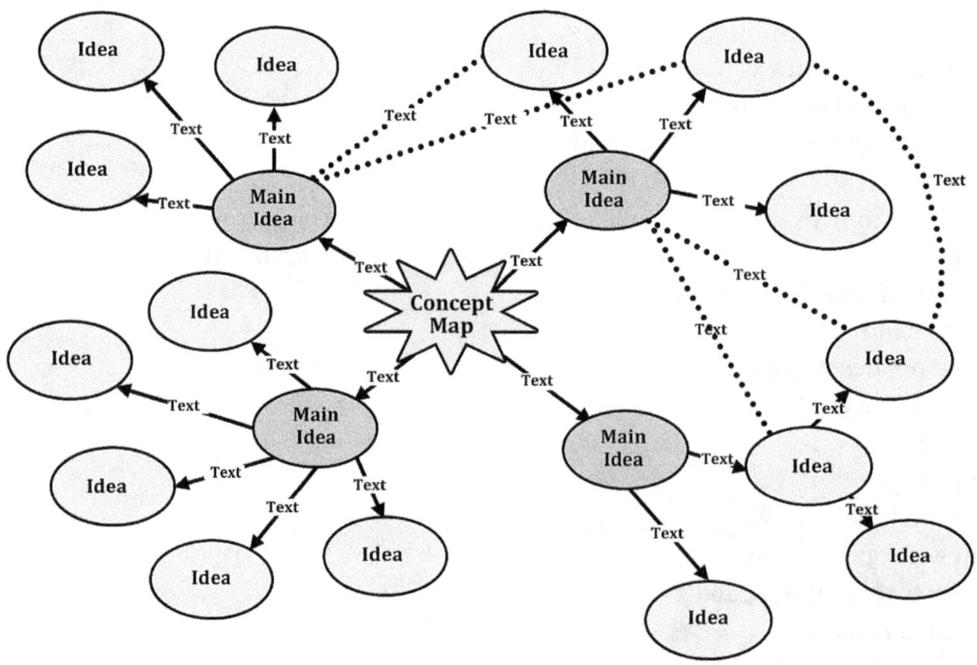

Table 2. Theme of parental involvement in school

Theme	Categories	Quotes
Parental involvement at school	Invited to meetings	"I receive invitations on slips of paper to come to meetings" "I get the announcements about meetings, I just don't go" "I have heard from the teacher about the meetings"
	Phone Calls	"I call and leave messages with his teacher" "The teacher calls me, she doesn't leave a very long message, then I have to call back and we play phone tag" "I don't have my cell phone anymore so it is hard to call me, but I use the phone at work to call"
	Emails	"I don't have the internet so I cannot get emails" "Email would be a good way to get ahold of me" "My son has an email account, I do not, but I suppose I could use his email, right?"
	They haven't involved me	"I do not think they try to involve me, I do not think they care" "I have not gotten any information from school" "I am sure the school does not want to hear what I have to say so I do not get the information about anything, anything at all."
	Need More Communication	"I would like to be involved, I do not know what to do, this school is different than the last one we were at". I think the teachers should try harder to get ahold of us parents" "I liked it better when he was little and I had to sign the assignment book, then I knew more what was going on"

Section 6

Data Analysis:

Examples of Research Studies using Mixed Methods

In this section three studies are used as innovative examples of how researcher use mixed methods to provide holistic interpretations of phenomena. Mixed methods researchers need to be experienced in three types of data analysis strategies. This chapter provides examples of innovative mixed methods research studies combining the use of inductive and deductive logics. The strength of a mixed methods approach is that it provides ample opportunities for researchers to be creative and eclectic in their approach, examining an issue from various angles using a myriad of data collection tools and data analysis techniques.

Chapter 12
Examining Online Communication:
A Method for the Quantitative Analysis of Qualitative Data

Michael G. Hughes
HumRRO, USA

Lauren Harkrider Beechly
IBM, USA

Jennifer A. Griffith
Alfred University, USA

Thomas A. Zeni
East Central University, USA

Cristina Byrne
University of Oklahoma, USA

Amanda Shipman
Kenexa, USA

Darin S. Nei
Hogan Assessment Systems, USA

Shane Connelly
University of Oklahoma, USA

Michael D. Mumford
University of Oklahoma, USA

ABSTRACT

Methods of individual communication continue to expand through online media. Given the dynamic nature of online communications, traditional methods for studying communications may not suffice. A hybridized content analytic approach that combines qualitative and quantitative methods offers a unique methodological tool to researchers who seek to better understand computer-mediated communications and the psychological characteristics of those who communicate online by evaluating qualitative information using quantitative methods. This means of measurement allows researchers to statistically evaluate whether investigated phenomena are occurring in combination with the richness that qualitative assessment provides. As with any approach to computer-mediated communication, various ethical considerations must be borne in mind, and, thus, are discussed in concert with this hybridized approach to content analysis.

DOI: 10.4018/978-1-5225-0007-0.ch012

INTRODUCTION

According to Internet World Stats (2011), the number of people who use the Internet worldwide is in the billions – to be more precise, 2.3 billion of the world's 7 billion people are Internet users. This number is staggering given that the Internet is a relatively new form of media. In fact, since 2000 this number has increased by over 528% (Internet World Stats, 2011). This exponential growth in Internet usage calls to mind a basic question: for what purposes are people using the Internet? According to research conducted by the PEW Internet and American Life Project (2012), the most common activities for adult users of the Internet are sending or reading email, using search engines, searching for maps or driving directions, checking the weather, and buying products. Other common activities include looking up information on politics, uploading photos to share with others, sending instant messages, and visiting social networking sites.

PEW (2010, 2012) reports that 68% of adults use social networking sites, and 93% of teenagers use these sites to communicate with their friends and family. Social networking is one of the top five Internet activities engaged in by teenagers and members of Gen Y (currently aged 21-35). These statistics suggest that among other uses like shopping and information searching, people are using the Internet to socialize and communicate at an ever-increasing level. Given the generational differences in these types of Internet activities, this trend is likely to become even more prevalent as online teenagers become adults. This broaches another, more important question: have communication research techniques been able to "keep up" with this ever-expanding channel of communication? While it may be a challenge for researchers to keep up with the pace of technology, this paper proposes a methodology for examining online communications that can be readily applied to a variety of formats. More specifically, this methodology is a form of content analysis that transforms qualitative data into quantitative data. Below we discuss the issues surrounding the nature of Internet communications that call for new methods of examination. Also, we provide a step-by-step procedural explanation as well as an example of this content analytic approach. Additionally, the advantages, disadvantages, and ethical issues surrounding this technique are discussed.

BACKGROUND

Nature of Online Communications

Over the past few decades, computer-mediated communication (CMC) has garnered a great deal of research focus and interest. With the advent of new online outlets for discussion arising at a remarkable pace, the breadth of communication research only continues to expand as each new medium brings with it unique intricacies and idiosyncrasies. Although, for the most part, previous approaches to the study of communication appear to have adapted well to the technologically driven new media of Internet communication and discourse, it seems reasonable for one to question if new methods of inquiry might be better-suited for capturing the wealth of information that such online contexts have to offer. In particular, Internet bulletin boards (or message board forums) have become an extremely popular means of modern discourse, but the online context and asynchronous nature of message board discussions entail unique concerns for communication researchers interested in studying these environments. Therefore, in this paper we propose a quantitative approach to analyzing qualitative content analytic data aimed specifically at studying the communications found on Internet message board forums.

Content analysis has been defined as the "research technique for the objective, systematic, and quantitative description of the manifest content of communication" (Berelson, 1952, p. 18). To be sure, the uses and applications of content analytic strategies have changed and developed since Berelson's initial examination of the subject. However, certain characteristics of content analysis continue to hold true. Typically, content analysis procedures center around examining what is said rather than trying to investigate latent intentions, reactions, or responses (Berelson, 1952). When inferences are made regarding the nature of communication, usually they are based upon the objective quantification of content rather than assumptions of intent made by the content analyst (Berelson, 1952). Indeed, conclusions are unambiguous and clear-cut when content analysis is applied to the manifest elements of communication (Rourke & Anderson, 2004). Still, sufficient empirical examples exist which have treated traditional content analytic methods as a means to arrive ultimately at inferred conclusions. However, once a researcher begins drawing inferences about underlying constructs such as attitudes, intentions, or affect, one exceeds the bounds within which content analysis was originally proposed (Rourke & Anderson, 2004). Further, because Internet outlets for communication can differ markedly from traditional face-to-face (FtF) contexts with respect to individuals' self-expression and sharing of personal details (McKenna, Green, & Gleason, 2002), we feel new considerations arise in applying content analytic methods to Internet communication in order to reliably and validly assess such latent characteristics of communication.

Perhaps the most apparent difference between online and FtF communication is the role of visual cues in interpersonal interaction. In FtF contexts, visual cues provide critical feedback that influence interactions (Deutschmann & Panichi, 2009). However, such visual feedback is entirely absent from most online message board discussions. Similarly, social cues and standards for communicative behavior in many online forums are severely limited when compared to FtF environments (Deutschmann & Panichi, 2009). In addition, even seemingly trivial factors such as one's dress or appearance can affect the role that perceived social status may have on a conversation (Kiesler, Siegel, & McGuire, 1984; Boucher, Hancock, & Dunham, 2008).

Another difference between online and FtF communication concerns the nature of the interaction taking place on each. Unlike FtF communication that is only synchronous, online communication can be either synchronous or asynchronous. This dual nature of online communication presents new challenges for researchers trying to apply traditional methods of content analysis in online settings. Even more, synchronous forms of CMC such as virtual chat or instant messaging can differ markedly from synchronous FtF communication (e.g., rapid shifts in topic, sub-conversations, disordered responses). However, differences are especially apparent when one compares FtF communication with the asynchronous and highly disorganized communication typically found on Internet message board forums. Whereas synchronous interaction allows participants to maintain the rhythm and progression of a conversation as they respond in turn to what has been stated previously, such mutual engagement and participatory understanding is virtually absent in most message board discussions as well as other asynchronous forms of online communication (Kalman, Ravid, Raban, & Rafaeli, 2006; Raban, Rafaeli, Ravid, & Kalman, 2006). Nevertheless, even the most rhythmically disordered types of CMC appear only to grow in popularity with some forms actually generating addictive usage (Morahan-Martin & Schumacher, 2000; Caplan, 2003).

One reason for the high levels of Internet CMC witnessed today may stem from certain individuals' motives for communication as a result of various personality traits. Researchers investigating online learning environments have suggested that certain students may prefer taking online courses because they may feel more comfortable interacting in a virtual classroom environment rather than a FtF setting (Zhu, 2006). Likewise, some individuals may seek out the anonymity that CMC contexts offer (Caplan,

2003). In turn, such anonymity may lead to more deceptive or aggrandized behavior when compared to FtF settings (Cornwell & Lundgren, 2001).

Additionally, although some online communication can be much more impersonal than in FtF contexts (Caplan, 2003), many types of CMC are even more personal when compared to FtF interaction (Walther, 1996; McKenna et al., 2002). According to Walther (1996), some online mediums are full of "hyperpersonal communication" characterized by impression management and carefully constructed social exchanges. Likewise, as a result of the anonymity found online, some individuals may much more readily disclose intimate details about themselves which they would not share in FtF settings (McKenna et al., 2002). Though clearly many Internet message boards are meant to serve the purpose of leisurely discourse or an informational resource, other forums such as political or ideological group forums exist to convey messages or justify important social values. These latter types of message board forums are especially interesting for researchers interested in examining the latent content of communication.

In the following sections, we detail typical online sources that can be used for communication research, the methodological procedures and considerations of applying content analysis to Internet communications, as well as some statistical considerations of this content analytic approach. Following the recommendations for content analysis proposed by Kaid and Wadsworth (1989), this approach adheres to the seven general steps in the process of data analysis: (1) formulation of a research question, (2) selection of a sample, (3) definition of categories, (4) outlining the coding process and training coders, (5) implementation, (6) evaluation of trustworthiness, and finally (7) analysis. Additionally, it is a directed approach in that the methodology involves the creation of a priori coding schemes based on theory or other relevant research concerns (Hsieh & Shannon, 2005). Furthermore, despite the methodological and scientific concerns that often arise when one uses content analysis to arrive at inferential conclusions (Rourke & Anderson, 2004), we intend to show that this quantitative approach to qualitative content analysis can be both reliable and valid when used for the observation of construct relationships and the drawing of causal inferences from online communication forums.

Common Sources of Online Communications

Internet Forums

Internet forums, or message boards, are interactive web-based forums where online posters are able to interact and the messages remain available for some period of time on the website for public viewing and additional responding. The basic unit of discussion in a message board is a post, which is one distinct response from one user or member on a given topic. Posts are contained within threads, which are distinct topics of discussion that are started by an individual member.

A wide range of information can be gathered and analyzed from Internet forums that pertain to both the individual members as well as the group. This information can help researchers make inferences about constructs pertaining to both individual members and the group. In regards to group-level constructs, almost all message boards have rules that govern content and access to discussions. However, depending on the nature of the group, some forums may be more restrictive than others. For instance, some message boards may only require the user to create a profile before they are allowed to access material, without any authentication procedures in place. On the other hand, some forums may have a formal application process requiring moderator approval before the user is allowed to access any of the material. Once on the site, members of Internet forums often need to create a profile before they are allowed to post or view threads. When creating a profile, users will often need to, at a minimum, create a user name or handle and select an image to accompany their posts, otherwise known as an avatar. Some forums may have rules governing the customizability of the website, such as the use of emotionally evocative symbols and pictures as avatars, whereas other sites may allow any type of symbol or photo to be used. When commenting on a thread, the user's handle and avatar often appear along with the post for other members to see. Some forums also include other information with posts, such as date of initial membership or length of tenure, and number of posts made on the message board forum. This information can help researchers make inferences about constructs at the group and individual levels, such as the status of the members, and the typical length of membership tenure for the group. In addition to creating a handle and avatar, users may choose to include more information about their personal profile including hometown, hobbies, occupation, personal photos, links to other websites, and other details about their interests.

While Internet forums are likely to provide a plethora of information pertaining to online communication, there are several issues that need to be considered when embarking on such a research endeavor. As was mentioned previously, message boards often differ in the degree of restrictive access and control over its members. As such, the researcher may not have access to all of the information on a given message board. Depending on the website, access to certain threads may be restricted. These same message boards may not allow members to post personal information, or they may place restrictions on who is allowed to view this information. At any rate, the level of restrictive access and control over the forum members will have a direct impact on the quality of data that can be obtained from the message boards, as well as the inferences that can be made based on the data.

Another issue with analyzing communication from message boards pertains to lurkers. Lurkers are visitors or registered users of online message boards that do not actively or publicly participate in the online community. Previous research has shown that people lurk for a variety of reasons (Nonnecke, Andrews, & Preece, 2006). Lurking may impact communications research because it is extremely difficult to determine what effect a given communication strategy may be having on a lurker since they, by definition, are not responding to the discussion. Making matters more difficult is the fact that the same person may be an active participant on one thread, but may be lurking on another thread. As such, it will be challenging for researchers to determine if respondents and lurkers are reacting similarly to the same communication strategy.

Blogs and Microblogs

Weblogs, or blogs as they are more commonly known, began to come into prominence in the late 1990s with the creation of blogging websites such as LiveJournal. Many of the first blogs were written as personal online diaries. They have since expanded to cover a variety of topics, including politics, religion, sports, and entertainment. Blogs are typically written and maintained by one person, although group blogs do exist. It is not uncommon for blogs to contain a combination of text, video, and links to other websites or blogs. With the advent of websites like Twitter, abbreviated versions of blogs, commonly referred to as microblogs, were introduced to the world (Yaping, 2011). Blogs and mircoblogs differ from Internet forums in several ways. First, most (micro)blogs may be accessible to anyone. Unlike many Internet forums, which require users to create an account, some (micro)blogs do not require such permission in order to view the contents. The creator of the (micro)blog is the only person that is able to post a new topic for discussion. However, other people are often able to make comments on a (micro) blog where the (micro)blog writer, as well as the general public, is able to view and respond to the comments. Although commenting on (micro)blogs may require membership or registration of some sort, most (micro)blogging sites give the authors control over which comments are shown and which comments are hidden to the public. Furthermore, other users are able to perpetuate a (micro)blog writer's posts(e.g., retweeting, linking, etc.).

Much of the information that is present with regard to individuals on message boards is also present on (micro)blogs. Many authors of (micro)blogs display information about their personal interests and lives, quite similarly to the personal profiles found on message boards. (Micro)blog authors may also display photos or avatars along with their blogs. Links to other (micro)blogs that the author follows is also a common feature of many profiles. In addition to these sources of information, the often personal nature of (micro)blogs allows researchers to make inferences about individual constructs such as personality of the (micro)blog author.

Like Internet forums, there are also considerations when conducting communication research on (micro)blogs. Personal information about viewers and commenters of (micro)blogs is generally limited, if at all present. Since many (micro)blogs are open to the public, an account or profile is often not needed to view the content. Some (micro)blogs may require a membership or account in order to post a comment, however in some instances comments can be posted to (micro)blogs anonymously. Depending upon the research question, (micro)blogs can be another source of qualitative Internet communication data that can be analyzed using the proposed quantitative approach.

Chat Rooms

Chat rooms are a prevalent form of online synchronous communication. Chat rooms are different than blogs and Internet forums because all parties are viewing and responding to the discussion simultaneously. Moreover, the primary focus of chat rooms tends to be for social interaction, rather than informational exchange. Similar to message boards, chat rooms require users to create an account and develop a handle. Participants in chat rooms often add personal information into their profile, which can provide insight into the personalities and other individual level constructs about the users.

There are several potential issues that will need to be considered when attempting to analyze chat room discussions. First, although some chat rooms are focused on a single or small set of topics or issues, the focus of the discussion can shift rapidly. As such, the quality and relevance of the data obtained will likely fluctuate considerably over time. In most chat rooms, members are free to enter and leave rooms at anytime. Depending on the research question, this may pose a problem to communication researchers as an ever-changing sample will likely make the interpretation of the data more difficult. Another difficulty in chat room research pertains to the nature in which member posts are displayed. Being a synchronous form of communication, there will likely be a lag in response time from one response to another. As such, it can be difficult to determine which responses correspond to one another when analyzing chat room discussions. Additionally, chat rooms can often have multiple sub-conversations present simultaneously which creates even more difficulty in determining the order of responses within a given chat room. In order to reduce this lag time, chat room participants often use short-hand or slang to convey information (e.g., emoticons). Therefore, in order to make sense of chat room communication, it will likely be necessary for the researcher to become familiar with the style and terminology of the site.

Social Networking

Social networking sites such as Reddit, Facebook, Instagram, Vine, and Twitter have become increasingly popular in past years. These sites allow individuals to create and customize their accounts and web pages through user preferences and shared text, pictures, and videos. While these sites do allow for individuals to communicate with others in several different ways, it is difficult to access these communications as security settings may require one party to grant access to the other party in order to view their account information. As such, content analysis of communication taking place on these sites is potentially more challenging, particularly on more closed systems such as Facebook. However, it might be possible to study more public profiles like those that are made for groups or organizations (e.g., activist groups, music bands). Groups typically grant access to the public without privacy restrictions. Accordingly, inferences stemming from data analyzed from these group profiles are likely to be different than data that is obtained from more individually oriented sources.

Using Content Analysis to Study Online Communication

While following the general framework of previous content analytic approaches, the proposed methodology provides sufficient structure for coherently investigating online message boards, while still being flexible enough to embrace the dynamic nature of the Internet. This method, as outlined below, addresses familiar concerns with qualitative analysis in general, while also shedding light on new considerations faced by researchers of the online world with respect to content analysis. As noted in a previous section, the development of any content analytic study involves the following steps: (1) research question formulation, (2) sample selection, (3) defining categories, (4) coding process outline, (5) implementation, (6) evaluation, and (7) analysis (Kaid & Wadsworth, 1989). These steps are illustrated in Figure 1 and further delineated in the following sections. Evaluation and analysis are not included in the following explanatory sections because these steps are dependent on the nature of the research questions and extend beyond the scope of this paper (i.e., an analysis plan is based on statistics and research methodology). However, these last two steps are important in drawing quantitative conclusions about qualitative data, and, as such, they are included in the example following the explanatory sections.

Figure 1. Kaid & Wadsworth (19889) approach to content analytic study

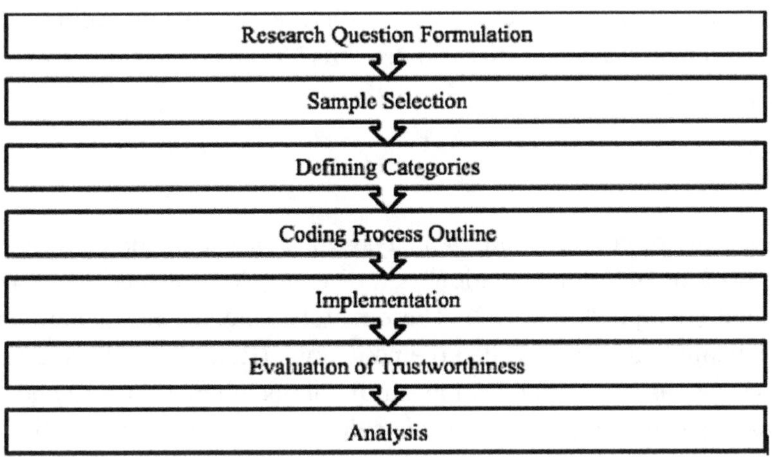

Research Question Formulation

Successful content analytic approaches begin, like all research, with the development of basic research questions. While easily overlooked during initial phases of research design, clearly defining the research questions and specifying constructs of interest will guide design efforts and inform every aspect of the study. For this reason, it is critical that researchers consider the core purpose of their research endeavors and work to fully operationalize and define constructs before moving on in the research process. As researchers move through the phases of designing an online communications study, there are multiple considerations to be had and choices to be made. Researchers must bear the purposes and goals of the researcher being conducted in mind as they make these decisions.

Sample Selection

Armed with clearly defined research questions, the next step will be to determine an appropriate sample for investigation. In some forms of content analytical research, such as historical data and psychobiography, the nature of the content analyzed provides a foundation for defining the sample (Simonton, 2003; Simonton, 1999; Elms, 1994). However, the sample is not generally as clearly defined when working with online communication content. The Internet changes constantly, and web sites pose unique challenges in that the size and nature of the population cannot often be determined (Schafer, 2002). For online communications analyses, Gerstenfeld, Grant, and Chiang (2003) utilize a technique referred to as purposive sampling. Web sites within the domain of interest were sampled using multiple existing sources tied to the domain. For example, in their study on extremist web sites (Gerstenfeld et al., 2003), the researchers compiled listings from a non-profit web site devoted to monitoring extremist sites, a published listing of extremist sites, sites categorized by a search engine within the domain of interest, as well as sites that were frequently linked to the above. This sample clearly does not include every possible extremist site, however it is representative of the domain in that a wide variety of major and minor sites are included (see also Angie, Davis, Allen, Byrne, Ruark, Cunningham, Hoang, Bernard, Hughes, Connelly, & Mumford, 2011; Byrne, Nei, Barrett, Hughes, Davis, Griffith, Harkrider, Hester, Angie, Robledo, Connelly, O"Hair, & Mumford, 2013; Dunbar, Connelly, Jensen, Adame, Rozzell, Griffith, & O' Hair, 2014). When such preexisting information is not available, other approaches must be used to select a sample. Simonton (1999) suggested utilizing eminence criteria in situations where samples lack clear definition. The use of eminence criteria is supported by the rationale that the most prominent cases within a domain of interest are generally the most representative. In online communications, this may included the selection of websites with the most "hits", or message board forums with the greatest number of members, posters, etc.

Defining Categories

After selecting an appropriate sample, a unit of analysis must be established to further define the study. A unit of analysis, in terms of communication, is a discrete element that is observed, recorded, and considered a data point (Krippendorf, 1980). These may be sentences or paragraphs, called syntactical units, or complete, single thoughts, called thematic units (Rourke, Anderson, Garrison, & Archer, 1999). When considering online communication, similar units of analysis exist and must be defined by the researcher. For example, is an entire web site considered a single unit? It may be when comparing its overall message or intent to other web sites. If considering communication on a message board within a web site, then the unit of analysis might be different. In that case, a researcher may examine the content of a specific post (a single message published by an individual), or even syntactical elements within a post. On a broader level, an entire thread (a series of inter-related posts by multiple individuals) may be examined as a thematic unit. Each unit of analysis answers different research questions, and a good content analytic technique examines an appropriate unit for answering the research questions at hand.

Coding Process Outline

Once the sample and unit of analysis have been selected, researchers must determine the most relevant constructs of interest for answering the proposed research questions, and identify the content to be included in the analysis. Content should be included that will inform the basic research questions, and can be of both "hard" and "soft" varieties. Hard content may be directly measured, such as the number of responses to a particular message, or the number of members in an online forum or more traditional means of content analysis (e.g., word counts, selected phrases, etc). Soft content, such as the intensity of affect within message, require skilled judgments to be made by expert raters. It is within the soft content that content analysis can be broadened and specified to answer more detailed questions about online behavior that hard content cannot answer. For instance, counting the number of times the word "terrorist" appears in a message board forum or on a social networking site does little to inform the context of the discussion or the implications of the discussion offline. Developing a complete "picture" or understanding of an online communication, requires integrating both hard and soft content. This is best accomplished by basing content decisions on prior theoretical work or modeling. In this way, qualitative communication content may be tied to psychological and communication theory and allow for inferences to be made from the analysis.

Whereas hard content can be fairly simple to measure, the ability to quantify and analyze soft content requires more work and planning, but offers the most fecund line of inquiry for online communication research. A system for measuring soft content must be created. Because measuring soft content entails some degree of subjective judgment, a rigorous process must be followed in developing the measurement system such that it incorporates a systematic approach to making subjective judgments. This involves developing metrics of the constructs of interest, determining the scale on which to measure these constructs, and finally rating, or assigning a quantity to the content. First, within each broad construct (e.g., bad attitudes), specific metrics must be defined (e.g., negative affect, rejection of argument) based on prior research and theory linking the metric to the construct of interest. Then each metric must be operationally defined (e.g., extent to which respondents use emotionally negative language in their posts). Next, a benchmark rating scale for each metric needs to be constructed, which typically involves using a 5-point Likert scale. These benchmarks provide concrete examples for low (1), moderate (3), and high (5) levels of the construct of interest (see Redmond, Mumford, & Teach, 1993 for more on constructing benchmark rating scales).

After developing metrics for all content of interest, a training program must be developed and implemented for the proper use of the metrics. Regardless of the number of researchers, all raters must complete the same training program as a group in order to develop congruent frames of reference (Bernardin & Buckley, 1981). Ideally, all researchers should participate in these training meetings simply as a check, but this is not necessary if constructs are defined clearly as all research personnel would have participated in that step in the process. Multiple raters must be trained on the correct procedures for content coding including developing an expert understanding, or shared mental model, of the constructs of interest, operational definitions, and use of benchmark scales. Rater training includes multiple practice sessions in order to develop each rater into an expert with the same understanding of the metrics as all other raters. This ensures that all raters have congruent understanding of the online communication phenomenon examined and can evaluate them in the same way fashion. In addition to multiple practices sessions, training should include discussions of differences in ratings until consensus is reached (see Bernardin & Buckley, 1981 for more on this method of rater training)

Implementation

After metrics are established, and training is completed, the process of content coding communications from the sample can begin. Multiple expert raters (3 to 4 raters are typically needed to insure reliable coding) are assigned to assess the same content. These raters will then code the information based on the metrics developed. In doing so, reliability of the metric scales can be assessed, and unbiased measures of the content can be obtained. Establishing reliability of the construct measures is an important part of assessing the validity of any conclusions the researcher wishes to draw from the measures and the study as a whole. In long-term studies where data is continually being coded, comparisons between raters over time can also provide critical indications of when training refreshers are needed in order to maintain the stability of the data.

The flexibility of content analysis paired with the range of opportunity presented by online communications studies requires additional considerations by the researcher. Simonton (2003) notes the massive potential of content analysis in longitudinal research. Online communication offers the choice between analyzing data in a single instance, or snapshot approach, or following communications over time thereby taking a longitudinal approach. However, the dynamic nature of online communication necessitates additional care when examining messages over time. For example, if communication is examined in real-time, there is the possibility that the nature of a message may change in between individual analyses by multiple raters thus reducing inter-rater reliabilities and the quality of the data. This can be a major concern when applying this content analysis approach to longitudinal studies. One proposed way to control for the dynamic nature of online communication is to consider the use of group ratings, rather than individual rating. When using a group approach, the specific time and context of ratings will remain constant within all raters, as they are all viewing the desired unit (e.g., thread, message board, etc.) at the same time. Generally, this involves group viewing of the desired websites, although this may also involve variations on this idea such as viewing the website for the same duration of time from remote locations. This ensures that all raters are viewing the same content, and, thus ratings are likely to be more reliable than individual ratings conducted at various intervals.

Illustration of Implementation and Future Research Guidelines

To further illustrate how this approach to content analysis can be applied to an online sample, this section describes a research study conducted using this methodology. In this project, the online communications of violent ideological, nonviolent ideological, and nonviolent non-ideological groups were assessed in online message board forums.

Ideological groups have clear, strongly held values that create a shared understanding of their environment (Mumford, Bedell-Avers, Hunter, Espejo, Eubanks, & Connelly, 2008). These groups typically have individuals with shared viewpoints, interests and goals (Van Dijk, 2006). As with many types of organizations, ideological groups are compelled to spread their message to others, and they must communicate their ideological values in order to recruit members and gain support. Because traditional media sources are typically highly regulated and expensive, ideological groups face difficulties disseminating information to wide audiences through these channels. New media such as online forums, however, allow ideological groups to more easily communicate their beliefs and values to both members and nonmembers (Allen, Angie, Davis, Byrne, O'Hair, Connelly, & Mumford,, 2008). Therefore, the purpose of this study was to determine how different communication strategies used by the initiator of an online forum impact the reactions of other participants responding on these sites. Specifically, the research question asked what communication strategies used by the initiator of a post lead to the most positive responses from other participants on violent ideological, nonviolent ideological, and nonviolent non-ideological forums. Based on these goals, this content analytic method was used to study what communication strategies are most successful in stimulating positive responses from members of different types of groups.

After developing the research question, the issue of sampling was considered. First, keyword searches were used to identify ten violent ideological, ten nonviolent ideological, and ten nonviolent non-ideological online forums. If the group articulated a shared mental model about goals, described actions to be taken to achieve these goals, and used this model to interpret relevant events, the group was classified as ideological; otherwise, the group was classified as non-ideological. Groups were considered violent if they condoned or if evidence was obtained that linked these groups to violent acts (e.g., through third party watch-group organizations). Next, only forums that had at least ten separate threads and had replies to the initial post were included in the study to ensure enough conversation existed for content analysis.

After identifying the thirty forums, we chose a thread as the unit of analysis because we were interested in studying *dialogue* between individuals. In other words, our research question necessitated gathering data about the communications of initiators and responders. For each forum, five threads were chosen. To address concerns regarding the constantly changing nature of online forums, threads were randomly chosen from the sample of forums for a given time period and then printed to capture an exact moment in time. A time frame was selected where threads were available from all forums sampled. These threads became the sample for content analysis of the impact of initial communication strategies on the responses from other forum participants.

Once the sample was identified, it was necessary to decide what constructs should be content analyzed. Relying on communications and psychological theory, we identified several communications strategies useful in developing persuasive arguments. For example, Hawkins (2004) suggests one effective strategy is fitting arguments within a universal understanding. Developing respect and credibility represents another strategy used to persuade others. Although we will not address all the communication strategies relevant to this particular example, it is important to recognize that all of the communication strategies that were content analyzed were selected based on theory and research. Not only were the communication strategies of the initial posters content analyzed but also the responses of the forum members. Again relying on psychological theory, we identified different response constructs to analyze including affective responses (e.g., good and bad attitudes) as well as other outcomes such as attitude change, questioning of the group, or degree of engagement in the group.

Once the communication strategies and response constructs were determined, specific benchmark scales were established in order to code for the presence of these elements. Each construct was operationally defined, and benchmark rating scales were developed for each construct using a 5-point Likert scale similar to the one shown in the Appendix. Sample threads from the online forums were read in order to help us create concrete examples for low (1), moderate (3), and high (5) responses for each construct. Giving raters examples from source material helps them to develop more similar understanding of how to appropriately rate each construct (Bernardin & Buckley, 1981). For a more detailed example of how to develop benchmark scales to content analyze the psychological processes and response patterns on online forums, see Angie et al. (2011).

The next step in this process was providing raters with extensive training on the constructs and benchmark rating scales. Training involved each rater completing a packet detailing the definitions and examples of each construct as well as practice threads to code. Ratings were compared among coders and disagreements were discussed until consensus was reached. Coders continued rating practice threads until high inter-rater agreement was reached on all constructs. Inter-rater agreement is an estimation based on the correlation of scores between raters who have rated the same items. Intra-class correlation (ICC) coefficient is used since it is appropriate for situations where items are rated by two or more raters (Shrout & Fleiss, 1979). Overall, the training helped ensure coders developed shared mental models and viewed each construct similarly.

Next, three researchers read the online discussion forum threads included in the sample and rated the content using the benchmark scales. Before analyzing this data, reliability statistics were computed to assess agreement among raters with high inter-rater reliabilities indicating raters viewed the constructs similarly. All constructs received coefficients greater than .70 which is considered an acceptable value for inter-rater reliability (Landis & Koch, 1977).

Additionally, some initial validation evidence was also obtained. Correlation coefficients between communication strategies and response constructs can be used to view the degree of relationships among variables. Constructs that are theoretically related should demonstrate a convergent pattern with high correlations whereas constructs that are not theoretically related should demonstrate a divergent pattern with low correlations (Campbell & Fiske, 1959). For example, negative affect and rejection of argument are both specific dimensions of the broad theoretical construct *bad attitudes*. High correlations were found between these variables showing further evidence that these variables are in fact related.

Finally, we used a variety of approaches to quantitatively analyze the data obtained from rating the online threads. Because this content analysis method quantitatively measures communication constructs, many statistical techniques can be used to address the specific research question. For our particular question, we first calculated basic correlations to see how different communication strategies (e.g., conditional agreement, rational persuasion, reframing arguments, refuting arguments) of the initial poster related to outcomes (e.g., group-level identity, attitude change, discussion stimulated or quelled, acknowledging group or individual errors) by the respondents. In this example, outcomes were measured using content analysis as well, but outcome metrics may come from entirely different sources (e.g., survey data, objective data, reader perceptions). Communication strategies that were strongly and positively associated with positive responses or strongly and negatively associated with negative responses suggest that these strategies are the most useful in stimulating successful responses from group members.

Further analyses were performed to address the question of what communication strategies characterize communications that are more successful. In other words, analyses investigated which communication strategies resulted in high ratings on positive response constructs (e.g., reconsidering position) and low ratings on negative response constructs (e.g., severe outgrouping). Positive responses in this case included metrics that reflected attitude and behavioral changes in a desired direction, and negative responses included metrics that measured attitude and behavioral changed in an undesired direction . To answer this question, we performed a series of discriminant function analyses (DFA). Based on the

Wilks' Lambda statistic, the DFA considers the predictive relationship of multiple variables operating together with respect to a given outcome variable. Therefore, this method can be used to show which communication strategies are associated with more successful and less successful communication outcomes for each response characteristic (e.g., bad attitudes, level of engagement). Additionally, the DFA shows what percentage of the communication strategies could be accurately classified into "more successful" and "less successful" categories.

In conclusion, this example demonstrates how the proposed content analysis method can be applied to communication in online forums and in particular how initial communication strategies can impact group member responses. In addition, it highlights the fact that conclusions regarding the effects of communication can be shown to be both statistically reliable and valid when the proposed content analysis method is used and proper techniques are followed. Developing sound research questions, determining the proper sample, defining constructs based on theory, using benchmark scales, training raters, and determining the appropriate plan of analysis are all necessary when using content analysis online.

Generalizability

As noted, this method of content analysis as a means of assessing attitudes and behavioral intentions has wide ranging applicability to numerous fields. Indeed, this approach provides a particularly useful framework for those concerned with public opinion, politics, marketing, and sociology. In terms of public opinion, content analyzing large social networks like Facebook or Twitter preceding or following major events, such as the Supreme Court ruing in favor of marriage equality, the spread of Ebola to the United States, or the increased prevalence of preventable diseases in areas of the country wherein parents opt out of vaccinating their children, can offer an interesting and enlightening look at how divided or united the public is in their attitudes. Those in the world of politics may find content analysis useful in evaluating presidential speeches and campaign commercials from the perspective of increasing or decreasing political contributions. Similarly, they may compare comment sections across various news sites to get an idea of the differences among the various voting blocks.

Marketing professionals have been utilizing the power of social networks for quite some time for customer outreach but pairing their current approach with content analysis of public response to commercials, online media, or other company-based events or incidents may illuminate how to better serve customer needs as well as how to turn lurkers into customers. Perhaps the field of study that may gain the most understanding from this means of gathering attitude and behavior information is sociology. Forums that allow member comments like Reddit, Voat, Newsvine, Fark, and 4Chan offer the unique ability to observe individuals in frank discussion with others on controversial topics without fear of observer effects. Assessing user comments on these sites may aid in understanding non-socially desirable attitudes (e.g., sexism, racism, homophobia) and behavior (e.g., death and rape threats) as ubiquitous online behavior.

Advantages and Disadvantages

When considering the use of the proposed approach of content analysis as a means for assessing online communication, it is beneficial to gain some understanding of the advantages and disadvantages associated with the content analytic process. Below, both advantages, such as lack of reliance on self-reported or survey data, greater objectivity and inferential power, and greater flexibility; and disadvantages, such as necessity of subjective judgments and large time and labor commitment, will be examined. While the following discussion of advantages and disadvantages of content analysis is not exhaustive, this list provides some indication of the problems better controlled with, and encountered during, this content analytic approach.

While much research depends upon self-reported data, content analysis approaches do not. Although self-reported data may provide some insight into the desired information, self-reported data from surveys tends to represent the individual's idealized self rather than the actual self. Individuals may feel prompted to respond in a way that is desirable to others, which can potentially lead to unreliable data due to the extensive use of impression management techniques, particularly when the topic is controversial (Ellison, Heino, & Gibbs, 2006). In such cases, self-reported data does not truly represent the individual providing it.

While it may seem that online administration of a self-report survey would offer more anonymity, and, thus, less socially desirable responding, Joinson (2001) offered that while individuals who received notice of his or her anonymity provided more information about themselves, the information was no more intimate than those individuals who did not receive any information regarding their anonymity. However, when researchers are absent from the process of study entirely, as in the case when examining interactions online, individuals may truly feel anonymous and divulge more intimate information about themselves (McKenna et al., 2002).

Rather than depending on self-reported data by way of a survey, content analysis of online communication relies on information an individual has willingly provided with no direct prompt from a researcher, hence this methods lacks pressure that may result from a researcher or laboratory setting. Therefore, the information retrieved online may be more likely to contain truthful information regarding individuals' attitudes, beliefs, and behaviors (e.g., Gerstenfeld et al., 2003; Cousineau, Rancourt, & Green, 2006). That this approach to content analysis does not rely on self-reported data is particularly advantageous for the study of online communication.

Specifically, this approach offers a look at real-time discussions that lacks editing, which is not found in speeches or books. Further, using the proposed method of content analysis allows researchers a look at the pure nature of CMC; it may supply researchers with more authentic and realistic information than methods that rely on videotaping or recording individuals. For instance, a researcher may ask individuals about the activities in which they engage while using the Internet, but individuals are unlikely to report that they engage in undesirable behaviors, such as cyber-bullying. However, with this content analytic approach, a researcher can examine cyber-bullying activities directly without prompting individuals for such information.

Generally, this approach to content analysis serves as a way to construct a more systematic form of qualitative data in order to identify patterns within a dataset (Hsieh & Shannon, 2005). By creating a set of criteria (e.g., benchmark rating scales) in which qualitative data is transformed into quantitative data, one can then make inferences based on statistical evidence, rather than inferring information based solely on qualitative data (Woodrum, 1984). The ability of researchers to statistically analyze data provides a way of exploring differences between different types of online communities and communications more objectively. This approach also allows for the acquisition of relevant statistical information, such as inter-rater reliability, which provides an important indicator of the extent to which the qualitative data is reliable and credible (Krippendorff, 2004; Lombard, Snyder-Duch, & Bracken, 2002).

In addition to greater inferential power, content analysis allows researchers to customize criteria and methodology to fit specific purposes. This is particularly relevant to researchers of online communication, as research criteria may vary depending on unit of analysis, sample, research question, or construct.

Benchmark rating scales may be developed to reflect nearly any research question. Aside from the necessity that the scales be theoretically relevant, there are no restrictions as to the flexibility of the content of the scale. Further, Clark and Watson (1995) note that a lack of substantial theoretical knowledge does not necessary deter the use of a content analysis approach; it merely requires that one be cautious during scale development not to overreach the scope of the research. This flexibility allows researchers to select criteria based on the both the construct of interest and the extent to which they possess theoretical knowledge about that construct (Krippendorff, 1980: Hsieh & Shannon, 2005; Simonton, 2003; Clark & Watson, 1995). This flexibility is particularly relevant when studying CMC, as forms and methods of communication vary extensively online. Pairing the dynamic nature of the Internet and the various research questions and constructs of interest make it a necessity that the chosen approach of content analysis be more tailored to CMC.

Another advantage to this approach to content analysis is its simplicity and straightforward nature. This simplicity can be better understood in comparison to computer software programs that offer content analytic capabilities. Computer software programs may seem particularly valuable to researchers with interest in studying online communications given the ease with which the data can be transferred to the software. However, these programs, while there is a substantial range, are generally resource intensive. Programs with less complex capabilities (e.g., word counts, frequencies) tend to be less expensive, yet sacrifice much of the richness of online data. More sophisticated programs offer greater content analytic capabilities (e.g., ontological themes, patterns of word usage) yet the cost, programming time, and expertise needed become more intensive. That is not to say that computer analytic techniques do not offer new and exciting methods for researchers to benefit from, however, at the present time a simpler and more straightforward approach to content analysis offers certain advantages over the available complex computer content analytic approaches.

However, content analysis is not without disadvantages. In the process of coding qualitative information so that it may be used for quantitative purposes, subjective judgments must be made. Information may be implied or mentioned in a subtle manner in online communication, not the direct and specific structure as dictated by the behavioral anchors in benchmark ratings scales, and it may be necessary to make an extrapolation based on the available information (Hsieh & Shannon, 2005). While this detracts from the objectivity of the content analysis approaches that do not strictly include word counts or other similar procedures, the use of subjective judgments in the content analysis process highlight the need to consider inter-rater reliabilities (e.g., Krippendorff, 1980). It is important to note that coding online communication with benchmark ratings scales is not synonymous with completing a self-reported survey. Although raters are completing documents formatted similarly to surveys, the subject of interest is the communicator or communication as whole, rather than the raters themselves. To summarize, raters are trained to code the unit of analysis (e.g., post, thread, etc.) for a specific set of criteria based on the research question, rather than responding to the items based on personal preferences or values, as one would do while completing a survey.

Content analysis also requires a large time commitment from multiple individuals. As noted previously, using content analysis procedures may be time consuming and highly labor intensive, both in the preparatory stage of research and the data collection phase. In order to engage in a content analysis approach effectively, individuals who will participate in the coding process must be trained extensively in order to both develop a complete understanding of the coding criteria and form a shared mental model of the concepts to be coded (Bernardin & Buckley, 1981). Further, content analyses that are conducted longitudinally may require a more long-term commitment than snapshot approaches (Weber, 1990).

While content coding of any media requires a large time commitment, content coding of online communication is particularly labor and time intensive. Because the Internet is constantly changing, this can make research somewhat difficult, depending on the research question. It may be necessary for individuals aiding in the coding of online communication to spend a large amount of time to merely locate the desired information before beginning the coding process (Krippendorff, 1980; Weber, 1990). Unlike stationary media, the location of necessary information may be altered at any time due to the dynamic nature of the Internet, which may make longitudinal coding particularly cumbersome. The shifting nature of the Internet also has implications in regards to synchronous communication media, such as chat rooms. As previously stated, it may be much more difficult to know exactly who is communicating with whom when coding synchronous media.

Ethical Considerations

While a variety of forms of online communication exist, there are several ethical considerations that need to be considered before conducting such research. To begin, it could be argued that people forfeit some degree of privacy when posting in public forums. Morio and Bucholz (2009) proposed a three level hierarchical model of anonymity used to explain online social behavior. The lowest level of anonymity proposed is visual anonymity, which occurs when individuals communicate online without the presence of their physical images. The Social Identity Model of Deindividuation Effects presented by Postmes, Spears, and Lea (1998) suggests that deindividuation is more likely to occur when people are visually anonymous. Many of the avatars that individuals use are salient examples of the visual anonymity that is afforded via online communication. The next level of anonymity in Morio and Bucholz's model is dissociation of identity. When individuals create online profiles that do not include their real names they should tend to become less responsible for their actions online as their new identity is not linked to their "true" identity. The online username or handle may represent an individual that is quite different from that of the person controlling the persona. The final level of anonymity in Morio and Bucholz's model is the lack of identifiability. At this level, the behavior of one individual cannot be discriminated from another individual's behavior. An example of this degree of anonymity would be a blog that allows viewers to comment under the name "anonymous." All three levels of anonymity are likely to be found in any given form of online communication. As a researcher, it is important to understand the perceived level of anonymity that is present on the site before conducting any research.

A second ethical consideration for online communication researchers is how to gain access to the material that is to be content analyzed. For blogs, this issue is not as crucial, as the information is almost always available to the general public. However, many message boards require an account to be established at a minimum, and others require more information from applying members, such as emailing the moderator explaining why one is interested in becoming a member of the online community. If the message board in consideration for analysis is open to the public, meaning that anyone is able to view the content on the site, then data can be collected without the consent of the participants, as long as personal identifiers are removed from the data. On the other hand, the ethical decision is not as clear in online communities that require new members to register, as the members of the community will likely perceive their exchanges as being more private than purely open forums.

A third ethical consideration is whether to inform the chosen online community that they are going to be observed. In instances where one must introduce his or herself to gain access to the community, it would almost certainly be unethical to withhold information pertaining to the true purpose of joining the online community. Additionally, it would be extremely difficult to gain consent from every member of an online community as their populations are likely to change frequently as new members are added. In fact, it may be difficult to obtain consent from the members even once they are all identified. Previous research on chat rooms found that when participants were told they were going to be observed, the researchers were kicked out nearly two-thirds of the time. However, when they did not inform the participants that there were going to be observed, the researchers were kicked out less than one-third of the time (Hudson & Bruckman, 2004). Also, when members know they are being observed, they might change their online behavior and communication patterns resulting in less realistic findings.

One option to remedy the issue of obtaining consent could be to change the names of the participants. However, this issue is not as straightforward as it may seem. In a study by Finn and Lavitt (1994), the researchers joined a group for sexual abuse survivors without the group members' permission, even though the group explicitly discouraged professionals from joining the group if they had not been victims themselves of sexual abuse. Finn and Lavitt then published the data using pseudonyms under the assumption that this would protect the identity of the group members. However, as King (1996) later noted, the exact dates and times of when the posts were made was also published which could have been linked to the original user login name. As such, this could have had the potential of eliminating the individuals' perceptions that the group was a safe place to discuss issues surrounding sexual abuse.

Given the above ethical considerations, it is clear that researchers need to carefully consider the characteristics of the target population before engaging in online communication research. In addition, some of the ethical issues that online communication researchers face could potentially be alleviated by using a quantitative analysis of qualitative data. With this method, the need to include personal identifiers in the research is reduced substantially as qualitative data is transformed into quantitative data via numeric rating scales.

CONCLUSION

Given the ever-changing landscape of the Internet and speed at which new media are introduced, communication researchers must be able to keep up with this plethora of information and data. We have proposed a method of content analysis that not only allows researchers to keep pace with technology but also grants a great deal of flexibility in its application. Specifically, this method can be tailored to almost any research question regarding online communication, and it can be applied to a wide variety of sources.

While this methodology is somewhat simple in nature compared to many other types of content analysis, its relative simplicity by no means hinders its wealth of capabilities. Though building benchmarks to examine the constructs of interest and training experts to agree are not inherently difficult or complex tasks, these techniques open up a huge array of opportunities for researchers to examine online communications nonetheless. Additionally, this methodology allows for the use of inferential statistics that cannot often be used with other forms of content analysis. At its core, this technique allows the researcher to capture "softer" data that might otherwise be unattainable. Furthermore, this technique allows the researcher to directly examine Internet communication behaviors instead of relying on self-report. This technique, however, is not without disadvantage, but we believe that its usefulness *far* outweighs its cost. While the dynamic landscape of the Internet might make a communications researcher cringe at the daunting tasks ahead, we hope that this methodology will provide a way to answer if only a few of the innumerable research questions regarding online communication that remain.

REFERENCES

Allen, M. T., Angie, A. D., Davis, J. L., Byrne, C. L., O'Hair, H. D., Connelly, S., & Mumford, M. D. (2008). Virtual risk: The role of new media in violent and nonviolent ideological groups. In R. L. Heath & H. D. O'Hair (Eds.), *The handbook of risk & crisis communication* (pp. 446–447). Mahwah, NJ: Erlbaum.

Angie, A. D., Davis, J. L., Allen, M. T., Byrne, C. L., Ruark, G., Cunningham, C. B., & Mumford, M. D. et al. (2011). Studying ideological groups online: Identification and assessment of risk factors for violence. *Journal of Applied Social Psychology*, *41*(3), 627–657. doi:10.1111/j.1559-1816.2011.00730.x

Berelson, B. (1952). *Content analysis in communication research.* New York: Free Press.

Bernardin, H. J., & Buckley, M. R. (1981). Strategies in rater training. *Academy of Management Review*, *6*, 205–212.

Boucher, E. M., Hancock, J. T., & Dunham, P. J. (2008). Interpersonal sensitivity in computer-mediated and face-to-face conversations. *Media Psychology*, *11*(2), 235–258. doi:10.1080/15213260801906471

Byrne, C. L., Nei, D. S., Barrett, J. D., Hughes, M. G., Davis, J. L., Griffith, J. A., & Mumford, M. D. et al. (2013). Online ideology: A comparison of website communication and media use. *Journal of Computer-Mediated Communication*, *18*, 25–39. doi:10.1111/jcc4.12003

Campbell, D. T., & Fiske, D. W. (1959). Convergent and discriminant validation by the multitrait-multimethod matrix. *Psychological Bulletin*, *56*(2), 81–105. doi:10.1037/h0046016 PMID:13634291

Caplan, S. E. (2003). Preference for online social interaction: A theory of problematic Internet use and psychosocial well-being. *Communication Research*, *30*(6), 625–648. doi:10.1177/0093650203257842

Clark, L. A., & Watson, D. (1995). Constructing validity: Basic issues in objective scale development. *Psychological Assessment*, *7*(3), 309–319. doi:10.1037/1040-3590.7.3.309

Cornwell, B., & Lundgren, D. C. (2001). Love on the Internet: Involvement and misrepresentation in romantic relationships in cyberspace vs. realspace. *Computers in Human Behavior*, *17*(2), 197–211. doi:10.1016/S0747-5632(00)00040-6

Cousineau, T., Rancourt, D., & Green, T. C. (2006). Web chatter before and after the women's health initiative results: A content analysis of in-line menopause message boards. *Journal of Health Communication*, *11*(2), 133–147. doi:10.1080/10810730500526661 PMID:16537284

Deutschmann, M., & Panichi, L. (2009). Talking into empty space? Signaling involvement in a virtual language classroom in Second Life. *Language Awareness*, *18*(3-4), 310–328. doi:10.1080/09658410903197306

Dunbar, N. E., Connelly, S., Jensen, M. L., Adame, B. J., Rozzell, B., Griffith, J. A., & O'Hair, H. D. (2014). Fear appeals, message processing cues, and credibility in the websites of violent, ideological, and nonideological groups. *Journal of Computer-Mediated Communication*, *19*(4), 871–899. doi:10.1111/jcc4.12083

Ellison, N., Heino, R., & Gibbs, J. (2006). Managing impressions online: Self-presentation processes in the online dating environment. *Journal of Computer-Mediated Communication, 11,* article 2.

Elms, A. C. (1994). *Uncovering lives: The uneasy alliance of biography and psychology*. New York, NY: Oxford University Press.

Finn, J., & Lavitt, M. (1994). Computer-based self help groups for sexual abuse survivors. *Social Work with Groups*, *17*(1-2), 21–46. doi:10.1300/J009v17n01_03

Gerstenfeld, P. B., Grant, D. R., & Chang, C. (2003). Hate online: A content analysis of extremist Internet sites. *Analyses of Social Issues and Public Policy (ASAP)*, *3*(1), 29–44. doi:10.1111/j.1530-2415.2003.00013.x

Hawkins, D. (2004). Explaining costly international institutions: Persuasion and enforceable human rights norms. *International Studies Quarterly*, *48*(4), 779–804. doi:10.1111/j.0020-8833.2004.00325.x

Hsieh, H. F., & Shannon, S. E. (2005). Three approaches to qualitative content analysis. *Qualitative Health Research*, *15*(9), 1277–1288. doi:10.1177/1049732305276687 PMID:16204405

Hudson, J. M., & Bruckman, A. (2004). "Go away": Participant objections to being studied and the ethics of chatroom research. *The Information Society*, *20*(2), 127–139. doi:10.1080/01972240490423030

Internet World Stats. (2011). Internet Usage Statistics: The Internet big picture -World Internet users and population stats. Retrieved from http://www.Internetworldstats.com/stats.htm

Joinson, A. N. (2001). Knowing me, knowing you: Reciprocal self-disclosure in Internet-based surveys. *Cyberpsychology & Behavior*, *4*(5), 587–591. doi:10.1089/109493101753235179 PMID:11725651

Kaid, L. L. & Wadsworth. (1989). Content analysis. In P. Emmert & L. L. Barker (Eds.), *Measurement of communication behavior* (pp. 197-217). New York, NY: Longman.

Kalman, Y. M., Ravid, G., Raban, D. R., & Rafaeli, S. (2006). Pauses and response latencies: A chronemic analysis of asynchronous CMC. *Journal of Computer-Mediated Communication*, *12*(1), 1–23. doi:10.1111/j.1083-6101.2006.00312.x

Kiesler, S., Siegel, J., & Mcguire, T. W. (1984). Social psychological aspects of computer-mediated communication. *The American Psychologist*, *39*(10), 1123–1134. doi:10.1037/0003-066X.39.10.1123

King, S. A. (1996). Researching Internet communities: Proposed ethical guidelines for the reporting of results. *The Information Society*, *12*(2), 119–128. doi:10.1080/713856145

Krippendorf, K. (1980). *Content analysis: An introduction to its methodology* (2nd ed.). London, GB: Sage.

Krippendorff, K. (2004). Reliability in content analysis: Some common misconceptions and recommendations. *Human Communication Research*, *30*, 411–433.

Landis, J. R., & Koch, G. G. (1977). The measurement of observer agreement for categorical data. *Biometrics*, *33*(1), 159–174. doi:10.2307/2529310 PMID:843571

Lombard, M., Snyder-Duch, J., & Bracken, C. C. (2002). Content analysis in mass communication: Assessment and reporting of intercoder reliability. *Human Communication Research*, *28*(4), 587–604. doi:10.1111/j.1468-2958.2002.tb00826.x

McKenna, K. Y. A., Green, A. S., & Gleason, M. E. J. (2002). Relationship formation on the Internet: What's the big attraction? *The Journal of Social Issues*, *58*(1), 9–31. doi:10.1111/1540-4560.00246

Morahan-Martin, J., & Schumacher, P. (2000). Incidence and correlates of pathological Internet use among college students. *Computers in Human Behavior*, *16*(1), 13–29. doi:10.1016/S0747-5632(99)00049-7

Morio, H., & Buchholz, C. (2009). How anonymous are you online? Examining online social behaviors from a cross-cultural perspective. *AI & Society*, *23*(2), 297–307. doi:10.1007/s00146-007-0143-0

Mumford, M. D., Bedell-Avers, K. E., Hunter, S. T., Espejo, J., Eubanks, D., & Connelly, M. S. (2008). Violence in ideological and non-ideological groups: A quantitative analysis of qualitative data. *Journal of Applied Social Psychology*, *38*(6), 1521–1561. doi:10.1111/j.1559-1816.2008.00358.x

Nonnecke, B., Andrews, D., & Preece, J. (2006). Non-public and public online community participation: Needs, attitudes, and behavior. *Electronic Commerce Research*, *6*(1), 7–20. doi:10.1007/s10660-006-5985-x

PEW Internet and American Life Project. (2010). *Teenage trend data*. Retrieved from http://pewInternet.org/Reports/2010/Social-Media-and-Young-Adults.aspx

PEW Internet and American Life Project. (2012). *Adult trend data*. Retrieved from http://pewInternet.org/Static-Pages/Trend-Data-%28Adults%29/Online-Activites-Total.aspx

Postmes, T., Spears, R., & Lea, M. (1998). Breaching or building social boundaries? SIDE-Effects of computer-mediated communication. *Communication Research*, *25*(6), 689–715. doi:10.1177/009365098025006006

Raban, D. R., Rafaeli, S., Ravid, G., & Kalman, Y. M. (2006). Speak *now* or forever hold your peace: Power law chronemics of turn-taking and response in asynchronous CMC. *Conference Papers – International Communication Association, 2006 Annual Meeting*, 1-24.

Redmond, M. R., Mumford, M. D., & Teach, R. (1993). Putting creativity to work: Effects of leader behavior on subordinate creativity. *Organizational Behavior and Human Decision Processes*, *55*(1), 120–151. doi:10.1006/obhd.1993.1027

Rourke, L., & Anderson, T. (2004). Validity in quantitative content analysis. *Educational Technology Research and Development*, *52*(1), 5–18. doi:10.1007/BF02504769

Rourke, L., Anderson, T., Garrison, D. R., & Archer, W. (1999). Assessing social presence in asynchronous text-based computer conferencing. *Journal of Distance Education*, *14*, 50–71.

Schafer, J. A. (2002). Spinning the web of hate: Web-based hate propagation by extremist organizations. *The Journal of Criminal Justice and Popular Culture*, *9*, 69–88.

Shrout, P. E., & Fleiss, J. L. (1979). Intraclass correlations: Uses in assessing rater reliability. *Psychological Bulletin*, *86*(2), 420–428. doi:10.1037/0033-2909.86.2.420 PMID:18839484

Simonton, D. K. (1999). Significant samples: The psychological study of eminent individuals. *Psychological Methods*, *4*(4), 425–451. doi:10.1037/1082-989X.4.4.425

Simonton, D. K. (2003). Qualitative and quantitative analyses of historical data. *Annual Review of Psychology*, *54*(1), 617–640. doi:10.1146/annurev.psych.54.101601.145034 PMID:12171999

Van Dijk, T. A. (2006). Ideology and discourse analysis. *Journal of Political Ideologies, 11*(2), 115–140. doi:10.1080/13569310600687908

Walther, J. B. (1996). Computer-mediated communication: Impersonal, interpersonal, and hyperpersonal interaction. *Communication Research, 23*(1), 3–43. doi:10.1177/009365096023001001

Weber, R. P. (1990). *Basic content analysis* (2nd ed.). London, GB: Sage.

Woodrum, E. (1984). "Mainstreaming" content analysis in social sciences: Methodological advantages, obstacles, and solution. *Social Science Research, 13*(1), 1–19. doi:10.1016/0049-089X(84)90001-2

Yaping, Y. (2011). A review on the research of microblog. *Journal of Intelligence.* Retrieved from: http://en.cnki.com.cn/Article_en/CJFDTOTAL-QBZZ201109013.htm

Zhu, E. (2006). Interaction and cognitive engagement: An analysis of four asynchronous online discussions. *Instructional Science, 34*(6), 451–480. doi:10.1007/s11251-006-0004-0

ADDITIONAL READING

Bender, J. L., Jimenez-Marroquin, M. C., & Jada, A. R. (2011). Seeking support on Facebook: A content analysis of break cancer groups. *Journal of Medical Internet Research, 13*(1), e16. doi:10.2196/jmir.1560 PMID:21371990

Creswell, J. W. (2013). *Research design: Qualitative, quantitative, and mixed methods approaches* (4th ed.). Thousand Oaks, CA: SAGE.

Herring, S. C. (2010). Web content analysis: Expanding the paradigm. In J. Lunsinger, L. Klastrup, & M. Allen (Eds.), *International handbook of Internet research* (pp. 223–249). Springer.

Hum, N. J., Chamberlin, P. E., Hambright, B. L., Portwood, A. C., Schat, A. C., & Bevan, J. L. (2011). A picture is worth a thousand words: A content analysis of Facebook profile photogrphs. *Computers in Human Behavior, 27*(5), 1828–1833. doi:10.1016/j.chb.2011.04.003

Jensen, K. B. (2013). *A handbook of media and communication research: Qualitative and quantitative methodologies*. NYC, NY: Routledge.

Kluver, R., Jankowski, N., Foot, K., & Schneider, S. M. (2014). *The Internet and national elections: A comparative study of web campaigning*. Routledge.

Lin, J. S., & Pena, J. (2011). Are you following me? A content analysis of TV networks' brand communication on Twitter. *Journal of Interactive Advertising, 12*(1), 17–29. doi:10.1080/15252019.2011.10722188

Lombard, M., Snyder-Duch, J., & Bracken, C. C. (2002). Content analysis in mass communication: Assessment and reporting of intercoder reliability. *Human Communication Research, 28*(4), 587–604. doi:10.1111/j.1468-2958.2002.tb00826.x

Neuendorf, K. A. (2002). *The content analysis guidebook.* Thousand Oaks, CA: SAGE.

Riff, D., Lacy, S., & Fico, F. (2014). *Analyzing media messages: Using quantitative content analysis research*. NYC, NY: Routledge.

Stemler, S. (2001). An overview of content analysis. *Practical Assessment, Research & Evaluation*, *7*(17), 137–146.

KEY TERMS AND DEFINITIONS

Content Analysis: Content analysis is a research technique, which analyzes communicated content in a systematic and quantitative way.

Hybridized Content Analysis: The analysis of qualitative information through quantitative means to allow for standardized comparisons of otherwise nonsystematic data. This typically requires the creation of evidence-based benchmarks in order to lessen subjective judgments.

Online Community: An online community is a community that exists primarily on the Internet and whose members interact virtually most often.

Qualitative Analysis: The analysis of information that is typically thought to be nonquantifiable or nonsystematic (e.g., historical context, body language, etc.).

Quantitative Analysis: The analysis of information through numerical measurement, which typically involves assessing objective criteria (e.g., number of votes, word counts, likes, retweets, etc.).

APPENDIX

Developing Respect/Credibility: A series of tactics designed to introduce oneself to discussion, to gain acceptance by others, and/or to position oneself as credible for future interaction and more advanced strategies. It includes the sub-strategies of acknowledgement, kindness, conditional agreement, rhetorical questions, and sharing values (shown in Table 1).

Table 1.

1	2	3	4	5
Initial poster never uses tactics to develop respect or credibility		Initial poster sometimes uses tactics to develop respect or credibility		Initial poster always uses tactics to develop respect or credibility
Never encourages other participants, agrees with another's point, or expresses shared values		May make a few encouraging remarks, agrees occasionally with others' points, or lends some support for shared values		Always acknowledges others' expertise, encourages new participants, and often expresses how they share others' values

Chapter 13
Mentoring and Support for the edTPA:
A Mixed Methods Program Review of edTPA Support Practices

Randa Suleiman
Cardinal Stritch University, USA

Clavon Byrd
Cardinal Stritch University, USA

ABSTRACT

The research question was: How effective is the current edTPA mentoring and support program for teacher candidates? This research utilized mixed method interactive program evaluation. An online survey collected teacher candidates' perceptions of mentoring and support for edTPA. The survey questions were organized around four constructs: Preparedness, support from instructor, support from university supervisors, and support from cooperating teacher. With N = 46, a comprehensive data analysis was conducted that identified areas of strength and need of the program. As a result, the researchers developed an edTPA mentoring and support program model.

INTRODUCTION

The Educator Teacher Performance Assessment (edTPA) is a performance assessment designed for pre-service teachers during the semester of student teaching. The edTPA was developed by the Stanford Center for Assessment, Learning, and Equity (SCALE). Formerly associated with the School Redesign Network at Stanford University, SCALE, the program provides technical consulting and support to schools and districts using performance-based assessment as part of a multiple-measures system for evaluating student learning and measuring school performance. According to edTPA's official website, this performance assessment was officially announced as a fully operating teacher - performance assessment in fall 2013 after two years field testing with 12,000 teacher candidates. In the United States, 35

DOI: 10.4018/978-1-5225-0007-0.ch013

states and the District of Colombia were implementing edTPA at different levels. It was a performance assessment "designed by the profession for the profession (edTPA, 2014)." The design of the edTPA was rooted in the methods and practices of National Board Teacher Certification (NBTC, 2014) and inclusive of numerous activities designed to demonstrate teaching competence. Among these activities were: Video taped lessons and various commentaries on teaching designed to extract deep knowledge of key aspects of teaching from NBTC candidates.

Likewise, the edTPA focuses on the central aspects of teaching practice, including: Planning, Instruction, and Assessment. In the edTPA, teacher candidates are charged with planning, implementing, and assessing a series of lessons known as a teaching segment. The common theme that streams across each of central aspects of teaching practice was the concept of academic language. Teacher candidates are charged with describing how they planned for, instructed and assessed their students' understandings of academic language. Each of the central aspects of edTPA, namely planning, instruction, and assessment are further described in the next section of this text.

Planning

This is the central element of the edTPA, as planning is the lynchpin of any lesson. All lessons begin with a plan, and the lesson plan's success or lack thereof is central to the success of the lesson. Within the edTPA Planning Task, teacher candidates are expected to submit a consecutive series of three to five lessons known as the lesson segment. Within the Planning Task, teacher candidates were expected to address the following elements of planning:

- Build content understanding.
- Support learner needs.
- Use knowledge of students in planning.
- Design assessments to monitor student learning.

Incorporating the above elements within the planning task enabled teacher candidates to justify their planning decisions via an exhaustive planning commentary where teacher candidates must support their planning practices with research-based theories and practices. The planning commentary was designed so that the teacher candidate must respond to various prompts in order to complete it.

Instruction

The edTPA Instruction Task is designed to coordinate the teacher candidate's implementation of the teaching segment. The Instruction Task was characterized by the following elements:

- Development of a learning environment conducive to effective teaching.
- Engaging students in learning activities by a variety of means.
- Facilitating deep thinking among students.
- Further exploration of subject specific pedagogy.

Teacher candidates were also encouraged to video record each lesson within the lesson segment in order to identify and submit a segment of video representative of the teacher candidates ability to teach an effective lesson inclusive of all of the above elements. Furthermore, teacher candidates were required to write an extensive instruction commentary that analyzed their teaching practices illustrative of teaching ability that addressed the above elements. A key element of the instruction task was that the teacher candidate must justify instructional strategies by citing applicably aligned research-based strategies. The instruction task was designed so that the teacher candidate must respond to specific prompts in order to complete it.

Assessment

The edTPA Assessment Task was designed to urge the teacher candidate to identify several key aspects of assessment within the lesson segment:

- Analysis of student learning.
- Providing feedback to students to advance their learning.
- Providing time and other supports to students in using the feedback from the teacher candidate.

The edTPA Assessment Task was highlighted by the assessment commentary in which the teacher candidate must write a descriptive account of assessment practices used throughout the lesson segment. The assessment task was designed so that the teacher candidate responded to numerous prompts. A key element of the assessment task was that the teacher candidate must justify next steps of instruction for the collective class as well as for specific students by citing applicably aligned research-based strategies.

With the increased consideration of edTPA as a pre-service teacher performance assessment and licensing requirement by many states including Wisconsin, institutions and schools of education faced increased pressure to effectively prepare teacher candidates to successfully complete a passing edTPA portfolio. The immediate facets of teacher preparation that affected Cardinal Stritch University were: In-class preparation, field experience preparation, portfolio assessment design, and edTPA implementation during student teaching. Prior to implementing the edTPA during the student teaching semester, the Cardinal Stritch University School of Education spent an immense amount of time preparing for the inevitability of the assessment.

BACKGROUND

The Cardinal Stritch University School of Education's (Stritch) first step in preparation for the edTPA was to learn as much about the assessment as possible. The faculty engaged in a deeply rooted mission to understand as much as possible about the edTPA including its purpose, design, and the research supporting the performance assessment. The faculty attended numerous conferences and presentations at the state and local level as well as engaged in local activities that meaningfully developed a shared understanding of the performance assessment. All of these efforts were completed a full three years prior to mandatory statewide implementation of the edTPA for higher education institutions (IHEs), and a full four years prior to implementation of the of the edTPA as consequential for licensure in the state of Wisconsin.

Armed with an understanding of the purpose and design of the edTPA, the Stritch faculty engaged in an intense curriculum mapping process that first analyzed the key components of the edTPA with an intention to embed activities and experiences within already established teacher preparation coursework.

The curriculum mapping process was constructed over several months, and it changed the nature of the teacher candidates' experiences within the teacher education program. Ultimately, several edTPA related activities and experiences were developed from existing experiences in order to maintain the teacher preparation program's identity. The most prevalent edTPA associated change was the re-design of the program's lesson plan template known as the Instructional Design Plan (IDP). The IDP was altered to include edTPA related skills and understandings. This was important because the IDP was woven throughout the fabric of the teacher education program. Teacher candidates were expected to use the IDP to design countless lessons throughout the course of the teacher education program, and teacher candidates would now have experience with edTPA-associated skills in virtually every course.

Stritch also recognized that a change to its periodic portfolio assessment process and design was essential to the curriculum mapping process as well. The portfolio and assessment process was designed to be administered to teacher candidates at strategic checkpoints throughout the course of the teacher preparation program through the assessment and verification of key assessments and other points of data acquired by teacher candidates over time. Overall, three portfolio checkpoints existed within Stritch's teacher preparation program. Each of these portfolio checkpoints was retrofitted to align with edTPA principles and activities. Within these various portfolio assessments, teacher candidates needed to demonstrate proficiency in skills, understandings, and experiences based upon coursework up to the point of the program when the portfolio was administered and assessed.

MAIN FOCUS OF THE CHAPTER

History of edTPA Implementation and Support at the Research Site

In terms of edTPA implementation, Stritch made a decision to implement the performance-based assessment as a program exit requirement a full two years in advance of mandatory statewide implementation of the edTPA for higher education institutions (IHEs), and a full three years prior to implementation of the of the edTPA being consequential for licensure in the state of Wisconsin. With this decision, Stritch made completion of the edTPA a mandatory program exit requirement for student teachers during the Fall 2013 semester. Armed with an understanding of the edTPA process, professors and teacher candidates set out to implement the edTPA with admittedly very few supports that were made available to teacher candidates during the process.

The Fall 2013 semester was the first step in implementing edTPA. It was a program exit requirement that all teacher candidates complete an edTPA entry during Professional Development Seminar. The implementation approach was holistic. Teacher candidates were exposed to three, four hour workshop sessions to learn about edTPA, receive support on edTPA tasks, ask questions about edTPA tasks, and to start working on the different tasks. This was the first exposure of that group of teacher candidates to edTPA. They had no background knowledge about edTPA requirements or expectations. A short pilot informal survey was conducted after the teacher candidates completed their edTPA entry in November to learn more about their experience. Based on initial feedback, teacher candidates strongly recommended spending more time unpacking each task and providing more support during each task. This recommendation prompted the researchers to design the edTPA mentoring and support program.

edTPA Implementation Design

The newly designed edTPA implementation for the Spring 2014 semester for teacher candidates was multi-faceted in its approach:

- Teacher candidates were required to attend four in-class workshops, held during traditional student teacher seminar class sessions, to complete the edTPA. Two sessions were devoted for the planning task, one for instruction task and one for the assessment task. Each workshop was three hours. Students were grouped by handbook when feasible or by grade level.
- Stritch provided edTPA training for university supervisors (US). There were two training sessions. The first training session focused on educating supervisors about edTPA and strategies that US could use to support teacher candidates in completing an entry. The second session focused on providing feedback to teacher candidates on the edTPA entry.

More details regarding the structure of Stritch's edTPA implementation efforts are discussed in the following two paragraphs.

During this first year of edTPA implementation, the Stritch faculty made the decision to introduce instruction of the edTPA through the framework of the Professional Development Seminar for Student Teachers course. This course traditionally met on periodic predetermined dates throughout the semester to discuss trending topics in education with a special emphasis on the student teaching experience. Additionally, this course was typically cross-enrolled with Portfolio III, which was designed as the exit portfolio for the teacher preparation programs. Student teachers used the key assessments administered during this course as artifacts for Portfolio III. Now, the content of these course meetings focused on instruction and completion of the edTPA. In all, course instructors devoted a total of three workshops to the edTPA; one workshop for each of the three edTPA tasks: Planning, Instruction, and Assessment. It is important to note that each workshop session was scheduled for a four-hour period of time. This format changed for the Spring 2014 semester. Four sessions were dedicated to edTPA with two sessions out of the four were focused on the planning task. The other change from the Fall 2013 semester was the frequency of the class meeting. Classes met for three hours bi-weekly. The total number of hours devoted to edTPA entry was the same 12 hours in total in both semesters. In the Fall 2013 semester the approach was holistic through providing the teacher candidates with the handbook and templates while allowing them to work on their own pace to complete each task. In the Spring 2014 semester the approach was changed to a scaffolded step-by-step approach. Each session was devoted to unpack a specific edTPA task and allow the teacher candidates to work in small groups to understand the task with the facilitation of the seminar instructor.

A key element of Stritch's early edTPA implementation efforts was the work of university supervisor (US). In their traditional roles working at Stritch, US mainly supervised student teachers in the field and supported course instructors with making connections to local schools and administrators for in-course experiences and career opportunities for teacher candidates. Under the edTPA-influenced model of course meetings, the role of US expanded, and this corps of individuals was expected to then provide edTPA support to teacher candidates. In fact, the role of university supervisor was catapulted into one of the most important support roles for teacher candidates in the new landscape molded by the edTPA. University supervisors saw their former background roles rise to the forefront because of the SCALE stipulation that course instructors and other faculty were not permitted to provide edTPA support to teacher candidates via feedback on their performance on the three tasks and associated commentaries.

Simply said, while faculty could not provide support to teacher candidates, university supervisors and cooperating teachers were free to do so. Stritch decided to focus its support efforts on university supervisors rather than cooperating teachers because the university had more direct control over the teacher candidates' interactions with University Supervisors than those that were held over Cooperating Teachers. Stritch was able to exercise flexibility in scheduling edTPA-geared professional development for University Supervisors that was difficult to establish and maintain for its school-based partners.

The professional development opportunities offered to university supervisors focused on each supervisors understanding of the edTPA in terms of purpose, design, and output. Also, a significant portion of this professional development focused on a constructive feedback model that US were expected to use to support teacher candidates throughout the edTPA process. This feedback model utilized active feedback for learning techniques that encouraged US to provide descriptive feedback to teacher candidates that questioned their work without providing specific instructions necessary for improvement. Teacher candidates were required to submit their completed work on each task to the US for feedback. This work was submitted in a prescribed manner with specific due dates for each of the three tasks. The platform utilized by US to review edTPA submissions from teacher candidates and provide feedback to guide their learning was Livetext, which was a digital assessment platform used by Stritch across all programs.

Statement of the Problem

The fall 2013 edTPA support model lacked variation and differentiation in supporting teacher candidates through the process of learning about and submitting an edTPA portfolio. The major issues associated with this lack of variation and differentiation included the following:

- Teacher candidates lacked experience with specific edTPA purpose, premises, and core concepts.
- In general, the teacher candidates' Cooperating Teachers possessed very little knowledge of specific edTPA purposes, premises, and core concepts.
- University Supervisors were not consistent in providing descriptive and constructive feedback to teacher candidates in a timely manner using the Livetext platform.

The primary issue was that the teacher candidates collectively had virtually no experience with specific edTPA premises and concepts. While the teacher candidates had engaged in best practices in pedagogy and experiences that were aligned with edTPA philosophies throughout the teacher preparation program, they had not had direct experience with the nuances of the edTPA—the teacher candidates were not privy to the intentionally embedded edTPA-aligned activities and experiences across the teacher education program. To put in kindly, the pioneer group of teacher candidates was not happy with Stritch's mandate that they complete the edTPA as a final program exit requirement. The teacher candidates often voiced their opposition to Professional Development Seminar for Student Teachers course instructors and to other faculty. The teacher candidates' collective dissatisfaction with the edTPA requirement was evident as one teacher candidate voiced his opinion: "This is not the student teaching seminar I was expecting. I feel like I have been robbed." Prior renditions of Professional Development Seminar at Stritch certainly lacked the rigor and high expectations that came along with the edTPA.

Another major issue facing Stritch's edTPA implementation design was the Cooperating Teachers' general lack of understanding about the edTPA in terms of its purpose, premises, and product. During many of the course sessions where edTPA was instructed, teacher candidates spoke candidly regarding their Cooperating Teachers' lack of awareness about the performance assessment. There seemed to be a great divide about how this major performance assessment that faced teacher candidates affected teaching practice for teacher candidates. In many cases, teacher candidates reported that their Cooperating Teachers had never heard of the edTPA.

One key issue that faced Stritch's edTPA implementation design was the University Supervisors' general lack of timeliness in reviewing teacher candidates' edTPA tasks. This led to a sense of frustration among the corps of teacher candidates who generally felt as though they were not receiving timely feedback necessary to improve their individual work samples. While the US were well-versed in the art of providing feedback necessary to guide the teacher candidates' learning, they were not as proficient in utilizing the Livetext digital assessment platform in order to put the feedback into the teacher candidates' hands. Many US preferred to either handwrite feedback to teacher candidates or to provide feedback to them via email. The inconsistency with which the US provided constructive feedback on edTPA tasks to teacher candidates helped to erode the teacher candidates' confidence in the edTPA implementation process.

LITERATURE REVIEW

Admittedly, there is limited research surrounding the edTPA, as it is a relatively new performance assessment for teacher preparation on the national scene. As multiple states have moved to adopt the edTPA as a viable assessment to measure the competence of teacher candidates with respect to their acumen in planning, instruction, and assessment the researchers are confident that edTPA related research and subsequent literature will move to the forefront.

According to SCALE website (2013), the edTPA was described as " a pre-service assessment process designed by educators to answer the essential question: "Is a new teacher ready for the job?" edTPA included a review of a teacher candidate's authentic teaching materials as the culmination of a teaching and learning process that demonstrates each candidate's ability to effectively teach his/her subject matter to all students." edTPA was described as a subject-specific national performance assessment that measured teaching behaviors and their impact on student learning. It was aligned to Common Core State Standards and the Interstate Teacher Assessment and Support Consortium (InTASC). At the time of this writing, five states were considering edTPA as a performance assessment for program completion and/ or as a licensure requirement including Wisconsin and 22 states are exploring its use as a performance assessment.

Using subject-specific evidence, edTPA focused on five dimensions of teaching. The first dimension was described as planning a learning segment in which the teacher candidates demonstrated their ability to plan standard aligned lessons that were differentiated to meet the needs of all students and were based upon students' background knowledge and previous academic learning. The second dimension of teaching was described by how the teacher candidates instructed and engaged students in the learning experience throughout the learning segment. Using unedited video clips, the teacher candidates completed a commentary to demonstrate their abilities to utilize effective subject-specific teaching strategies to deepen content understanding through eliciting and monitoring students' response. The third dimension of teaching was described as assessing students' learning. By analyzing patterns in students learning using classroom based assessment and showing evidence of teacher candidate feedback to students, teacher candidates completed an exhaustive commentary to demonstrate their ability to summarize a whole class performance, identify areas of strength and areas of growth of three focus students, and explain how their feedback was designed to guide their students' learning experiences from the point of the feedback moving forward in time.

The fourth and fifth dimensions of teaching were described as intertwined with the previous three dimensions. The fourth dimension was analysis to teaching effectiveness. Through writing commentaries in the previous three dimensions (planning, instruction, and assessment), teacher candidates demonstrated their abilities to justify their plans to meet the needs of diverse learners based on research and theory. Also, in the instruction commentary, the teacher candidates demonstrated their individual ability to justify the effectiveness of their teaching and what they would change. While in the assessment commentary, the teacher candidates demonstrated their ability to analyze students' work to inform their future planning and instruction. The last dimension of teaching was defined as the teacher candidates' ability to introduce and foster an understanding of academic language among their students. Teacher candidates uses their various commentaries, lesson planning, video clips, and other artifacts to demonstrate their ability to deepen content understanding through the support of the students' oral and written academic language (ACCTE & Stanford University, 2013).

A critical component of this mixed methods research study is the concept of teacher mentoring. In a study of mentoring practices in an urban school context, Yendol-Hoppey, Jacobs, & Dana (2009) studied the unique challenges that mentors faced in supporting the retention of new teachers in an urban setting. The first assertion declared within this particular study was that mentors must support mentees in navigating the first years of teaching in a challenging context to ensure that the mentees actually focused on their learning and the learning of their students. This assertion holds particular relevance to this research study in that one of the most important roles of the instructor for Professional Development Seminar is to help the teacher candidates understand and articulate their rationale for engaging in specific student learning experiences. Another assertion declared by the Yendol-Hoppey et. al study was that mentors must work with school leadership to actively nurture a learning culture within the school. In the case of the edTPA, the mentors (instructors of Professional Development Seminar) must work with the college administration to procure the resources necessary to support teacher candidates in preparation of successful edTPA portfolios. A major component of the mentoring and support program resulting from this research focuses on how the instructors aligned resources to support teacher candidates through the process of developing an edTPA portfolio.

Smith & Ingersoll (2004) noted that the most salient factors affecting the reduction of new teacher turnover was having a mentor from the same field, having common planning time with other teachers in the same subject, or collaboration with other teachers, and being part of an external network of teachers. Those researchers cautioned that each of the above-mentioned factors were not statistically significant in impacting teacher turnover, but the combination of multiple factors listed had an affect on teacher retention. The authors of this researcher study understood that a combination of mentoring practices would be a key component of the resulting edTPA support program. The findings of this research study suggested that successful edTPA support should be structured around common planning time with teacher candidates with the same or similar edTPA handbooks on specific, edTPA focused full day sessions, in concert with findings from Smith & Ingersoll (2004).

RESEARCH PURPOSE AND DESIGN

The research purpose was to discover and describe the effectiveness of edTPA mentoring and support program designed by the researchers measured by teacher candidates' perceptions. Within that research purpose, a program model was constructed to describe a framework for effective mentoring and candidates' support program, which would lead to a successful completion and submission of an edTPA portfolio. The by-product of this research was that the resulting edTPA support model would potentially increase the teacher candidates' probability of attaining a state teaching license as a result of passing edTPA, which was consequential for teacher licensure in the state of Wisconsin beginning with the Fall 2016 semester.

This design and results of this research study are useful to individuals learning about mixed methods research methodology. More specifically, this research study is an example of how the topic of edTPA Support was researched from a mixed methods perspective; both qualitative and quantitative data collection and analysis were used throughout this study. As a result of this study's use of mixed methods design, the researchers are more confident in the triangulation of data that resulted from this study.

Description of Research Site and Sample

The site for this research study was Cardinal Stritch University, a Franciscan-centered university in Milwaukee, WI., Cardinal Stritch University is a Franciscan, private, small, liberal-arts based university offering a myriad of undergraduate and graduate degrees in multiple disciplines. The specific college used within this research study was the College of Education and Leadership, which consisted of the following schools: School of Education, School of Leadership, and the School of Urban Initiatives. The specific school used for this research study was the School of Education.

The Cardinal Stritch University School of Education consists of two initial certification teacher preparation programs: The Undergraduate Teacher Education Program (UGTE) and the Master of Arts—Teaching Program (MAT). The MAT program offered courses leading toward initial licensure in Milwaukee, WI and Madison, WI. Both the UGTE and the MAT programs were considered initial licensure programs in that they offered initial licensure to teacher candidates. The site and sample of this study consisted of students who sought initial licensure in either the UGTE or MAT program, including the MAT program's Madison location. It is important to note that the researchers served dual roles as Professional Development Seminar instructors within this study.

The research online survey was provided to all teacher candidates in the Master of Arts and Teaching program (MAT) in Milwaukee and Madison, Undergraduate Teacher Education Program (UG) in Milwaukee at the end of the edTPA mentoring and support program in April and after submitting their completed edTPA portfolio for the Spring semester 2014. The total enrollment in the student teaching seminar at the research site was 53 teacher candidates, of these candidates; 43 were MAT teacher candidates in Milwaukee and Madison, 10 UG teacher candidates in Milwaukee. The total number of completed surveys was n = 45 (84.90% of total enrollment), of these surveys; n = 35 (81.40% of total MAT student teaching seminar) surveys were completed by MAT teacher candidates and n = 10 (100.00% of total UG student teaching seminar) surveys were completed by UG teacher candidates. Of the total number of surveys completed, there were 20 (44.44%) males and 25 (55.55%) females. The percent of teacher candidates who completed edTPA portfolio in the different programs and gender are shown in Figure 1. The number and percent of edTPA portfolio completions were illustrated in Table 1

Figure 1. Distribution of teacher candidates who completed the survey by program and gender

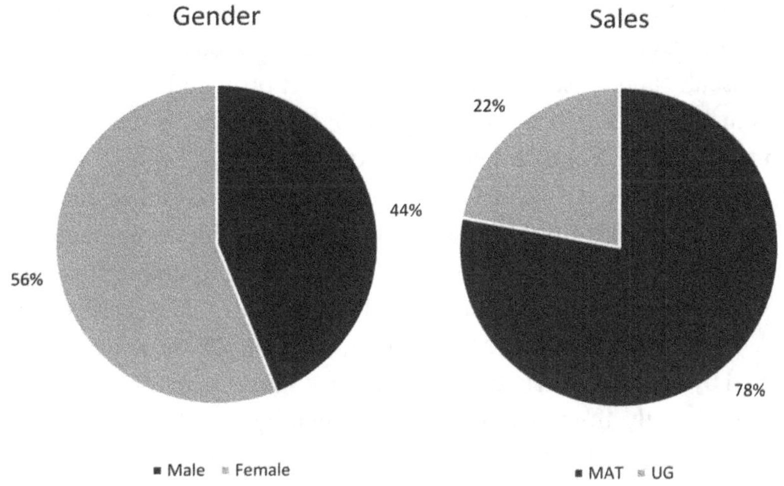

Table 1. Distribution of students by edTPA portfolio license area

		Percent	Response Count
Valid	Elementary Literacy	31.1	14
	MC Science	2.2	1
	MC Math	2.2	1
	MC English	2.2	1
	Secondary Science	4.4	2
	Secondary Math	4.4	2
	Secondary English	6.7	3
	Secondary History	15.6	7
	Special Education	28.9	13
	Performance Arts	2.2	1
	Total	100.0	

Methodology

A mixed methods interactive program evaluation was utilized to determine the effectiveness of the edTPA support model initiated by Cardinal Stritch University throughout its initial licensure teacher preparation programs. This research was conducted as the procedures were developed and implemented by the researchers. An online descriptive survey was used to collect teacher candidate perception of the edTPA mentoring and support program utilized during the Spring 2014 semester to complete and submit an edTPA portfolio. The survey collected a satisfaction rating from teacher candidates engaged in preparing an edTPA portfolio using a six point Likert scale and qualitative data to expand and to provide more details regarding the teacher candidate experience.

For the satisfaction rating, there were four subscales: edTPA preparation subscale, understating edTPA subscale, edTPA submission subscale, and support provided subscale. There were 4 statements in edTPa preparation subscale, 6 statements in the understanding edTPA subscale, 6 statements in the edTPA submission subscale, and 11 statements in the support provided subscale. The support provided subscales measured four areas of support: Support provided by seminar instructor, by university supervisors, by cooperation teacher, and by resources provided.

The purpose of the qualitative data was to collect participants' (teacher candidates') opinions and attitudes about the procedures and support embedded within edTPA implementation for this group of students. Participants' responses were compiled and analyzed using qualitative coding strategies. For the qualitative data open-ended questions there were 11 statements. The focus of these statements was on two areas: edTPA preparation and support provided. The support provided subscale looked deeply on the support provided by seminar instructor, by university supervisors, by cooperation teacher, and by resources provided. Additionally, the researchers utilized member checking (Cresswell, 2003) to confirm the coded qualitative themes unearthed through the survey instrument.

Program Description

The edTPA mentoring and support program was constructed as a differentiated workshop to meet the needs of all teacher candidates. The program was built around four areas: Mini lectures, resources, work time, and Q&A forums.

The mini lectures were used to discuss and explain each task in the following terms:

- What do I need to know?
- What do I need to do?
- What will I write?
- *What will I submit*?

The resources used were:

- Detailed edTPA introduction at student teaching kickoff.
- Cooperating teacher edTPA resource packet.
- edTPA weekly calendar.
- University supervisors expertise and visits.
- Task checklist, weekly task summary email, edTPA and making good choices handbook.

The work time was planned within the sessions where teacher candidates experienced working with peers across multiple areas of licensure, and working individually. The Q&A forums were offered regularly during class sessions or via Canvas, the university's learning management system.

PRESENTATION AND SUMMARY OF DATA

SPSS data analysis was used to test for the reliability of the descriptive survey subscales (construct). How I feel about my preparation subscale of the edTPA mentoring and support survey had a high reliability,

Cronbach's α = 0.832. Understanding edTPA subscale of the edTPA mentoring and support survey had a high reliability, Cronbach's α = 0.828. How I feel about my completed edTPA submission subscale of the edTPA mentoring and support survey had a medium reliability, Cronbach's α = 0.675. How I feel about the support provided subscale of the edTPA mentoring and support survey had a high reliability, Cronbach's α = 0.793. The overall reliability of all subscales combined of the edTPA mentoring and support survey had a high reliability, Cronbach's α = 0.860.

A principal axis factor analysis was conducted on the 28 items with oblique rotaBon (direct oblimin). The Kaiser-Meyer-Olkin measure verified the sampling adequacy for the analysis, KMO = 0.475 and the lowest KMO values for individual items was greater than 0.37, which is below the acceptable limit of 0.5 (Field, 2013). This was caused by the small sample size (n = 45). An initial analysis was run to obtain Eigenvalues for each factor in the data. Six factors had Eigenvalues over Kaiser's criterion of 1 and in combination explained 64.56% of the variance. The scree plot was ambiguous and showed inflexions that would justify retaining either 4 or 6 factors. The researchers retained 6 factors because of the convergence of the scree plot and Kaiser's criterion on this value. The items that cluster on the same factor suggest that factor 1 represents edTPA preparation, factor 2 represent support by university supervisor, factor 3 represent support by the cooperating teacher, factor 4 represents how teacher candidates feel about the submitted edTPA portfolio, factor 5 represent support by the instructor, and factor 6 represent edTPA resources. These factors correspond with the subscale initially used to construct the survey with slight variations. For example, support is divided into three separate factors. Most survey items loaded under factors 1 and 5. Factor 1 has a relationship with factors 2, 4, 5, and 6. Factors 2 and 3 have fairly small relationship with the other factors. Overall, the findings of the reliability and factor analysis findings confirmed that the survey is reliable and valid. Factor analysis is exploratory in nature. It will be used to guide future hypothesis. It is used to inform the researchers about patterns within the data sets for future construct of the survey.

Descriptive Statistics

Descriptive statistics were used to describe the characteristics of the sample and test for assumptions. These statistics include the mean, standard deviation, and range of scores. The mean is a hypothetical value used to summarize the data. Variance is the average error between the mean and the observation; as a result, it measures how well the mean represents the data. But, it measures the variation squared; therefore, the standard deviation is used to make sure that the average error is in the same units as the data set. Standard deviation is the square root of the variance (Field, 2013). A smaller standard deviation compared to the mean, shows that the data points are close to the mean; therefore, the mean is representative of the data. Larger standard deviation compared to the mean, shows that the data points are distant from the mean; therefore, the mean does not represent the data (Field, 2013).

Table 2 shows the means and standard deviations for survey statements under school culture items for all respondents. A six point Likert scale was used as stated in Chapter Three with the response scale ranging from 1 = (strongly disagree) to 6 = (strongly agree). These results are reported to show the relative strength of individual items, and to present the full data-set before disaggregated comparisons are made. Table 3 shows descriptive statistics summary for "How I feel about my Preparation" Subscale.

In regard of "how I feel about my preparation" subscale, four items are considered. A review of Table 2 shows the individual mean and standard deviation of each item. Table 4 shows the overall mean for all the items under this subscale is M = 4.56 and relatively medium size standard deviation, value of

Table 2. Descriptive statistics summary for all survey items

	N	Minimum	Maximum	Mean	Std. Deviation
I feel well informed about the edTPA.	45	1	6	4.60	1.074
I feel academically prepared for the edTPA.	45	2	6	4.60	1.074
I feel I was prepared to complete the edTPA on time.	45	1	6	4.44	1.253
I feel the student teaching seminar adequately prepared me to complete the edTPA.	45	1	6	4.67	1.108
I read the edTPA handbook and all other provided resources.	45	2	6	4.89	1.005
I understood the edTPA handbook.	45	2	6	4.53	.919
I understood the edTPA requirements.	45	2	6	4.58	.812
I understood the format of edTPA templates.	45	1	6	4.69	1.184
I understood each edTPA task.	45	1	6	4.42	1.215
I understood what I need to submit for each task.	45	1	6	4.62	1.093
I fulfilled all my responsibilities as a teacher candidate.	45	4	6	5.44	.624
I believe that I put forth my best effort to complete the edTPA entry.	45	1	6	4.84	.928
I am very proud of my work. I did my best.	45	1	6	4.53	.944
I was motivated to complete and do well on the edTPA.	45	1	6	4.04	1.278
The people most responsible for what I learn are my instructors.	45	1	6	3.58	1.158
The person most responsible for what I learn is myself.	45	4	6	5.13	.786
My Instructor supported me during the edTPA.	45	3	6	5.53	.726
My cooperating teacher supported me during the edTPA.	45	1	6	4.53	1.546
My university supervisor supported me during the edTPA.	45	1	6	4.53	1.531
The edTPA task checklists helped me with my submission.	45	2	6	4.98	.988
The edTPA event calendar helped me stay on track.	45	2	6	4.62	1.093
My instructor fulfilled his/her responsibilities toward preparing me.	45	3	6	5.29	.843
My cooperating teacher fulfilled his/her responsibilities toward preparing me.	45	1	6	4.38	1.655
My university supervisor fulfilled his/her responsibilities toward preparing me.	45	1	6	4.60	1.543
My university supervisor provided me with valuable constructive feedback.	45	1	6	4.53	1.517
My university supervisor provided me with timely constructive feedback.	45	1	6	4.58	1.685
My cooperating teacher provided me with timely constructive feedback.	45	1	6	4.44	1.423

$SD = .92$. Therefore, the mean is representative of the "how I feel about my preparation" subscale. The teacher candidates felt prepared to complete an edTPA portfolio.

Table 5 shows descriptive statistics summary for "Understanding edTPA" Subscale. In regard of "Understanding edTPA" subscale, six items are considered. A review of Table 5 shows the individual mean and standard deviation of each item. Table 6 shows the overall mean for all the items under this subscale is $M = 4.62$ and relatively medium size standard deviation, value of $SD = .77$. Therefore, the mean is representative of the "Understanding edTPA" subscale. The teacher candidates felt that they understand edTPA requirements, task, and expectations.

Table 3. Descriptive Statistics summary for "how I feel about my preparation" subscale

	N	Minimum	Maximum	Mean	Std. Deviation
I feel well informed about the edTPA.	45	1	6	4.60	1.074
I feel academically prepared for the edTPA.	45	2	6	4.60	1.074
I feel I was prepared to complete the edTPA on time.	45	1	6	4.44	1.253
I feel the student teaching seminar adequately prepared me to complete the edTPA.	45	1	6	4.67	1.108
Valid N (listwise)	45				

Table 4. Overall descriptive statistics summary for "How I feel about my preparation" subscale

	N	Minimum	Maximum	Mean	Std. Deviation
How I feel about my preparation	45	2.25	6.00	4.5778	.92137
Valid N (listwise)	45				

Table 5. Descriptive statistics summary for "Understanding edTPA" subscale

	N	Minimum	Maximum	Mean	Std. Deviation
I read the edTPA handbook and all other provided resources.	45	2	6	4.89	1.005
I understood the edTPA handbook.	45	2	6	4.53	.919
I understood the edTPA requirements.	45	2	6	4.58	.812
I understood the format of edTPA templates.	45	1	6	4.69	1.184
I understood each edTPA task.	45	1	6	4.42	1.215
I understood what I need to submit for each task.	45	1	6	4.62	1.093
Valid N (listwise)	45				

Table 6. Overall descriptive statistics summary for "Understanding edTPA" subscale

	N	Minimum	Maximum	Mean	Std. Deviation
Understanding edTPA	45	2.33	5.83	4.6222	.76821
Valid N (listwise)	45				

Table 7 shows descriptive statistics summary for "How I feel about my completed edTPA submission" Subscale. In regard of "How I feel about my completed edTPA submission" subscale, six items are considered. A review of Table 7 shows the individual mean and standard deviation of each item. Table 8 shows the overall mean for all the items under this subscale is M = 4.60 and relatively small size standard deviation, value of SD = .60. Therefore, the mean is representative of the "How I feel

Table 7. Descriptive statistics summary for "How I feel about my completed edTPA submission" subscale

	N	Minimum	Maximum	Mean	Std. Deviation
I fulfilled all my responsibilities as a teacher candidate.	45	4	6	5.44	.624
I believe that I put forth my best effort to complete the edTPA entry.	45	1	6	4.84	.928
I am very proud of my work. I did my best.	45	1	6	4.53	.944
I was motivated to complete and do well on the edTPA.	45	1	6	4.04	1.278
The people most responsible for what I learn are my instructors.	45	1	6	3.58	1.158
The person most responsible for what I learn is myself.	45	4	6	5.13	.786
Valid N (listwise)	45				

about my completed edTPA submission" subscale. The teacher candidates felt proud motivated and had ownership of their submitted portfolio.

Table 9 shows descriptive statistics summary for "How I feel about the support provided" Subscale. In regard of "How I feel about the support provided" subscale, eleven items are considered. A review of Table 10 shows the individual mean and standard deviation of each item. Table 8 shows the overall mean for all the items under this subscale is $M = 4.73$ and relatively medium size standard deviation, value of $SD = .76$. Therefore, the mean is representative of the "How I feel about the support provided" subscale. The teacher candidates felt supported in the process of completing an edTPA portfolio.

In order to understand support at a deeper level, the researcher divided the subscale into four sub-categories: support by the student teaching seminar instructor, university supervisor, the cooperating teacher and the resources provided. For the support provided by the student teaching seminar instructor subcategory, two items are used. The teacher candidates felt very supported by the instructor. Table 11 shows the overall mean for all the items under this subcategory is $M = 5.41$ and relatively medium size standard deviation, value of $SD = .70$. Therefore, the mean is representative of the "How I feel about the support provided by the instructor" subcategory.

A look at the support provided by the cooperating teacher and university supervisors subcategories reveals a mean, value of $M = 4.45$ and 4.56 respectively with relatively large size standard deviation, value of $SD = 1.37$ and 1.46 respectively. Four items are used under the support by the cooperating teacher subcategory and six items are used under the support by the university supervisor subcategory. The mean is not a representative of these subcategories. This shows a variation in the support provided by the cooperating teacher and university supervisor. This could be identified as an area of need. Table 12 shows the overall mean for all the items under these subcategories.

Table 8. Overall descriptive statistics summary for "How I feel about my completed edTPA Submission" subscale

	N	Minimum	Maximum	Mean	Std. Deviation
How I feel about my completed edTPA submission	45	2.83	5.83	4.5963	.60353
Valid N (listwise)	45				

Table 9. Descriptive statistics summary for "How I feel about the support provided" subscale

	N		Maximum	Mean	Std. Deviation
My Instructor supported me during the edTPA.	45	3	6	5.53	.726
My cooperating teacher supported me during the edTPA.	45	1	6	4.53	1.546
My university supervisor supported me during the edTPA.	45	1	6	4.53	1.531
The edTPA task checklists helped me with my submission.	45	2	6	4.98	.988
The edTPA event calendar helped me stay on track.	45	2	6	4.62	1.093
My instructor fulfilled his/her responsibilities toward preparing me.	45	3	6	5.29	.843
My cooperating teacher fulfilled his/her responsibilities toward preparing me.	45	1	6	4.38	1.655
My university supervisor fulfilled his/her responsibilities toward preparing me.	45	1	6	4.60	1.543
My university supervisor provided me with valuable constructive feedback.	45	1	6	4.53	1.517
My university supervisor provided me with timely constructive feedback.	45	1	6	4.58	1.685
My cooperating teacher provided me with timely constructive feedback.	45	1	6	4.44	1.423
Valid N (listwise)	45				

Table 10. Overall descriptive statistics summary for "How I feel about the support provided" subscale

	N	Minimum	Maximum	Mean	Std. Deviation
How I feel about the support provided	45	2.91	6.00	4.7293	.76442
Valid N (listwise)	45				

Table 11. Overall descriptive statistics summary for "How I feel about the support provided by instructor" subcategory

	N	Minimum	Maximum	Mean	Std. Deviation
How I feel about the support provided by Instructor	45	3.00	6.00	5.4111	.70137
Valid N (listwise)	45				

Table 12. Overall descriptive statistics summary for "how I feel about the support provided by cooperating teacher and university supervisor' subcategory

	N	Minimum	Maximum	Mean	Std. Deviation
How I feel about the support provided by CT	45	1.00	6.00	4.4519	1.36544
How I feel about the support provided by US	45	1.00	6.00	4.5611	1.45691
Valid N (listwise)	45				

Table 13. Overall descriptive statistics summary for "How I feel about the support provided by resources" subcategory

	N	Minimum	Maximum	Mean	Std. Deviation
How I feel about the support provided by resources	45	2.00	6.00	4.8000	.89443
Valid N (listwise)	45				

In regard of the support provided by the resources subcategory, the teacher candidates felt supported. The overall mean for all the items under this subcategory is $M = 4.80$ and relatively medium size standard deviation, value of $SD = .89$. Two items are used to calculate the mean. Table 13 shows the overall mean for all the items under this subcategory.

Differences in Survey Subscales

A Paired Sample test is used to measure the differences between the four survey subscales. Table 14 presents the results of Paired Sample test for the four subscales respectively. The finding in Table 14 shows that the t value for pair 1: how I feel about my preparation and understanding edTPA is -0.376 with a significance level of $p = 0.708$. The probability value (p) is more than .05; therefore the result is not significant. There is no statistically significant difference in the mean value of how I feel about my preparation subscale and the mean of understanding edTPA subscale mean value ($p < .05$). Table 14 also shows the results of the Paired Sample test for pair 2: how I feel about my precreation subscale and how I feel about my completed submission subscale as well as pair 3: how I feel about my preparation subscale and how I feel about the support provided subscale. The t values for pairs 2 and 3 are -0.152 and -1.02 respectively with a significance level of $p = 0.880$ and 0.312 respectively. The probability value (p) is more than .05; therefore the result is not significant. There is no statistically significant difference in the mean value of how I feel about my preparation subscale and the mean of how I feel about my submission and how I feel about the support provided subscales mean values ($p < .05$).

Table 14. Paired sample test difference in mean values

		Paired Differences					t	df	Sig. (2-tailed)
		Mean	Std. Deviation	Std. Error Mean	95% Confidence Interval of the Difference Lower	Upper			
Pair 1	How I feel about my preparation - Understanding edTPA	-.04444	.79189	.11805	-.28235	.19346	-.376	44	.708
Pair 2	How I feel about my preparation - How I feel about my submission	-.01852	.81841	.12200	-.26439	.22736	-.152	44	.880
Pair 3	How I feel about my preparation – How I feel about the support provided	-.15152	.99311	.14804	-.44988	.14685	-1.023	44	.312

Similarly, a Paired Sample Test is used to measure the difference in the mean values of understanding edTPA subscale and how I feel about my submission subscale and how I feel about the support provided subscale respectively. Table 15 shows that the *t* value for pair 1: understanding edTPA and how I feel about my completed submission subscale and pair 2: understanding edTPA and how I feel about the support provided are 0.209 and -0.852 respectively with a significance level of $p = 0.836$ and 0.399 respectively. The probability value (p) is more than .05; therefore the result is not significant. There is no statistically significant difference in the mean value of how I feel about my preparation or support subscales and the mean of understanding edTPA subscale mean value ($p < .05$).

Lastly, a Paired Sample Test is used to measure the difference in the mean values of how I feel about my submission subscale and how I feel about the support provided subscale. Table 16 shows that the *t* value for pair 1 is -0.952 with a significance level of $p = 0.346$. The probability value (p) is more than .05; therefore the result is not significant. There is no statistically significant difference in the mean value of how I feel about my submission subscale and how feel about the support provided subscale mean value ($p < .05$).

A review of the previous data analysis, one can conclude that there is no significant difference in the mean value between the survey subscales. Therefore, the overall mean of all the items on the survey is used to represent teacher candidates' perception of the effectiveness of edTPA mentoring and candidate support program designed by the researchers to successfully complete edTPA entry. The overall mean of all items on the survey is $M = 4.63$ and relatively small size standard deviation, value of $SD = .56$. Twenty-eight items were used to calculate the mean. Table 17 shows the overall mean for all the items in the survey.

Table 15. Paired sample test difference in mean values

		Paired Differences							
					95% Confidence Interval of the Difference				
		Mean	Std. Deviation	Std. Error Mean	Lower	Upper	t	df	Sig. (2-tailed)
Pair 1	Understanding edTPA - How I feel about my preparation	.02593	.83330	.12422	-.22443	.27628	.209	44	.836
Pair 3	Understanding edTPA - How I feel about the support provided	-.10707	.84268	.12562	-.36024	.14610	-.852	44	.399

Table 16. Paired sample test difference in mean values

		Paired Differences							
					95% Confidence Interval of the Difference				
		Mean	Std. Deviation	Std. Error Mean	Lower	Upper	t	df	Sig. (2-tailed)
Pair 2	How I feel about my completed submission - How I feel about the support provided	-.13300	.93667	.13963	-.41440	.14841	-.952	44	.346

Table 17. Overall descriptive statistics summary for all the survey items

	N	Minimum	Maximum	Mean	Std. Deviation
Survey overall mean	45	3.34	5.77	4.6314	.55821
Valid N (listwise)	45				

Summary of Findings

A Paired Sample Test was used to find if there was a difference between the four subscales' mean values. There was no statistically significant difference between the mean values between the four subscaless. Overall mean of all items on the survey is $M = 4.63$ and relatively small size standard deviation, value of $SD = 0.56$. A six point Likert scale was used, the response scale ranged from 1 = (strongly disagree) to 6 = (strongly agree). Teacher candidates felt prepared and understood the edTPA tasks. Also, they felt satisfied about their submitted edTPA portfolio and they felt that they did their best. Teacher candidates also felt supported by their seminar instructors, university supervisors and cooperating teacher as well as the resources provided.

Open Ended Questions with Qualitative Data

The response rate of 93.33% represent the 43 teacher candidates completed the open-ended questions describing their edTPA experience. As teacher candidates responded to a question asking them to describe what went well in their edTPA experience, the following themes were identified:

- All tasks were completed on time following the edTPA submission timeline (calendar).
- Allowed teacher candidates to reflect deeply on their own teaching practices and identify areas of strength and areas of growth.
- Teacher candidates felt prepared and supported by their seminar instructor, CT, US and resources.
- Deepened teacher candidates understanding of best practices and effective teaching strategies.

The main two themes identified from teacher candidates response to a questions asking them to describe what could be improved in their edTPA experience were need more work time and need more support from cooperating teachers. Other themes were need more support with technology, more interactive review of edTPA task with peers and university supervisors to get more feedback and lastly, start the process earlier to gain better understanding of edTPA.

In order to get a deeper understanding of the mentoring and support provided, teacher candidates were asked to describe how they were supported by the seminar instructor, university supervisor and cooperating teacher. 98.00% of teacher candidates felt positively supported by their seminar instructor. The main themes identified were clear and detailed walkthrough each task, prompt responses, available for questions, and well organized. The three areas of need that were identified by teacher candidates to get more support from the seminar instructor are more class time, more technology support and use sample edTPA entries.

The percent of teacher candidates who felt supported by their cooperating teacher was 71.42%. Most teacher candidates felt that their CT mainly supported them during the planning task, video recording, answered questions, and provided feedback. The two areas of need that were identified to provide more support from CT were to learn more about edTPA and provide more feedback (be more involved in the edTPA process). 83.33% of teacher candidates felt positively supported by their university supervisor are providing descriptive feedback, answered questions, and supported to plan the edTPA experience. Areas of need identified were more thoughtful and specific feedback, clear communication, and timely responses.

The researchers utilized member-checking (Cresswell, 2003) with the teacher candidates during one of the Professional Development Seminar sessions. The researchers posted the prevailing themes on the board, one at a time, and asked the teacher candidates to discuss the relevance of each theme. The teacher candidates corroborated the notion that they needed more work time to do well with edTPA tasks. One student stated, "I really want to do well on the edTPA, but I do not have enough time to do the work I know I can do. Student teaching, itself, keeps me so busy." Also, while the teacher candidates felt supported in the edTPA process, they noted that Stritch should focus more efforts in the area of CT development of edTPA understanding.

Summary of Findings

Teacher candidates were asked open-ended questions to describe their experience in completing an edTPA entry and the support provided to them through this experience. The overall strengths identified were the support provided by the seminar instructor, the support provided by the use of resources. The overall areas of need identified were more work time, more technology support, specific and clear feedback, and educate the CT and US about edTPA.

External Data to Support Findings: Pearson Reports

Six completed edTPA entries were randomly chosen to be submitted to Pearson as an external assessment to score and provide a performance report. Three elementary literacy entries, one secondary science, one secondary history/social studies, and one secondary English entry were submitted. Due to the small sample size and to protect students' identity, the data are used anonymously without identifying the edTPA handbook. Table 18 shows the teacher candidates scores as reported by Pearson as well as the mean score for each task.

According to the 2013 edTPA Field Test Summary Report published by Scale published November 2013, the percentage of candidates who would have "passed" edTPA along this recommended cut-score band ranged from 78 percent (score of 37) to 58 percent (score of 42). A review of table 18 shows that 100% of the teacher candidates would pass at cut scores 37 and 38. With a cut scores ranging from 39-42, 83.33% of the teacher candidates would pass. The lowest mean score was 2.53 and the highest mean score was 4.00 with an average mean score value of $M = 3.21$. According to the 2014 edTPA Field Test Summary Report published by Scale published July 2014 the average mean score value for secondary history/social studies is $M = 2.99$, for secondary science is $M = 3.05$, for secondary English $M = 3.09$ and the average mean score value for elementary literacy is $M = 2.88$. The teacher candidates' average mean score is higher than the national average mean score, which confirms the survey result.

Table 18. edTPA teacher candidates total scores and mean scores

Candidate	Planning Total Score	Planning Mean Score	Instruction Total Score	Instruction Mean Score	Assessment Total Score	Assessment Mean Score	Total Score	Total Mean Score
Candidate 1	15.5	3.1	13	2.6	16	3.2	44.5	2.97
Candidate 2	14	2.8	11	2.2	13	2.6	38	2.53
Candidate 3	15	3	19	3.8	18	3.6	52	3.47
Candidate 4	17	3.4	14	2.8	15	3	46	3.07
Candidate 5	19	3.8	12.5	2.5	16.5	3.3	48	3.20
Candidate 6	17	3.4	22	4.4	21	4.2	60	4.00
Mean	16.25	3.25	15.25	3.05	16.58	3.32	48.08	3.21

External Data to Support Findings: CT and US Evaluations

Another source of data used to verify the survey findings were the end of student teaching university supervisor and cooperating teacher performance evaluation completed by teacher candidates in June 2014. Teacher candidates were asked to evaluate the university supervisor performance; a total of 41 supervisors and cooperating teachers were evaluated. Teacher candidates responded positively to a question asking them "The supervisor was very helpful and supportive;" only one supervisor out of the 41 got a negative response. The three themes identified from the open-end question were: provided feedback, answered my questions, and provided great advise. Two of these themes align with the themes identified earlier. The one area of need identified by some teacher candidates was provide more constructive feedback, which aligns with the earlier findings.

Summary of Findings

Six edTPA entries were randomly selected to be scored by Pearson. The overall average mean score $M = 3.21$ is higher than the national average mean score. This confirms the survey finding that teacher candidates felt prepared, confident about their edTPA submission, understood the edTPA task and felt supported by their seminar instructor, university supervisor, cooperating teacher, and resources provided.

University supervisor and cooperating teacher performance evaluation at the end of student teaching were used to verify the findings of the research survey. The researchers focused on the question asking teacher candidates about the supervisor or the cooperating teacher being "helpful and supportive." The responses from the teacher candidates aligned with the research findings and similar themes that were identified in the online survey analysis.

RESEARCH MODEL

Based on the descriptive survey findings and the data analysis, the researcher chose the following model to represent the support and mentoring program (Figure 2). Based on the quantitative data analysis supported by the qualitative data, the mentoring and support program consists of three modules, structural support, pre-seminar, and seminar module. The structural support module is constructed based on three cores: edTPA handbooks, filed placement and learning management system. Theses cores needed to happen before the seminar to ensure a successful early start. First, field placements needed to be done earlier in the process at least one month before seminar. Second, the placement needs to align to the edTPA handbooks. For example, a teacher candidate doing the elementary literacy handbook needs to be in an elementary classroom teaching literacy. This sounds very obvious, but as the field placement recruitment officer might not be aware of all the edTPA requirements. Lastly, all edTPA handbooks and templates need to be uploaded to a secured online learning management system that can be accessed by all teacher candidates. Handbooks and templates needs to be assigned to respective teacher candidates based on teaching license requirements.

The pre-seminar module is constructed based on three cores: contact with cooperating teacher, training the university supervisor and edTPA kickoff session. The research finding strongly emphasized the importance of educating cooperating teachers about edTPA early in the process in order to support teacher candidates.

Contact the CT core is built around establishing early communication with the cooperating teachers through sending a letter explaining the edTPA process and expectations, providing literature about edTPA, and providing edTPA submission calendar. The purpose of this core was to build background knowledge about edTPA, to establish understanding of the edTPA expectations from all teacher candidates, to solicit support and to provide examples of how the CT can support the teacher candidate. The communication had to be established as early as possible. The researchers suggested multiple attempts of communications through email, mail and face meeting if possible. The literature explaining edTPA was retrieved from the edTPA resource library hosted by American Association of Colleges for Teacher Education (AACTE). The CT handouts were adapted from the PowerPoint presentation hosted in the library. Finally, the researchers developed the letter and the edTPA calendar.

Figure 2. Research model

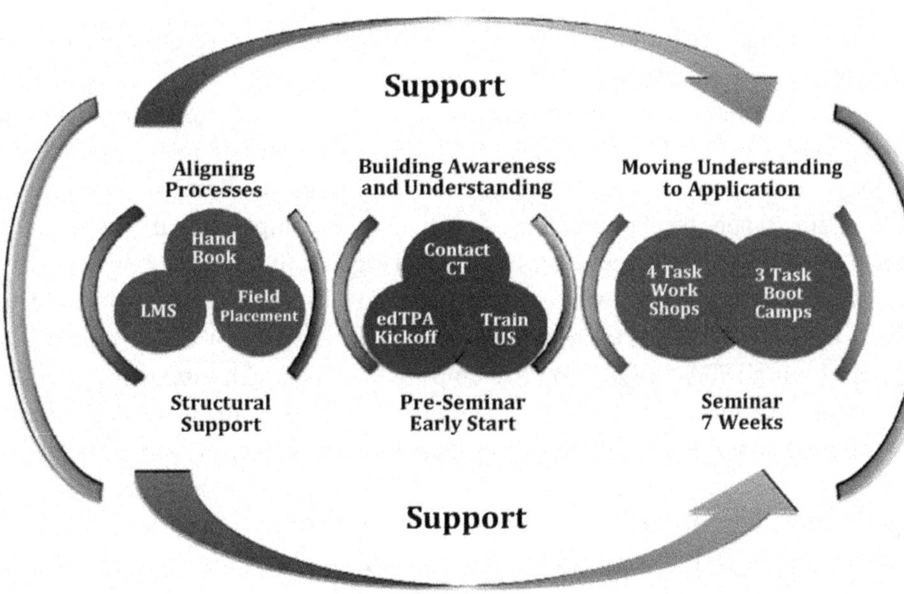

The second core was training university supervisors. This was a crucial core to be started earlier in the process and then to continue through out the process. The main two goals of the training were to educate supervisor about the edTPA and train them on providing descriptive feedback. The supervisor was the primary contact person with the CTs in the field. Part of their responsibilities is to train the CTs. The teacher candidates identified the need for feedback from both CT and US. The goal of the initial training before the seminar was to educate US about edTPA using literature and to communicate edTPA expectations. This training enabled US to answer teacher candidates' questions as well as the CT. the second training goal was to train US on giving descriptive feedback, to discuss the different types of feedback and finally to share allowable support and feedback under edTPA regulations. The researchers developed the training sessions material.

The third core was the edTPA kickoff session. All teacher candidates were invited to a kickoff session three weeks before the start of the student teaching seminar. The goal of this session was to provide teacher candidates with an overview of edTPA, explain the three tasks briefly, share the edTPA weekly calendar, and the CT handout. The key part of the kickoff session was to give teacher candidates access to their repsective edTPA handbooks and Making Good Choices document. Teacher candidates were instructed to carefully read both handbooks as well as share the edTPA calendar and handouts with their respective CT in the first meeting. The more the teacher candidates read about the edTPA and gained familiarity with the expectations, the more they felt prepared to make sound decisions related to the learning segment.

The seminar module was constructed based on two cores; task workshops and tasks boot camps. The four tasks workshops are three hours each. At the beginning, allow each teacher candidate to select an edTPA buddy from the same license area. During the workshop the teacher candidates with the facilitation of the seminar instructors, deconstructed each task. The workshop began with an overview of the task highlighting "what do I need to do" section of the handbook while referring to the evidence chart. The second part of the workshop was dedicated to deconstruction of the commentary prompts and the task rubrics with reference to the Making Good Choices document. It was essential that teacher candidates develop an understanding of the task, commentary, and artifact. This was accomplished through small group discussion and pair discussion with the edTPA buddy. The first workshop was designated for the learning segment and lesson plans, the next three workshops were designated to the planning task, instruction task and assessment task, respectively.

There were three edTPA boot camp sessions of six hours each, which were designated for task work time and feedback from peers. In the research study findings, teacher candidates' comments showed a great emphases on the need for more work time and more feedback. The first half of each boot camp was designated for work time where teacher candidates were able to complete all the required writing (task commentaries) and prepared the task-related artifacts. During this time, support was provided from the seminar instructors, volunteer teachers who completed an edTPA entry previously, and from technical support personnel. The second half of each boot camp was designated for peer feedback. Teacher candidates worked with their edTPA buddies to provide peer feedback using the task checklist developed by the researchers and adapted from the Making Good Choices document. In addition, each teacher candidate met with his/her US to review the edTPA task at hand, discuss any concerns, answer any questions, monitor progress, and provide oral feedback. The three boot camps session were designated to support the planning, instruction, and assessment tasks respectively.

The two arrows wrapping the three modules represented support. Support was critical for the success of this program. Teacher candidates' comments in the research survey highlighted to importance of support received from seminar instructor, cooperating teacher, and university supervisor. Different forms of support can include but limited to availability to answer questions personally or through email. Build background knowledge and fully understand the edTPA tasks to answer any questions, dismiss any concerns, and provide advice. The seminar instructors needed to make sure to be available through phone, email or personally to support not only the teacher candidates, but also the cooperating teacher and university supervisor as well. The instructor needed to designate office hours to meet with anyone who needs support.

Summary of Findings

The research purpose was to discover and describe the effectiveness of edTPA mentoring and support program measured by teacher candidates' perceptions. An online survey instrument was used to measure teacher candidates' perception. The analysis of the quantitative and qualitative data collected by survey instrument revealed themes that supported the model. Teacher candidates felt very prepared, they felt that they have a deep understanding of the edTPA, they felt confident about their submission and felt supported by their seminar instructors and resources provided. These findings were supported through a triangulation of data from external resources such as Pearson official scoring reports and end of student teaching university supervisor and cooperating teacher evaluation. Based on these findings, the researchers developed an edTPA mentoring and support model. The model was based on three modules: structural module, pre-seminar module and seminar module. All three modules were constructed based on cores all wrapped with support.

REFERENCES

American Association of Colleges for Teacher Education & Stanford University. (2013). *edTPA General Information*. Retrieved from http://edtpa.aacte.org

American Association of Colleges for Teacher Education & Stanford University. (2013). *Using edTPA Handout*. Retrieved from http://edtpa.aacte.org

Charmaz, K. (2006). *Constructing grounded theory: A practical guide through qualitative analysis*. Thousand Oaks, CA: SAGE Publications.

Crawford, S. D., Couper, M. P., & Lamias, M. J. (2001). Web surveys: Perception of burden. *Social Science Computer Review*, *19*(2), 146–162. doi:10.1177/089443930101900202

Creswell, J. W. (2003). *Research Design: Qualitative, Quantitative, and Mixed Methods Approaches* (2nd ed.). Thousand Oaks, CA: Sage Publications.

Department of Public Instruction. (2014). *Teacher Evaluation Process Manual. Version 4*. Retrieved from http://dpi.gov

Dillman, D. A., Tortora, R. D., & Bowker, D. (2001). *Principles for constructing web surveys*. Retrieved December 12, 2013, from Social and Economic Science Research Center.http://survey.sesrc.wsu.edu/dillman/papers.htm

Fetterman, D. M. (1994). Empowerment evaluation. *Evaluation Practice, 15*(1), 1–15. doi:10.1016/0886-1633(94)90055-8

Field, A. (2013). *Discovering statistics using SPSS* (4th ed.). London: Sage Publication.

Kaye, B. K., & Johnson, T. J. (1999). Research methodology: Taming the cyber frontier. *Social Science Computer Review, 17*(3), 323–337. doi:10.1177/089443939901700307

Owens, J. M., & Rogers, P. J. (2003). *Program evaluation forms and approaches* (3rd ed.). London: Sage Publication.

Patton, M. Q. (1996). A world larger than formative and summative. *Evaluation Practice, 17*(2), 131–144. doi:10.1016/S0886-1633(96)90018-5

Posavac, E. J., & Carey, R. G. (2003). *Program evaluation methods and case studies* (6th ed.). Upper Saddle River, NJ: Prentice Hall.

Salkind, N. J. (2006). *Statistics for people who think they hate statistics*. Thousand Oaks, CA: Sage Publications.

Smith, T. M., & Ingersoll, R. M. (2004). What are the effects of induction and mentoring on beginning teacher turnover? *American Educational Research Journal, 41*(3), 681–714. doi:10.3102/00028312041003681

Solomon, D. J. (2001). Conducting web surveys. *Practical Assessment, Research & Evaluation, 7*(19).

Yendol-Hoppey, D., Jacobs, J., & Dana, N. F. (2009). Critical concepts of mentoring in the urban context. *New Educator, 5*(1), 25–44. doi:10.1080/1547688X.2009.10399562

KEY TERMS AND DEFINITIONS

Artifacts: Represent authentic work completed by the teacher candidates and their students. These include lesson plans, copies of instructional and assessment materials, video clips of your teaching, and student work samples.

Commentaries: Written documents to describe the artifacts, explain the rationale behind their choice, and analyze what the teacher candidates have learned about their teaching practice and their students' learning.

Cronbach's Alpha: A numerical coefficient of reliability. Computation of alpha is based on the reliability of a test relative to other tests with same number of items, and measuring the same construct of interest. In general, Cronbach's alpha factor will increase when the correlation between the items increase. The reliability coefficient of 0.7 or higher is considered "acceptable" in most social science research.

Likert Scale: A rating scale useful when a behavior, attitude, or other phenomenon of interest needs to be evaluated on a continuum. For the purposes of this study, the scale of strongly disagree, mostly disagree, disagree, agree, mostly agree, strongly agree were utilized.

Mean: A type of average where scores are summed and divided by the number of observations.

Pearson: The evaluation system. It is used to provide technical platform to collect, score, and deliver results to teacher candidates and preparation programs.

Rubrics: Subject-specific evaluation criteria used to score your performance on edTPA. These rubrics are included in the handbook following the directions for each task. The descriptors in the five-level rubrics address a wide range of performance, beginning with the knowledge and skills of a novice not ready to teach (Level 1) and extending to the advanced practices of a highly accomplished beginner (Level 5).

Standard Deviation: Represents the average amount of variability in a set of scores. In practical terms, it is the average distance from the mean. The larger the standard deviation, the larger the average distance each data point is from the mean of the distribution.

Chapter 14
Morphological Ontology Design Engineering:
A Methodology to Model Ill-Structured Problems

Joey Jansen van Vuuren
Council for Scientific and Industrial Research, South Africa

Marthie M. Grobler
Council for Scientific and Industrial Research, South Africa

Louise Leenen
Council for Scientific and Industrial Research, South Africa

Ka Fai Peter Chan
Council for Scientific and Industrial Research, South Africa

Zubeida C. Khan
Council for Scientific and Industrial Research, South Africa

ABSTRACT

In the Social-technical domain scientists are often confronted with a class of problems that are termed messy, ill-structured or wicked. These problems address complex issues that not well-defined, contain unresolvable uncertainties, and are characterized by a lack of common agreement on problem definition. This chapter proposes a new mixed methods research technique, Morphological Ontology Design Engineering (MODE), which can be applied to develop models for ill-structured problems. MODE combines three different research methodologies into a single, methodology. MODE draws from research paradigms that include exploratory and descriptive research approaches to develop models. General morphological analysis offers a systematic method to extract meaningful information from domain experts, while ontology based representation is used to logically represent domain knowledge. The design science methodology guides the entire process. MODE is applied to a case study where an ontological model is developed to support the implementation of a South African national cybersecurity policy.

DOI: 10.4018/978-1-5225-0007-0.ch014

INTRODUCTION

The purpose of this chapter is to show how the sum of various research methodologies can be combined to provide a more powerful mixed methods research methodology (Peffers, Tuunanen, Rothenberger, & Chatterjee, 2007) to model ill-structured problems. An ill-structured problem is characterised by a high degree of uncertainty and conflicts about values. This research methodology focuses on understanding the messiness of real world problems resulting in the development of constructs, models, methods or instantiations. Fritz Zwicky, the father of General Morphological Analysis (GMA) stated that "... within the final and true world image everything is related to everything, and nothing can be discarded a priori as being unimportant" (Ritchey, 1998).

The proposed hybrid methodology, referred to as Morphological Ontology Design Engineering (MODE), will combine two qualitative research techniques (design science and general morphological analysis) with a quantitative methodology (ontology based representation) to present a holistic artefact. MODE draws on the strengths of different research paradigms and is used to develop models for difficult domains. Domains are areas of expertise that can be used to create models. Domain knowledge in this chapter refers to knowledge of specialists or focus groups in an area of human endeavour or a specialised discipline. These models capture and represent domain knowledge, and support the development of solutions to problems areas in the domain. The authors apply the MODE methodology to the cybersecurity environment to develop a multi-layered, multi-dimensional domain model. Ontologies are used to present knowledge in such a way that computers can interpret and do some reasoning with this knowledge. An ontology normally consists of a set of entities and relations that describes a particular knowledge domain.

On a high level, the design science methodology guides the entire MODE process. This methodology involves a series of approaches with the intent of producing new theories, artefacts, and practices (Peffers, et al, 2007). Constructs conceptually describe the problem and solution domain and can be refined throughout the design cycle. Models describe the relationships among the constructs of the artefact, and therefore can be regarded as the problem and solution statements. Methods provide an algorithm or guideline on how to perform a task in solving a particular problem. Instantiations are the technical implementation of a construct, model and method within the operational environment.

The extraction of information from subject experts is a challenge that is addressed in MODE via a variation of the classic GMA method developed by Fritz Zwicky (Ritchey, 1998). GMA is a facilitated process where subject experts work together to collect and classify domain information. In order to represent the extracted knowledge in a meaningful format for the design of an artefact, an ontology is developed. An ontology is a semantic technology that enables a rich representation of domain knowledge. It provides a formal, encoded description of domain knowledge: all the entities, their attributes and their inter-relationships are represented. The output of the MODE methodology is a single shareable model of the environment (developed from the ontology component), agreed-upon by subject experts (developed from the GMA component).

MODE can be applied by any researcher to solve problems in a multi-layered, multi-dimensional problem space where there are uncertainties. The focus of this mixed methods methodology is to model problems that are ill-defined and difficult to model with normal qualitative and quantitative techniques. The uniqueness of MODE lies in the physical model derived from qualitative information that is presented in a quantitative methodology. The research team requires the inclusion of a GMA facilitator and an ontology engineer in the research team. In small research projects, this role can be done by one researcher with facilitation and ontology engineering skills, but it will be challenging to capture all the knowledge and facilitate at the same time {Martins, 2012 #226}. The authors of the chapter used specialised software, the CARMA software package, in their case study but access to this software is not a requirement. Spreadsheet applications using statistical formulas or simple programming applications can be developed to compute the solution spaces.

The methodologies paragraph that follows gives an overview of the application of mixed methods research methodology as well as the three core methodologies that are combined to form the MODE methodology. In addition, the combination of these techniques is discussed and the new integrated methodology is presented. A case study on the South African cybersecurity environment is then presented. The book chapter is concluded with an overview of the benefits and challenges of the MODE.

Miller (Miller, 2012) states that GMA is not a suitable mechanism for design but rather imposes structure on a method that is purposely structurally amorphous. In his Colombia-FARC case study Miller uses GMA to structure a military operational environment and show relationships graphically similar to what is used in the case study in this chapter. As such, the authors are of the opinion that GMA is a valuable component of the proposed MODE methodology.

METHODOLOGIES

This section provides an overview of the Mixed Methods Research Methodology as well as the three core methodologies that are combined to form the MODE methodology: design science, general morphological analysis, and ontology based development. The section concludes with a description of the MODE methodology.

Mixed Method Methodology

Mixed Method Methodology, or integration of qualitative and quantitative research as a technique of research, was initially discussed in 1998 by Tashakkori and Teddlie (Tashakkori & Teddlie, 1998). Mingers motivates the use of mixed methods due to the fact that research is usually not a single, discrete event but a process that typically proceeds through a number of phases (Mingers, 2001). Mingers goes on to state that in some cases the adoption of a single research methodology can provide a limited view of a particular research situation. A mixed methodology approach can address the constraint that a single research methodology often imposes: a selected methodology may be useful in certain phases of a research process but less appropriate in subsequent phases (see Miller's earlier critique of GMA as example). Even in the case where two or more methods perform similar functions, combining a range of approaches may yield a better result (Mingers, 2001). Mingers also argues that methodologies from different paradigms can be applied to solve a problem because it is possible to detach research methodologies from one paradigm and use them, critically and knowledgeably, within a different context and with different assumptions. The advantages of the multimedia method as discussed by Teddlie and Tashakkori (2003) are the option to validate data and results by combining a range of data sources, methods, or observers (triangulation); the discovery of fresh or paradoxical factors that stimulate further work, and the widening of the study to incorporate broader aspects of the situation (Mingers, 2001). Lewis and Grimes have shown in a range of surveys that a richer and more holistic result can be found due to the often implicit background assumptions of theories (Lewis & Grimes, 1999).

Design Science Methodology

The design science based research methodology focuses on understanding the messiness of real world problems with the "context being a core part of the story and not an extraneous variable to be trivialized" (Barab & Squire, 2004). Although this methodology is regularly used in research on education, its application as a methodology for information systems research is discussed by Peffers et al. (Peffers, et al., 2007).

Peffers et al. (2007 p. 86-91) provides the following steps as a depiction of the reasoning or approaches in the design cycle methodology:

- Identify the research problem and motivations for solving the problem.
- Define the objectives and the contribution of a feasible solution artefact within the restrictions by analysing the problem.
- Design and develop the artefact to include constructs, models, methods or instantiations. This step can be conducted before a complete articulation of the supporting concepts, models or theories exist.
- Demonstrate the artefact's ability to solve the problem. This can be conducted through simulation, experimentation, case studies or other accepted research methods in a practical environment.
- Measure or evaluate the artefact to determine its ability to solve the problem using various quantitative and qualitative analysis techniques and knowledge of relevant metrics. The hypothesis can be confirmed or contradicted. The hypothesis can also be amended as a result of the additional information gained in the process.
- Results of the problem solution, analysis design, evolution and its utility are communicated.

The designed solution must be aligned with the state of the problem and can include different configurations of either existing, or new and existing elements that determine the artefact's desired functionality, architecture and construction. The method used for the implementation need not be novel but must serve the purpose of the project. New knowledge is also gained from the environment and techniques applied. These steps are shown in Figure 1.

Modified General Morphological Analysis

The MODE methodology makes use of the Modified General Morphological Analysis (MGMA) approach that is based on the classical GMA.

Figure 1. General model for reasoning in the design cycle (Adapted from: Peffers, et al., 2007)

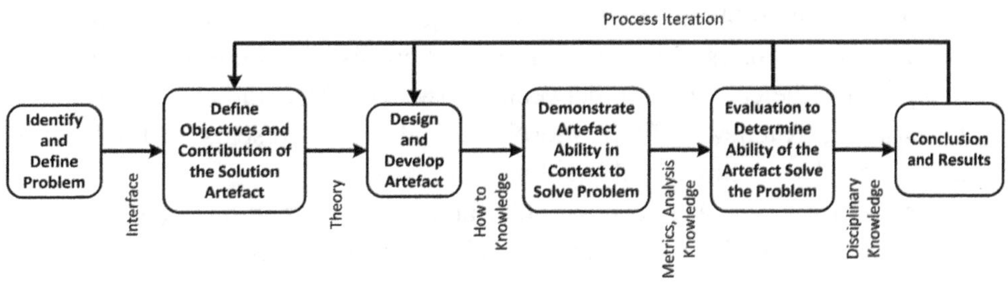

GMA is a well-known research technique aimed at solving complex problems. This form of non-quantified modelling relies on the judgmental processes of subject matter experts. GMA uses facilitated workshops (pre-workshop and workshop) with the group of subject matter experts that can address the specific problem complex. Each of these subject experts' view of the problem may differ due to differences in frame of reference. The challenge is to discuss these problems until a good consensus is achieved {Miller, 2012 #446}.

During the preparation (pre-workshop) phase, the facilitators meet with the client to investigate the nature of the problem and to determine whether GMA would be a suitable methodology to apply. Once agreed, the workshop takes place over three or more days during which the invited domain experts contribute by providing insight into the complex problem identifying the sum of relationships contained in multi-dimensional environment of the complex problem. If the problem space is small one facilitator can do both the facilitation and modeling. It is, however, difficult to capture all the information and facilitate at the same time, and therefore it is recommended that at least two facilitators form the MGMA team.

One of the principles of GMA is to identify the relationships and given uncertainties inherent in such multi-dimensional problem spaces and present this in a structured, reduced format, called a morphological field. Subject experts identify dimensions and relationships in the complex problem space with every dimension consisting of one or more variables. The same variable can exist in more than one dimension. Due to complexity when there are more than three dimensions and overlapping variables, the variables of all the dimensions are displayed in columns besides each other in multi-dimensional configuration space called a morphological field. The basis of GMA is to make sense of these non-quantifiable variables within a given messy problem by providing a formal structure to the multi-dimensional problem complex (Ritchey, 1998, 2006, 2011).

GMA is inherently qualitative, and do not interchange well with traditional quantitative methods, such as causal modelling and simulation (Ritchey, 1998). Although consensus should be reached by the group of subject matter experts participating in the GMA's cross-consistency assessment, it does allow for group discussion and move away from the accepted group think that often occurs during group discussions. Opinions are recorded with specially developed GMA software or with another suitable application. The solution space is computed by calculating combinations of value pairs and can be done by a customised programming application or a spreadsheet application. One of the advantages of the GMA methodology is that it caters for possibility rather than probability. In addition, the resulting morphological field presents a manner of traceability that enables reproducibility, which is necessary for scientific study.

A morphological analysis is carried out in two phases. The Analysis phase breaks the problem down into components that define the problem space. During the analysis phase, the most important or variables of the problem are identified and defined. By grouping similar elements identified by the focus group together with a heading, the facilitator guides the group into stratifying the problem-complex into variables (headings). Each variable is then given a range of values or conditions. The morphological field is populated by placing each variable in a column with its values or conditions in the rows below. One state (or solution) of the problem is found by selecting one value from each column and computing the combination of consistent pairs of values. An internal consistency between two values of two different variables represents the co-existence between the two variables, rather than the cause and effect between the two variables (Ritchey, 1998). A morphological field represents the total solution space of the problem and thus can have many possible solutions or scenarios.

The Synthesis phase finds links between variables and synthesizes an outcome space to support the development of a functioning model. In a synthesis cycle, the participants reduce the number of possible solutions by doing a Cross-Consistency Assessment (CCA): the relationships of every pair of values in the morphological field are checked for consistency. These exclusive relationships can be classified as logic contradictions, logical constraints or empirical constraints. Logical constraints represent only true or false possibilities, therefore consistent or not consistent. Empirical constraints use statistic probabilities to determine if it is likely or unlikely to be consistent. Two factors can be logically consistent but if the probability of them existing together is unlikely, then the two factors are empirically exclusive or inconsistent. Normative constraints are relationships that society will consider the norm but must be used with caution {Miller, 2012 #446}. The set of possible solutions is reduced to contain only internally consistent configurations.

The GMA methodology comprises a number of iterative steps, in which a subject specialist or focus group iterates through a number of analysis and synthesis cycles. The five iterative steps are described below.

Step 1: Identifies the variables which best define the essential nature of the problem complex or scenario. It is a good practice to work with no more than six or seven variables at a time.

Step 2: Defines a range of relevant, discrete *values or conditions for each variable.* The variable and variable-condition matrix is the morphological field: a multi-dimensional coordinate system that implicitly contains an outcome space for the problem complex thus defined.

Step 3: Assesses the internal consistency of all pairs of variable conditions and weeds out all inconsistent or contradictory pairs. It is usually at this point that participants begin to understand what the variables and variable conditions actually represent. Often, poorly defined variables and variable conditions are adjusted or removed completely. Steps 1 and 2 can now be streamlined and all participants are comfortable with the cross-consistency.

Step 4: Synthesizes an internally consistent outcome space. The software application does this by running through all of the possible solutions in the morphological field and "reducing" the field by restricting all outcomes containing internal contradictions. The surviving configurations represent the total output or solution space (excluding unknown values) of the given problem complex

Step 5: Involves the iteration of the process, if necessary. The solution space is scrutinised and Steps 1, 2 and 3 are carried out again, if necessary, to adjust variables, alternatives and consistency measures, and Steps 4 and 5 should be run again.

The iterative nature of both the GMA and MGMA methodology is witnessed better by developing scenarios.

The descriptions and examples of the dimensions and values, as agreed upon by the workshop participants, are recorded in the GMA software for easy future reference. This ensures that all future viewers of the particular model are guided in their interpretation of the dimensions and variables based on the initial understanding thereof by the original group of participants.

In the authors' experience, some of the requirements of the classical GMA method are problematic. The modified GMA process addresses the following challenges identified during the classical GMA process:

- Subject matter experts, who are often senior professionals, do not want to commit to attend a GMA session for periods of three days or longer. If they do commit to attend these longer workshops, they are often absent for periods during the session. The absences create disruptions because the facilitator has to update the absent participant upon his/her return to the session.

- The GMA facilitators often have knowledge regarding the problem area that is being addressed. The classical GMA process requires a facilitator not to influence the process nor add to add any information or subject related contribution.

The MGMA follows the same steps as the GMA but in the MGMA process, facilitators are allowed to contribute their knowledge during the preparation phase: they assist with the pre-selection of one or two variables and associated values. However, during the sessions facilitators adhere to the rule of not influencing the proceedings. It is important to note that the predetermined variables should not be forced on the participants. If they do not agree with these variables or their values they are allowed to change them. Pre-determined information should not restrict the exploration of the question, but rather serve as a catalyst to promote a faster formulation of the morphological field. Using the technique of cross-consistency checking in the MGMA, the system identifies possible solutions and eliminates inconsistent combinations in a grid-box structure. This cross-consistency matrix also results in the exclusion of inconsistent relationships in the ontology model as the inconsistent relationships are not included in the solutions of the GMA.

The models produced as a result of the MGMA workshop are interactive models; their value lies in the what-if scenario play. The case study was done using the Computer Aided Resource for Morphological Analysis (CARMA) software application that has a feature to perform what-if analysis. Some examples of the what-if scenario play feature are shown, but due to space restrictions, a comprehensive overview cannot be provided. It is important to note that the MGMA is a structuring process for a complex problem and does not provide a solution to a problem.

Ontology Based Representation

A fundamental aspect of research is to efficiently obtain, represent and manage new knowledge. Knowledge representation systems are aimed at designing systems that machines can interpret and understand. In addition to representing and managing knowledge, ontologies can be used to infer new knowledge and detect inconsistencies in the knowledge by means of powerful reasoning capabilities. A such, it is important to note that a research team using MODE requires the inclusion of an ontology engineer.

Ontologies provide a common vocabulary for domain knowledge and provide a shared source of information. The most cited definition of an ontology was proposed by Gruber (Gruber, 1993) who defined an ontology, in the context of computer science, as: "a formal specification of a conceptualisation". Since then, this definition has been refined by Guarino (Guarino, 1998) to elucidate what is meant by a formal specification of a conceptualization: "… a logical theory accounting for the intended meaning of a formal vocabulary, i.e. its ontological commitment to a particular conceptualization of the world. The intended models of a logical language using such a vocabulary are constrained by its ontological commitment. An ontology indirectly reflects this commitment (and the underlying conceptualization) by approximating these intended models". Conceptually, an ontology consists of a set of entities and relations that describes a particular domain.

Ontologies are used for Semantic Web applications because they formally and logically represent a domain of interest such that machines are able to classify and understand the domain. The Semantic Web is an extension of the World Wide Web; the vision of the Semantic Web is to allow for machines to process web content thereby improving the general web experience for users.

Ontology engineering refers to the tasks, methods, and tools that are required to construct ontologies. There are a number of existing methodologies that provide guidelines and assist with the ontology engineering process, from knowledge acquisition to ontology evaluation. Widely-used methodologies to facilitate ontology engineering include MOKI (Ghidini, Kump, Lindstaedt, Mahbub, Pammer, Rospocher, & Serafini, 2009), the NeON Suarez-Figueroa, de Cea Buil, Dellschaft, Fernandez-Lopez, Garcia, Gomez-Perez, Herrero, Montiel-Ponsoda. Sabou, 2008), and OntoSpec (Kassel, 2005). Typically, most ontology development methodologies differ in terms of tools, languages, activities, and tasks. Simperl et al. provide a general summary of the main activities used in most methodologies (Simperl, Mochol, & Bürger, 2010). This is displayed in Figure 2.

An ontology language is required to describe the ontology. The Web Ontology Language (OWL) (Horrocks, Patel-Schneider, & Van Harmelen, 2003) is a knowledge representation language used for creating ontologies in a richer, semantic format such that they facilitate machine processing. OWL has been endorsed as a W3C (World Wide Web Consortium) recommendation, and is based on Description logics (DL) languages. Resource Description Framework (RDF) is used to describe an ontology by using triples. In a RDF ontology, each triple consists of a subject, predicate and object. OWL is on a higher level

Figure 2. The main activities in ontology engineering methodologies (Adapted from: Simperl, et al., 2010)

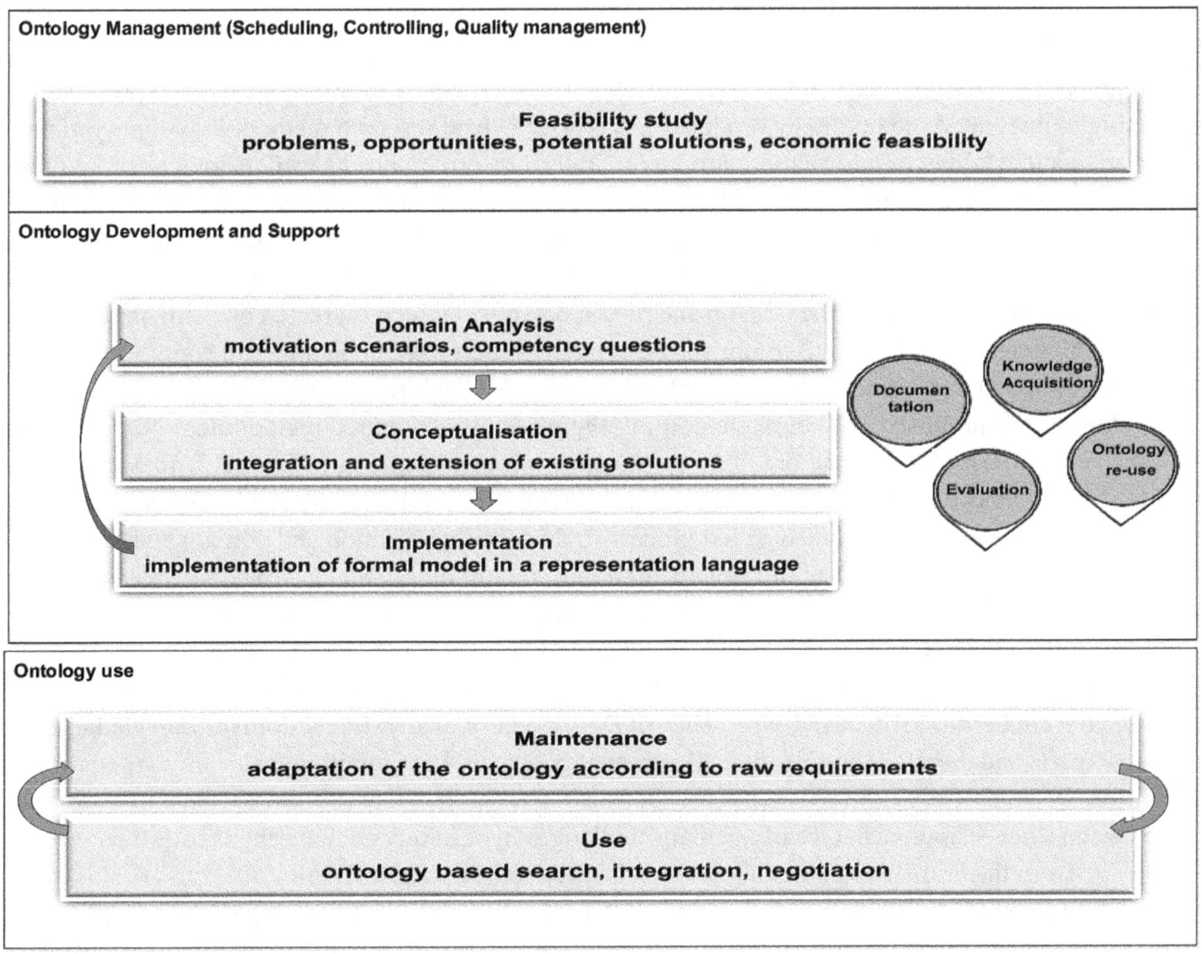

than syntactic RDF, which allows a developer to add semantics or meaning to such triples. In OWL, an ontology developer is also able to specify cardinality values of relations and logical operations between entities. Furthermore, with OWL ontologies it is possible for machines to draw new conclusions or infer new knowledge from the existing domain knowledge by using built-in automated reasoners. In terms of ontology engineering tools, Protégé (Stanford University, 2013) is the most widely-used tool for creating and editing domain ontologies in OWL. Protégé provides a graphical interface of the ontology to developers, automated reasoners that validate the consistency of the ontology and infer new knowledge, and other useful plugins that aid ontology development.

The use of ontologies has increased considerably over the years. Ontologies have been used as a formal mechanism as part of scientific discourse and advancing research. Several researchers have considered developing ontologies for usage in a scientific discipline. For instance, in the microbiology domain, the Gene ontology is used to unify the descriptions of gene terminologies among various databases (Ashburner, Ball, Blake, Botstein, Butler, Cherry, Davis, Dolinski, Dwight, Eppig, Harris, Hill, IsselTarver, Kasarskis, Lewis, Matese, Richardson, Ringwald, Rubin, Sherlock, 2000). To assist with national security research in South Africa, an ontology-based methodology is proposed (Jansen van Vuuren, Leenen, & Zaaiman, 2014).

MORPHOLOGICAL ONTOLOGY DESIGN ENGINEERING

The combination of the research methodologies of design science, GMA and ontology based representation into a single, integrated methodology produces a powerful hybrid research methodology that combines quantitative and qualitative information, long term and short term goals, as well as logical and empirical evidence. The MODE methodology draws on the positive aspects of all the research paradigms to enable improved models. The methodology consists of an overarching iterative process which is derived from the design science principles, with each cycle consisting of two stages. Stage 1 is the exploratory research (which is derived from the GMA principles and a model which is created by using the ontology engineering methodology). Stage 2 comprises descriptive research which includes the demonstration of the MGMA and ontology models and the evaluation of the MGMA results and the ontology.

In addition, the proposed MODE methodology (Figure 3) is a structured methodology that complies with the required activities of the ontology engineering process outlined in Figure 2. In Stage 1, the MODE methodology uses a literature survey that provides for the requirements of a feasibility study. The MGMA process is a structured and detailed process to do conceptualisation and domain analysis. Stage 1 is completed by the development or implementation of the ontology. Stage 2 makes provision for the evaluation, use and updates that are similar to the maintenance and evaluation processes of an ontology.

The benefit of using GMA, and therefore also the MGMA, is that it assists the researcher in discovering new, non-obvious relationships or configurations, and encourages the identification and investigation of boundary conditions (Ritchey, 1998). The MGMA process also extracts domain knowledge from domain experts and therefore ensures that all parameters, values and conditions relevant to the problem domain are systematically explored and clearly defined for the purpose of the development of an artefact in the design science approach. One of the main challenges in ontology engineering is to extract domain knowledge from the domain experts: this process is often tedious and difficult.

On completion of the MGMA in Stage 1, the MGMA results are used to set up the ontology model. MGMA models are made up of variables and a cross-consistency matrix depicting the relationship of

Figure 3. Morphological ontology design engineering methodology

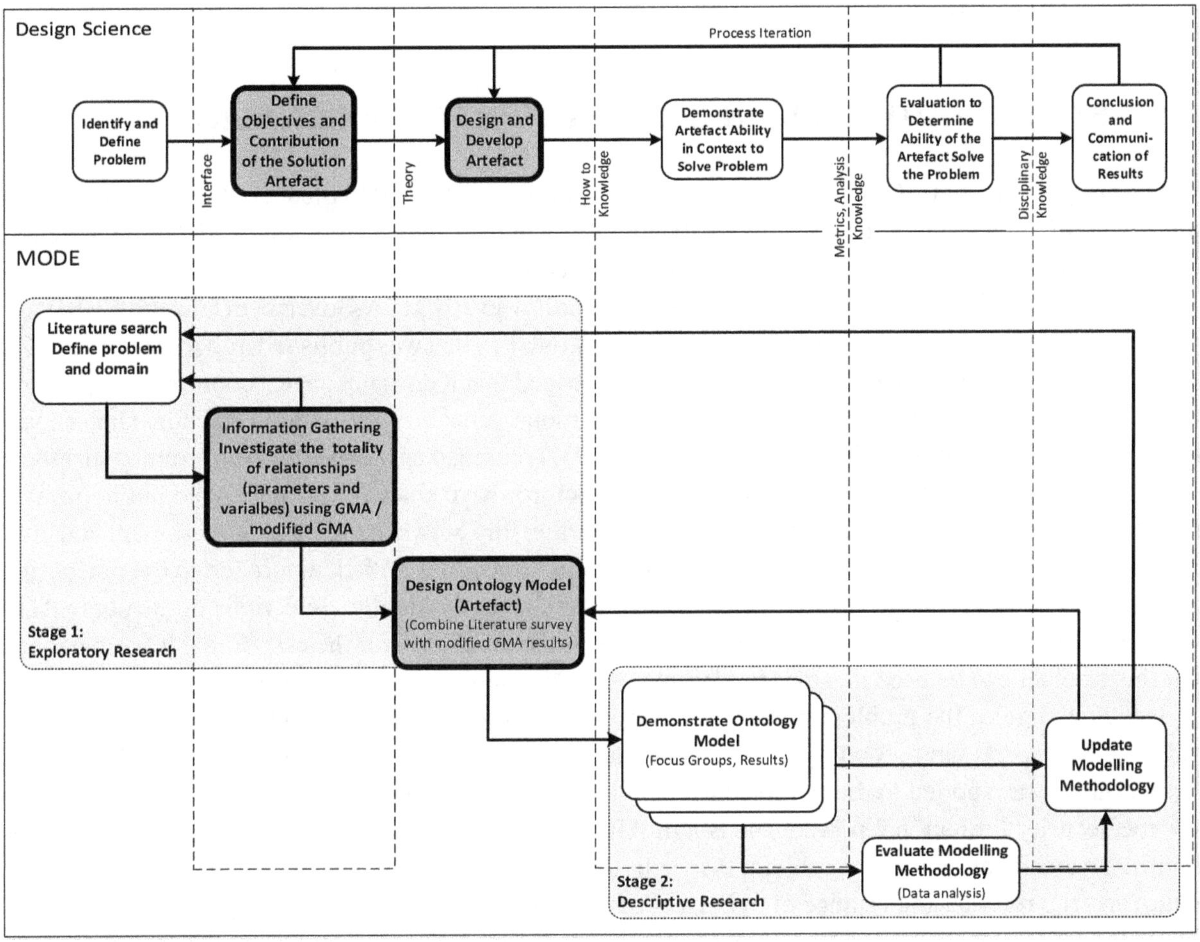

each variable to another variable; consistent, unknown or inconsistent. These variables will eventually become the entities in the ontology model with the cross-consistency matrix information used as the basis to model relations between entities in an ontology model. The strength of Stage 2 is the demonstration of the models to experts for feedback and the adjustment of the models accordingly. After completion of Stage 2, any of the loops can be followed to improve the artefact.

Although the new methodology is time consuming due to the number of iterations required (refer to the overarching process iteration), the MGMA model and artefact provide a valuable contribution to the information domain. The MGMA results relate directly to the information that is required as input to the ontology (artefact, as defined by the design science methodology).

This use of this new methodology will be illustrated in the next section by means of a case study on the South African cybersecurity environment. Although the case study is the only project in which the full MODE was applied, the authors have used parts of MODE to complete several other projects:

- MGMA has been applied to create a model for the development of a cybersecurity culture as well as models for the development of a Cybersecurity Incident Response Team (CSIRT).

- Ontologies have been developed for Network Attack Prediction and Command and Control in Network Warfare.

CASE STUDY: SOUTH AFRICAN CYBERSECURITY ENVIRONMENT

A case study of the South African cybersecurity environment is selected to illustrate the implementation of the MODE methodology. The cybersecurity environment is not well understood and the implementation of a national cybersecurity policy is an ill-structured problem.

South Africa published its first draft cybersecurity policy in 2010 (SA Goverment Gazette, 2010) followed by the National Cybersecurity Policy Framework (NCPF) that was published in 2012 (Dube, 2012). With the approval of the NCPF, South Africa acknowledged that it does not have a coordinated approach in dealing with cybersecurity and that access to a common source of information is lacking (Jansen van Vuuren, Leenen, & Zaaiman, 2014). As such, the NCPF is regarded as the overarching strategy that must guide cybersecurity within South Africa. Various structures have since been established but to have an efficient cybersecurity strategy, there is a need for a partnership between business, government and civil society. Despite the common understanding of the importance of a holistic approach to cybersecurity, at the time of writing the NCPF was still not made available to the public. Selections of the content are made available through media statements and were used as foundation for the case study. It is envisioned that this artefact can be used in support of implementing a South African cybersecurity policy.

The challenge of the problem modelling task was to extract accurate knowledge from the policies and domain experts, and secondly to represent such knowledge in a meaningful format. The MODE methodology was applied to the cybersecurity environment to develop a model for the implementation of cybersecurity for national power. The South African cybersecurity environment can be regarded as a multi-dimensional, complex problem. As such, this case study was selected due to its ill-structured nature and the relevant importance of cybersecurity. The case study was done in 3 phases. The subsection below discusses Phase 1, which includes Stage 1 and Stage 2 of the MODE methodology (refer to *Figure 3*). A discussion of Phase 2 then follows.

Phase 1

Stage 1: Exploratory Research

This stage is used to define the problem and capture the knowledge needed to build the model or artefact as shown in Figure 4.

As stated in the previous section, Stage 1 of the MODE methodology is in line with the required activities of the ontology engineering process outlined in Figure 2. The case study was also used to test the correlation. The following summary depicts the correlation between the MODE methodology and the well-known ontology engineering process using the case study example.

- The domain in question, the cybersecurity environment, was analysed and found to be an ill-structured problem with key issues such as questions about the role players and their responsibilities, objectives and functions of the cybersecurity environment. A combination of both quantitative and qualitative methods was deemed to be necessary for model construction, such as depicted

Figure 4. Stage 1: Exploratory research

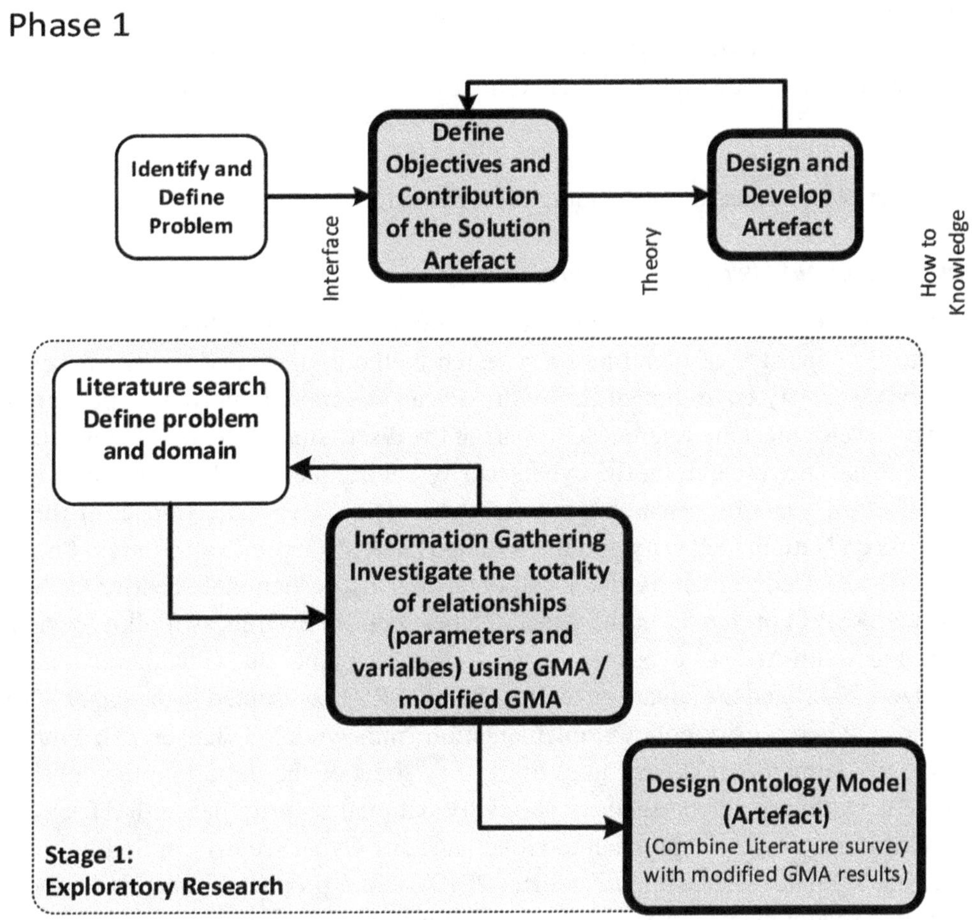

by the proposed MODE research methodology. The GMA method was considered for acquiring domain information because it offers a method for the extraction of meaningful information from subject matter experts. Ontology based representation was considered for accurate domain knowledge representation because ontologies enable a rich semantic representation of such information.

- The cybersecurity environment is complex in nature with many variables. To begin the analysis of the domain, a focus question for GMA workshop delegates was formulated: *"What are the most important functions and stakeholder groupings that define the South African Cybersecurity Environment for national power?"*
- To answer the multi-dimensional focus question, variables were identified within the GMA session. These variables were defined, populated with conditions, and redefined by the participants of the GMA session so as to conceptualise a solution model. After comparing different values of the model for consistency, an initial GMA model was obtained.
- An ontology was then constructed by using the output of the GMA model; the variables correspond to the ontology classes while consistent value pairs correspond more or less to the object properties of the ontology. The ontology and GMA model were evaluated by the GMA participants to assess its representation of the cybersecurity environment domain. The feedback provided

by the participants resulted in an improved ontology. As the cybersecurity environment domain changes with new trends and threats; continued maintenance will be required.

- The ontology is to be used by cybersecurity domain experts from the government, business, or civilian sector, as a unified and coordinated approach in dealing with the national cybersecurity of the country.

A detailed description of the implementation of Stage 1 follows.

Definition of the Domain Using a Literature Survey

A literature survey is the first step in the exploratory research that is part of Stage 1 of the MODE methodology. The survey consisted of literature on research methodologies and papers discussing several aspects of the cybersecurity environment in South Africa. The most important papers considered for the methodology development are referenced as part of the discussion of the new MODE technology in the previous sections. The research on the cybersecurity environment in South Africa was performed over a period of a few years by research team members. The paper "The Impact of the Increase in Broadband Access on National Security and the Average citizen" (Jansen van Vuuren, Phahlamohlaka, & Brazzoli, 2010) was used to identify the social determinants for national security for South Africa. The toolkit (CYBERSAT) presented in the paper "Cyber security awareness toolkit for national security: an approach to South Africa's cyber security policy implementation" (Phahlamohlaka, Jansen van Vuuren, & Coetzee, 2011) and the advanced toolkit (XCYBERST) presented in the paper "Development of a South African cybersecurity policy implementation framework" (J. Jansen van Vuuren, Leenen, Phahlamohlaka, & Zaaiman, 2013) were used to identify some of the important factors and role players for the implementation of a cybersecurity policy for national security in South Africa. Governance structures were discussed in "An Approach to Governance of CyberSecurity in South Africa" (Jansen van Vuuren, Leenen, Phahlamohlaka, & Zaaiman, 2012), and a proposed implementation framework was discussed in "Development of a South African cybersecurity policy implementation framework" (Jansen van Vuuren, et al., 2013). These papers form the basis of the literature review on the cybersecurity environment in South Africa. Other papers that were used as background include "Implementation of a Cyber Security Policy in South Africa: Reflection on Progress and the Way Forward" (Grobler, van Vuuren, & Leenen, 2012), "Using an ontology as a model for the implementation of the National Cybersecurity Policy Framework for South Africa" (Jansen van Vuuren, et al., 2014) and "Governance of cybersecurity in South Africa" (Jansen van Vuuren, Phahlamohlaka, & Leenen, 2012).

Information Gathering Using MGMA

Taking into consideration the problems experienced with the traditional GMA process, this case study employed the MGMA methodology to extract domain knowledge from subject matter experts. This was done during a two-day workshop with the aim of extracting domain knowledge from experts for structuring the relations between stakeholders and functions in the South African cybersecurity environment. The MGMA process was initiated by presenting the subject matter experts with some predetermined variables and values to facilitate a shortened variable brainstorming phase. The facilitators are also researchers in the cybersecurity field and therefore could contribute some knowledge.

During a pre-workshop conducted by the facilitation team, the focus question for the workshop was identified as "What are the most important functions and stakeholder groupings that define the South

African Cybersecurity Environment for national power?" The team selected the following variables prior to the MGMA session as a proposed kick-off point for discussion:

- National power elements.
- Policy elements.
- High level stakeholders.
- Lower level stakeholders.
- Functions.
- Implementation process.

The variables and some corresponding proposed values were sourced from the Extended Cyber Security Toolkit for National Security (XCyberST); the toolbox was also provided to the MGMA workshop participants as additional background information. Although the participants discussed and changed these pre-selected variables, their inclusion helped considerably to shorten the initial phase during which the morphological field was created. This resulted in the participants focusing on the selection of relevant variables and values from early on the first workshop day, whereas past experiences has shown that the participants tended to start with a rather large number of possible variables before narrowing them down into a workable number.

The workshop progression is detailed below.

- **Step 1:** The expert group identified several variables. These variables were eventually reduced to six (Figure 5). It should be noted that another variable, *High level Role Players,* was initially identified by the group of MGMA participants, but that the group agreed that this variable should be excluded from the morphological field because it can be regarded as a higher level categorisation of the other two role player variables.
- For each of the variables identified in Step 1, **Step 2** defines a range of relevant, discrete *values or conditions or parameters.* Figure 6 shows the values or conditions for each of the variables in Figure 5.
- **Step 3:** The cross-consistencies for all the variables were agreed upon and entered into the cross-consistency matrix. Due to the large size of the matrix only a partial consistency assessment is shown in Figure 7.
- **Steps 4 and 5:** The fully consistent solution space was calculated by the CARMA software application and is shown in Figure 8 in the shaded blocks. Many of the values were not included in the solution space and the group analysed the models to detect why this was the case. It can be seen that that the solutions space is relatively small as a large number of inconsistencies in value pairs resulted in the excluding of many of the values. The solution space shows definite groupings of values that are inherently linked (blue shaded blocks) and definite groupings of values that are inherently not linked (white blocks).

Figure 5. MGMA cybersecurity environment variables

Low level role players - Lead departments	Low level role players - Support departments	Structures	Governance	Functions	Capacity / Capability

Figure 6. Variables and values for the cybersecurity environment

Low level role players - Lead departments	Low level role players - Support departments	Structures	Governance	Functions	Capacity / Capability
	Department of International Relations and Cooperation (DIRCO)	Cybersecurity Response Committee (CRC)	International and regional conventions	Enforcement	Infrastructure
Department of Justice and Constitutional Development (DOJ & CD)	Department of Trade and Industry (DTI)	Cybersecurity Centre	Policy	Set general cyber measures	Research and development
Department of Defence (DOD)	Department of Education (DOE)	Government Security Incident Response Teams (CSIRTs)	Standards	Human capacity building (excess / national)	Qualified manpower
South African Police Service (SAPS)	State Information Technology Agency (SITA)	Cyber Command	Legislative measures	Threat identification	Cryptology
Department of Safety and Security (DOSS, SSA, National Communication)	Department of Provincial and Local Government (DPLG)	National Cybercrime Centre (SAPS)	Best practice	Collaboration (Cooperation)	Netizens
Department of Telecommunications and Postal Services (DTPS), including DOC	Department of Public Service and Administration (DPSA)	24/7 cyber contact point (SAPS)		Coordination	Offensive and defensive capacity
Department of Science and Technology (DST)	National Treasury, including Financial Intelligence Centre (FIC) and SARS	International structures		Combat cybercrime	
Financial sector				Cultivate cybersecurity culture	Cyber qualification framework (including certifications)
Basic Education service providers		Cybersecurity Hub		Cyberwarfare	
Higher Education service providers		Private Sector Security Incident Response Teams (CSIRTs)		Protect Information Communication Systems (ICS)	
Training providers		South Africa Banking Risk Information Centre (SABRIC)		Protect National Critical Information Infrastructures Systems (NCIIS)	
Software and end user device providers				Conduct research and development	
Infrastructure providers					
Access providers					
Private security cluster					

The resulting model was also used to explore different scenarios. *If-then* scenario creation shows which values are relevant in relation to the chosen drivers and indicates where a variable may need unpacking or require a different categorisation of its values.

MGMA model scenario play was done with the selection of a single driver or more than one driver selected. For example, in Figure 9 the *Department of Justice and Constitutional Development (DOJ&CD)* is selected as single driver (red shaded block). The blue shaded blocks show all the values that are consistent with this driver. Therefore, the *State Information Technology Agency (SITA)*, the *Cybersecurity Response Committee (CRC)*, *Government Security Incident Response Teams (CSIRTs)*, *Standards*, *Conduct research and development* and *Infrastructure* values can all be regarded as being in some relationship with the DOJ&CD.

Similarly, scenario play was also done with the selection of more than one driver. For example, in Figure 10 the *Research and development* and *Protect National Critical Information Infrastructures Systems (NCIIS)* are selected as the two drivers (red shaded blocks). The blue shaded blocks show all the values that are consistent with both these drivers. Therefore, the *Department of Telecommunications and*

Figure 7. Cross-consistency matrix for the cybersecurity environment

		ional Development (DOJ & CD)	Department of Defence (DOD)	h African Police Service (SAPS)	SSA, National Communication)	Services (DTPS), including DOC	Science and Technology (DST)	Financial sector	asic Education service providers	her Education service providers	Training providers	e and end user device providers	Infrastructure providers	Access providers	Private security cluster	ations and Cooperation (DIRCO)	ment of Trade and Industry (DTI)	Department of Education (DOE)	ation Technology Agency (SITA)	and Local Government (DPLG)	vice and Administration (DPSA)	lligence Centre (FIC) and SARS	ity Response Committee (CRC)	Cybersecurity Centre	ent Response Teams (CSIRTs)	Cyber Command	onal Cybercrime Centre (SAPS)	24/7 cyber contact point (SAPS)	International structures	Cybersecurity Hub	ent Response Teams (CSIRTs)	sk Information Centre (SABRIC)	ational and regional conventions	Policy	Standards	Legislative measures	Best practice	Enforcement
		Low level role players - Lead departments														Low level role player							Structures										Governance					Fun
Low level role players - Support departments	Department of International Relations and Cooperation (DIRCO)	X	X	X	X	X	X	X	X	X	X	-	-	X	X																							
	Department of Trade and Industry (DTI)	X	X	X	X	X	X	-	X	X	K	-	-	-	K																							
	Department of Education (DOE)	X	X	X	X	X	K	X	-	-	-	X	X	X	X																							
	State Information Technology Agency (SITA)	-	-	-	-	-	-	X	-	-	X	-	-	-	K																							
	Department of Provincial and Local Government (DPLG)	X	X	X	X	X	X	K	X	K	K	K	K	K	K																							
	Department of Public Service and Administration (DPSA)	X	X	X	X	X	X	X	X	X	X	X	X	X	X																							
	National Treasury, including Financial Intelligence Centre (FIC) and SARS	-	-	-	-	-	-	X	K	K	X	X	X	X	X																							
Structures	Cybersecurity Response Committee (CRC)	-	-	-	-	-	-	X	K	-	X	X	X	X	X	X	X	-	-	X	K	X																
	Cybersecurity Centre	-	-	-	-	-	-	X	-	-	X	X	X	X	X	X	X	K	-	-	X	X																
	Government Security Incident Response Teams (CSIRTs)	-	-	-	-	-	-	X	-	-	X	X	X	X	X	X	X	-	-	-	X	K																
	Cyber Command	X	-	X	X	X	X	X	X	X	X	X	X	X	X	X	X	X	-	K	X	X																
	National Cybercrime Centre (SAPS)	K	K	K	K	K	K	-	K	K	K	K	K	-	-	X	X	X	-	X	X	K																
	24/7 cyber contact point (SAPS)	K	K	K	K	K	K	-	K	K	K	K	K	-	-	X	X	X	X	X	X	X																
	International structures	X	X	X	-	-	X	X	X	X	X	K	X	X	K	-	-	X	X	X	X	X																
	Cybersecurity Hub	X	X	X	X	X	X	-	X	X	K	-	-	-	-	X	X	X	X	X	X	K																
	Private Sector Security Incident Response Teams (CSIRTs)	X	X	X	X	X	X	-	X	X	K	X	-	-	-	X	X	X	X	X	X	X																
	South Africa Banking Risk Information Centre (SABRIC)	X	X	X	X	X	X	-	X	X	X	X	X	K	X	X	X	X	X	X	X	X																
Governance	International and regional conventions	-	-	-	-	-	K	X	X	X	X	X	X	X	X	-	-	X	X	X	X	X	-	K	X	K	X	X	-	-	X	X						
	Policy	X	X	-	-	-	-	X	X	X	X	X	X	X	X	X	X	-	-	-	X	X	-	-	-	-	X	X	X	-	X	X						
	Standards	-	X	X	-	-	-	X	X	X	X	X	X	X	X	X	X	-	-	-	X	X	-	-	-	-	X	X	-	-	-	-						
	Legislative measures	-	X	X	X	-	X	X	X	X	X	X	X	X	X	X	X	X	X	X	X	X	-	X	X	X	-	-	X	X	X	X						
	Best practice	X	X	X	X	-	-	X	X	X	X	X	-	X	X	X	X	X	-	-	X	X	-	-	-	-	X	X	-	-	-	-						
Functions	Enforcement	-	K	-	K	-	X	K	X	X	X	X	K	K	K	X	X	X	X	X	X	X	K	-	-	-	-	X	X	K	-	-	K	K	X	-	X	
	Set general cyber measures	-	-	K	-	K	X	K	X	X	X	-	K	K	X	X	X	X	K	X	X	X	-	K	X	-	K	X	K	-	X	-	-	-	-	K	-	
	Human capacity building (excess / national)	K	K	K	X	-	-	K	-	-	-	X	X	X	K	X	-	-	X	X	X	-	K	X	X	K	K	X	K	-	K	-	X	-	X	X	X	
	Threat identification	X	-	-	-	-	-	-	X	K	K	K	K	X	-	X	X	X	-	X	X	X	K	-	-	-	-	K	K	-	-	-	X	X	K	X	-	
	Collaboration (Cooperation)	X	-	-	-	-	-	K	X	K	K	K	K	K	-	-	-	X	K	K	X	X	-	-	K	-	-	-	-	-	-	-	-	K	-	X	X	
	Coordination	-	K	-	-	-	-	K	X	-	X	K	X	K	X	X	X	-	X	X	X	X	-	-	X	K	-	-	K	-	-	-	X	-	K	-	X	

Ready Comb 411600 41 230 NUM

Postal Services (DTPS), including DOC, SITA, CRC, Policy and *Best Practice* values can all be regarded to participate in some relationship with the two initially selected driver values.

Due to the systematic exploration of pairwise consistencies during a GMA process, the problem area is covered in depth. GMA has been referred to as *totality research* due to its ability to derive all the possible solutions to any given problem space. Accordingly, this type of knowledge extraction serves as excellent input to an ontology relationships between concepts are represented semantically.

Design of the Initial Ontology Model

The output of the MGMA was used to create an ontology that models the role players, functions and inter-relationships in the South African cybersecurity environment (shown in Figure 11). The authors could formally structure the captured knowledge in the cybersecurity ontology by presenting variables and values from the MGMA model as classes and object properties. The consistency between value pairs is represented as relations between classes, also known as object properties. As such, the central class is called *RolePlayers*. This class is subdivided into two disjoint subclasses representing high level and low level role players. Due to certain structures having leading roles over other structures or departments, the latter subclass was divided into two disjoint subclasses, *LeadDepartments* and *SupportDepartments*. The members of the high level role player class include the different sectors in the business, education,

Figure 8. Solution space calculated by CARMA

Low level role players - Lead departments	Low level role players - Support departments	Structures	Governance	Functions	Capacity / Capability
	Department of International Relations and Cooperation (DIRCO)	Cybersecurity Response Committee (CRC)	International and regional conventions	Enforcement	Infrastructure
Department of Justice and Constitutional Development (DOJ & CD)	Department of Trade and Industry (DTI)	Cybersecurity Centre	Policy	Set general cyber measures	Research and development
Department of Defence (DOD)	Department of Education (DOE)	Government Security Incident Response Teams (CSIRTs)	Standards	Human capacity building (excess / national)	Qualified manpower
South African Police Service (SAPS)	State Information Technology Agency (SITA)	Cyber Command	Legislative measures	Threat identification	Cryptology
Department of Safety and Security (DOSS, SSA, National Communication)	Department of Provincial and Local Government (DPLG)	National Cybercrime Centre (SAPS)	Best practice	Collaboration (Cooperation)	Netizens
Department of Telecommunications and Postal Services (DTPS), including DOC	Department of Public Service and Administration (DPSA)	24/7 cyber contact point (SAPS)		Coordination	Offensive and defensive capacity
Department of Science and Technology (DST)	National Treasury, including Financial Intelligence Centre (FIC) and SARS	International structures		Combat cybercrime	
Financial sector				Cultivate cybersecurity culture	Cyber qualification framework (including certifications)
Basic Education service providers		Cybersecurity Hub		Cyberwarfare	
Higher Education service providers		Private Sector Security Incident Response Teams (CSIRTs)		Protect Information Communication Systems (ICS)	
Training providers		South Africa Banking Risk Information Centre (SABRIC)		Protect National Critical Information Infrastructures Systems (NCIIS)	
Software and end user device providers				Conduct research and development	
Infrastructure providers					
Access providers					
Private security cluster					

government, non-government and private sectors. The high and low level role player classes are then bound by a lead/support relation. For example, in order for a department to lead a certain activity, it must have some supporting departments and vice versa. Figure 11 shows some of the relations of the high level classes.

In this model, individuals in the class represent specific entities, for example the State Security Agency (SSA) is an individual in the government subclass. It is important to note that these entities are represented as subclasses of the *RolePlayers* class in the ontology that resulted from the next iteration of the MODE process.

The *Function* class (refer to Figure 12) aids in developing policies and procedures. Each role player has some functions that they should provide. *Capability* refers to the role players' ability to perform certain tasks. This class can predetermine or limit some of the functions that a specific role player may have. *Structure* represents grouping of entities. These structures are dynamic and can be formed in an ad hoc manner. The *Governance* class refers to the rulebook of the Cybersecurity environment. Policy statements, security standards and legislation rules are subclasses of this class. Each governance function is governed or overseen by a certain role player. Certain functions, such as enforcement, can be

Figure 9. MGMA single driver scenario

Low level role players - Lead departments	Low level role players - Support departments	Structures	Governance	Functions	Capacity / Capability
	Department of International Relations and Cooperation (DIRCO)	Cybersecurity Response Committee (CRC)	International and regional conventions	Enforcement	Infrastructure
Department of Justice and Constitutional Development (DOJ & CD)	Department of Trade and Industry (DTI)	Cybersecurity Centre	Policy	Set general cyber measures	Research and development
Department of Defence (DOD)	Department of Education (DOE)	Government Security Incident Response Teams (CSIRTs)	Standards	Human capacity building (excess / national)	Qualified manpower
South African Police Service (SAPS)	State Information Technology Agency (SITA)	Cyber Command	Legislative measures	Threat identification	Cryptology
Department of Safety and Security (DOSS, SSA, National Communication)	Department of Provincial and Local Government (DPLG)	National Cybercrime Centre (SAPS)	Best practice	Collaboration (Cooperation)	Netizens
Department of Telecommunications and Postal Services (DTPS), including DOC	Department of Public Service and Administration (DPSA)	24/7 cyber contact point (SAPS)		Coordination	Offensive and defensive capacity
Department of Science and Technology (DST)	National Treasury, including Financial Intelligence Centre (FIC) and SARS	International structures		Combat cybercrime	
Financial sector				Cultivate cybersecurity culture	Cyber qualification framework (including certifications)
Basic Education service providers		Cybersecurity Hub		Cyberwarfare	
Higher Education service providers		Private Sector Security Incident Response Teams (CSIRTs)		Protect Information Communication Systems (ICS)	
Training providers		South Africa Banking Risk Information Centre (SABRIC)		Protect National Critical Information Infrastructures Systems (NCIIS)	
Software and end user device providers				Conduct research and development	
Infrastructure providers					
Access providers					
Private security cluster					

carried out by a governing role player to ensure certain standards are met. For example, the role player *Department of Education* has the capability of *Research and Development* and the function of *Human Capacity Building*. Equally, role player *Department of Justice* has some relationship with the structure *the Cyber Contact Point*, whilst *Governance of Standards* has a relationship with the role player *Department of Science and Technology*.

The input from the MGMA helped immensely with the structuring of the entities within the ontology. However, the consistency of value pairs was difficult to look up using the GMA software. By implementing the structure into an ontology, relations can be clearly defined and the reasoner can make inferences across the body of knowledge. Questions can also be posed to the ontology. In addition, this provides the support group with a shareable knowledge base that can be easily updated without the need of doing manually consistency checks between all the variables.

Stage 2

After completion of the first version of the ontology, some of the experts valuated the implementation models and changes were made on these models to have the first updated version of the ontology (figure

Figure 10. MGMA multiple driver scenario

Low level role players - Lead departments	Low level role players - Support departments	Structures	Governance	Functions	Capacity / Capability
	Department of International Relations and Cooperation (DIRCO)	Cybersecurity Response Committee (CRC)	International and regional conventions	Enforcement	Infrastructure
Department of Justice and Constitutional Development (DOJ & CD)	Department of Trade and Industry (DTI)	Cybersecurity Centre	Policy	Set general cyber measures	Research and development
Department of Defence (DOD)	Department of Education (DOE)	Government Security Incident Response Teams (CSIRTs)	Standards	Human capacity building (excess / national)	Qualified manpower
South African Police Service (SAPS)	State Information Technology Agency (SITA)	Cyber Command	Legislative measures	Threat identification	Cryptology
Department of Safety and Security (DOSS, SSA, National Communication)	Department of Provincial and Local Government (DPLG)	National Cybercrime Centre (SAPS)	Best practice	Collaboration (Cooperation)	Netizens
Department of Telecommunications and Postal Services (DTPS), including DOC	Department of Public Service and Administration (DPSA)	24/7 cyber contact point (SAPS)		Coordination	Offensive and defensive capacity
Department of Science and Technology (DST)	National Treasury, including Financial Intelligence Centre (FIC) and SARS	International structures		Combat cybercrime	
Financial sector				Cultivate cybersecurity culture	Cyber qualification framework (including certifications)
Basic Education service providers		Cybersecurity Hub		Cyberwarfare	
Higher Education service providers		Private Sector Security Incident Response Teams (CSIRTs)		Protect Information Communication Systems (ICS)	
Training providers		South Africa Banking Risk Information Centre (SABRIC)		Protect National Critical Information Infrastructures Systems (NCIIS)	
Software and end user device providers				Conduct research and development	
Infrastructure providers					
Access providers					
Private security cluster					

13). Questions were posed to the ontology to check relationships. The changes mainly consisted of correcting information that was incorrectly transferred from the MMGA model to the ontology.

Phase 2

In this phase, stages 1 and 2 will be combined due to the iterative process of feedback and adjustment due to a round table discussion with two members from the original MGMA workshop. During this phase (Figure 14), the experts used literature surveys and expert knowledge to adjust both the MGMA and ontology models.

In the expert feedback session of Phase 1, it was identified that the initial model did not satisfy the query requirements and did not include the full domain. To address the identified shortcomings, the modellers and experts made use of the literature survey results to check and adjust the ontology model. There were also some variables and values that were not modelled correctly due to ambiguous definitions and thus resulted in incorrect interpretations of aspects of the MGMA by the ontology engineers.

Figure 11. Depiction of concepts and relations in the initial ontology model

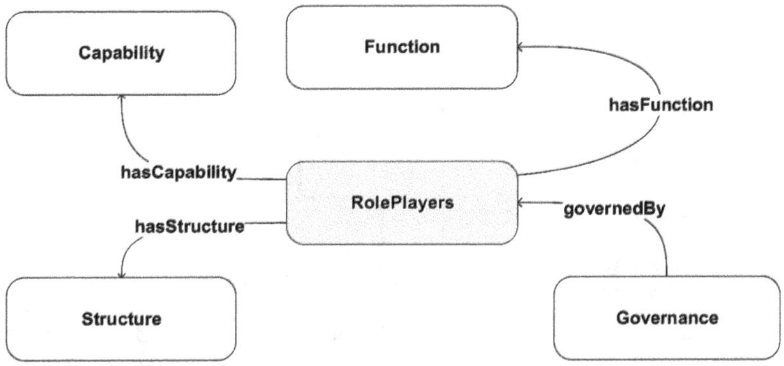

Figure 12. Role players and functions in the cybersecurity ontology

Figure 13. Stage 2: Descriptive research

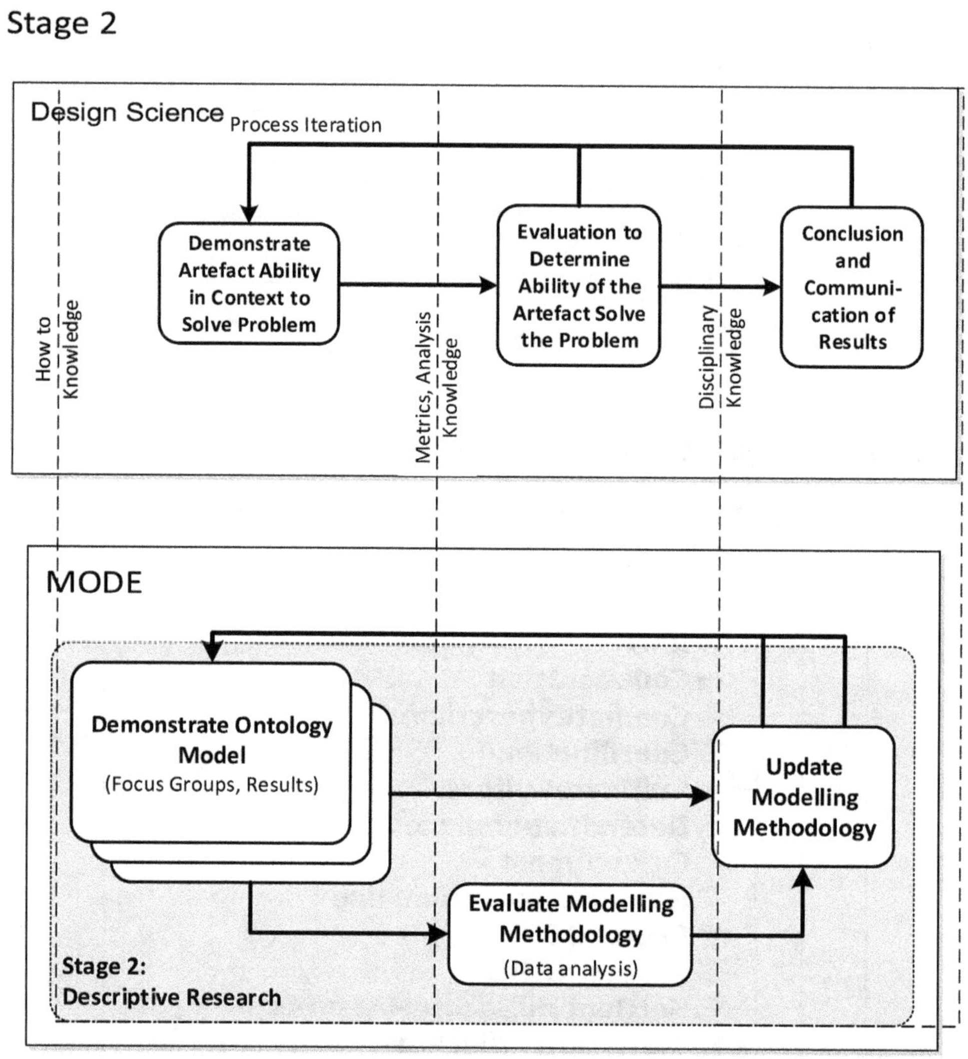

These corrections required an iterative process because each change in the ontology or MGMA model influenced the other model as well.

Based on the additional research done and the expert feedback received, the ontology model was adjusted to the new model depicted in Figure 15.

The modelling team realised that the core class is actually the *Functions class* and not the *RolePlayers* class. The definition of the Functions variables and some of its variables values were tightened to ensure correct interpretations. Additional values were also identified (Figure 16).

Figure 17 shows the Functions and Social Determinants (called National Cyber Power class in the ontology) classes and their subclasses in the ontology. The latter class was included due to the information extracted from the literature that was revisited during this iteration. Another class, *Strategic Objectives* was also added to the ontology.

Figure 14. Phase 2

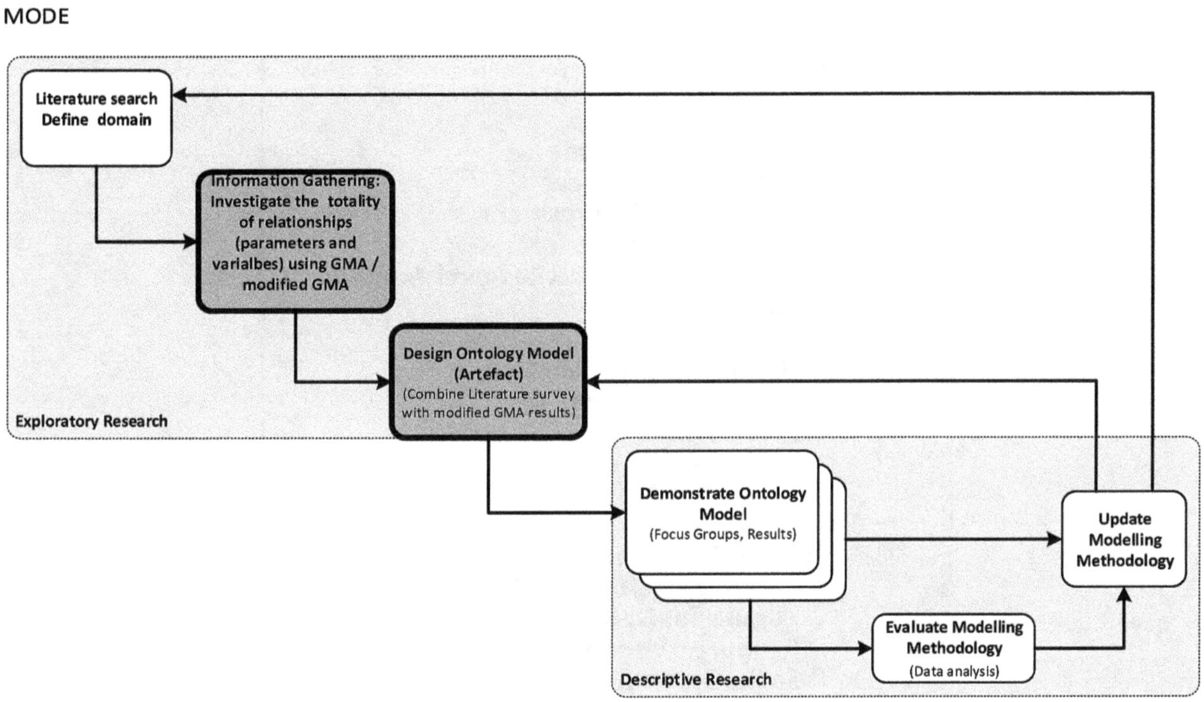

Figure 15. High level classes and relations in the ontology

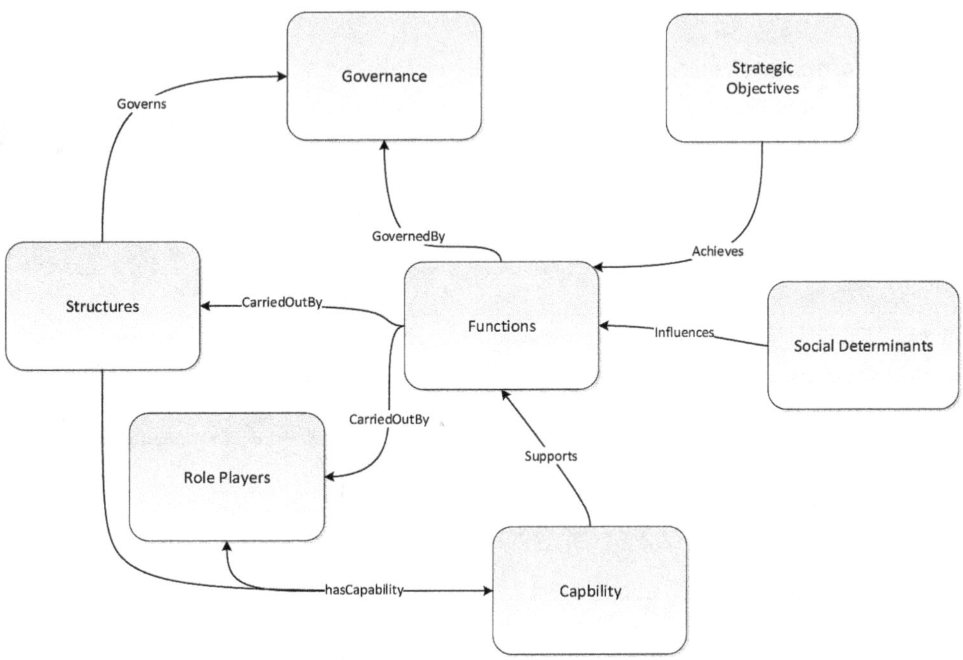

Figure 16. National cyber class in ontology

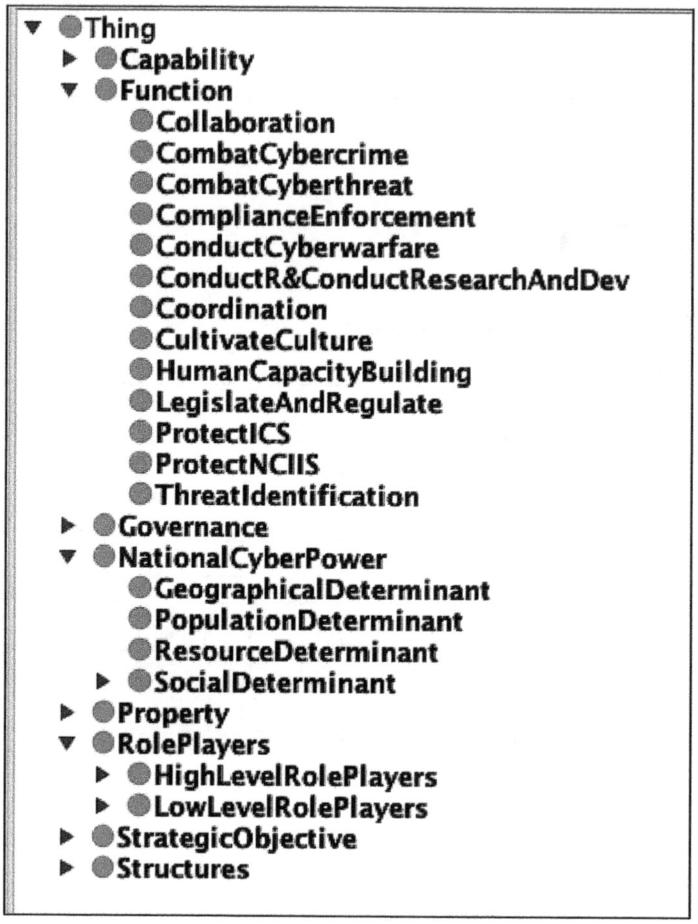

Figure 17. Social determinant relations with functions in Protégé

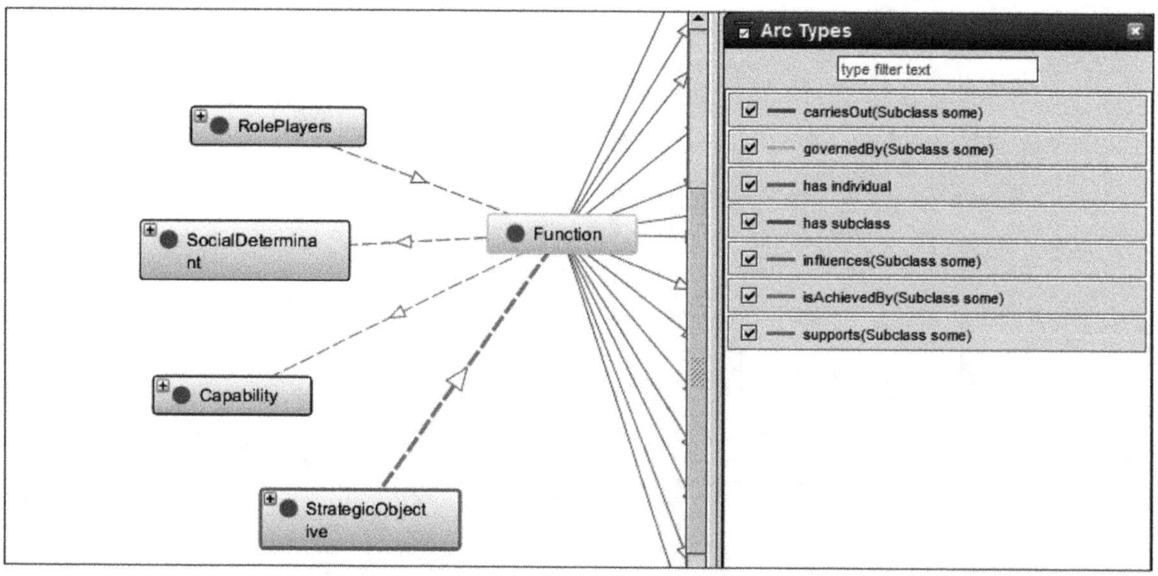

Figure 18. MGMA solution space Phase 2

Low level role players - Lead departments	Low level role players - Support departments	Structures	Governance	Functions	Capacity / Capability
	Department of International Relations and Cooperation (DIRCO)	Cybersecurity Response Committee (CRC)	International and regional conventions	Compliance enforcement	Infrastructure
Department of Justice and Constitutional Development (DOJ & CD)	Department of Trade and Industry (DTI)	Cybersecurity Centre	Policy	Legislate and regulate	Research and development
Department of Defence (DOD)	Department of Education (DOE)	Government Security Incident Response Teams (CSIRTs)	Standards	Human capacity building (excess / national)	Qualified manpower
South African Police Service (SAPS)	State Information Technology Agency (SITA)	Cyber Command	Legislative measures	Threat identification	Cryptology
Department of Safety and Security (DOSS, SSA, National Communication)	Department of Provincial and Local Government (DPLG)	National Cybercrime Centre (SAPS)	Best practice	Collaboration (Cooperation)	Netizens
Department of Telecommunications and Postal Services (DTPS), including DOC	Department of Public Service and Administration (DPSA)	24/7 cyber contact point (SAPS)		Coordination	Offensive and defensive capacity
Department of Science and Technology (DST)	National Treasury, including Financial Intelligence Centre (FIC) and SARS	International structures		Combat Cyber Threats	
Financial sector				Cultivate cybersecurity culture	Cyber qualification framework (including certifications)
Basic Education service providers		Cybersecurity Hub		Cyberwarfare	
Higher Education service providers		Private Sector Security Incident Response Teams (CSIRTs)		Protect Information Communication Systems (ICS)	
Training providers		South Africa Banking Risk Information Centre (SABRIC)		Protect National Critical Information Infrastructures Systems (NCIIS)	
Software and end user device providers				Conduct research and development	
Infrastructure providers				Combat Cybercrime	
Access providers					
Private security cluster					

Figure 18 shows the solution space of the Phase 2 model after the MGMA model was also updated. The blue shaded blocks show all the values that from part of the solution space, whilst the white coloured blocks show all the values that are not part of any solutions. Similarly to Phase 1, scenario play can be performed with single and multiple drivers.

Phase 3

During this phase another expert of the original MGMA panel was given an opportunity to evaluate the updated models. After a session during which the second and first experts jointly discussed and modified weaknesses in the updated models, the second expert also viewed and commented on the ontology (Figure 19).

This expert's main concern was a particular role player that was not part of the complete solution in the GMA model. The consistency table was explored and the two experts subsequently changed several consistency values. This resulted in the role player becoming part of the solution. A few other minor changes were made to other value pair combinations' consistency values. These changes were accommodated in the MGMA model Figure 20. The resulting MGMA model's solution space is now much larger than the initial model. Similarly to Phase 1 and Phase 2, scenario play can be performed with single and multiple drivers.

Figure 19. Evaluation process Phase 3

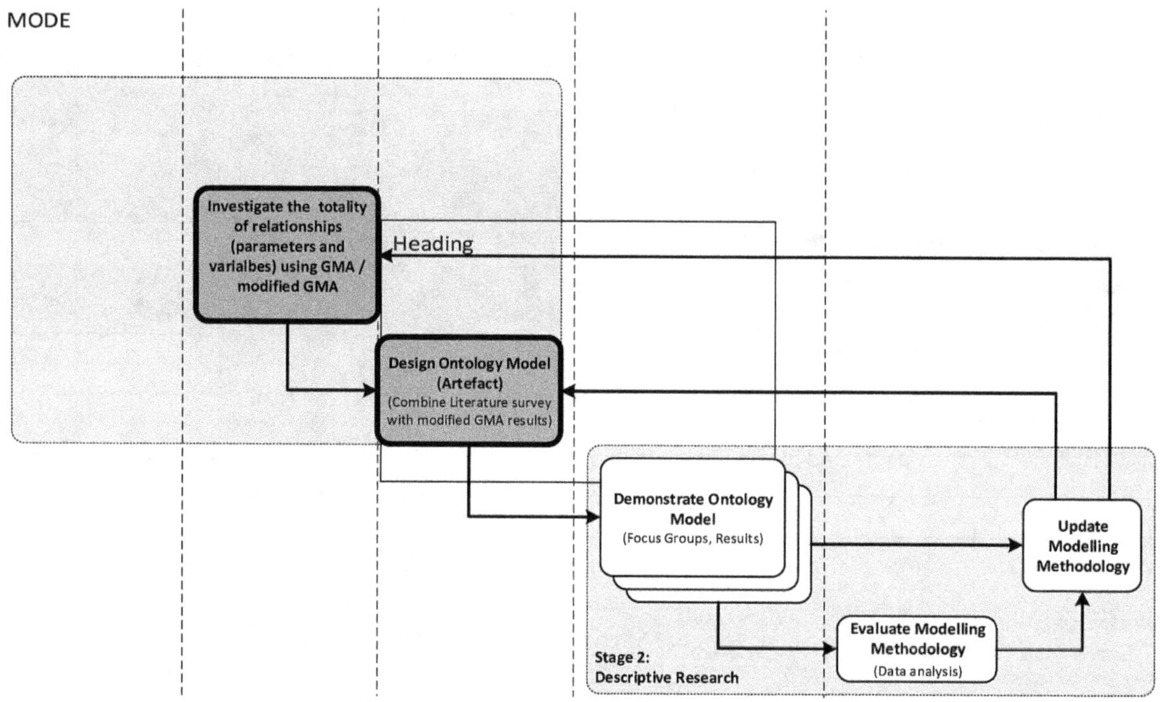

The ontology was also updated to reflect the above changes. The two experts then posed questions to the ontology to check relationships, for example, "Who is responsible to perform a specific function?" Several omissions were noted and updated.

The overall impressions of the second expert was very positive and he expressed appreciation for the value that the ontology in conjunction with the MGMA model bring to the cybersecurity domain: He answered that these two artefacts were "...on the right track in terms of who should have been invited to the party" when the team asked him whether the models would be useful for implementing a cyber-security policy in South Africa. He also stated that this approach provided a better solution and that an ontology will support the development Object Orientated and Agent-based solution systems which may be required during the implementation of the policy.

Limitation of the Case Study

There are many role players in the cybersecurity environment in South Africa and it was impossible to include subject experts to represent the whole environment. This shortcoming may have influenced the resulting artefact but future refinements are planned which will include more subject experts.

EVALUATION OF THE MODE METHODOLOGY

By combining the three core research methodologies to propose the MODE methodology, many of the individual benefits of the core methodologies are inherited and some of the challenges are addressed.

Figure 20. Solution Phase 3

Low level role players - Lead departments	Low level role players - Support departments	Structures	Governance	Functions	Capacity / Capability
	Department of International Relations and Cooperation (DIRCO)	Cybersecurity Response Committee (CRC)	International and regional conventions	Compliance enforcement	Infrastructure
Department of Justice and Constitutional Development (DOJ & CD)	Department of Trade and Industry (DTI)	Cybersecurity Centre	Policy	Legislate and regulate	Research and development
Department of Defence (DOD)	Department of Education (DOE)	Government Security Incident Response Teams (CSIRTs)	Standards	Human capacity building (excess / national)	Qualified manpower
South African Police Service (SAPS)	State Information Technology Agency (SITA)	Cyber Command	Legislative measures	Threat identification	Cryptology
Department of Safety and Security (DOSS, SSA, National Communication)	Department of Provincial and Local Government (DPLG)	National Cybercrime Centre (SAPS)	Best practice	Collaboration (Cooperation)	Netizens
Department of Telecommunications and Postal Services (DTPS), including DOC	Department of Public Service and Administration (DPSA)	24/7 cyber contact point (SAPS)		Coordination	Offensive and defensive capacity
Department of Science and Technology (DST)	National Treasury, including Financial Intelligence Centre (FIC) and SARS	International structures		Combat Cyber Threats	
Financial sector				Cultivate cybersecurity culture	Cyber qualification framework (including certifications)
Basic Education service providers		Cybersecurity Hub		Cyberwarfare	
Higher Education service providers		Private Sector Security Incident Response Teams (CSIRTs)		Protect Information Communication Systems (ICS)	
Training providers		South Africa Banking Risk Information Centre (SABRIC)		Protect National Critical Information Infrastructures Systems (NCIIS)	
Software and end user device providers				Conduct research and development	
Infrastructure providers				Combat Cybercrime	
Access providers					
Private security cluster					

The MGMA as the methodology to extract knowledge from domain experts resulted in a systematic analysis of the cybersecurity environment, and the new methodology contributed to the discovery of new relationships or configurations, which may not have been obvious as cybersecurity is a relatively new environment. It provided multiple combinations or permutations that might have been overlooked in other circumstances.

Benefits

By combining the core methodologies, the following benefits can be derived from applying the MODE methodology to the South African cybersecurity environment:

- The ability to share a common understanding of domain knowledge among practitioners by providing a single formal structured model.

- Reuse of domain knowledge; the ontology for the South African cybersecurity environment will allow others to expand and add to this knowledge base.
- Making domain assumptions explicit; miscommunication can arise from implicit assumptions. The responsibility of the stakeholders is depicted.
- Analysis of domain knowledge; by using a logic reasoner, inferences can be made from the capture knowledge base.
- The *what-if* analysis capability of the MGMA model can show discrepancies in the gathered domain information.
- A rigorous process that focuses an expert group to consider/argue pairwise comparisons; this leads to the identification of relationships to be used in the ontology model.
- Repeated design science iterative loops result in better understanding of the environment and an improved model.
- A traditional problem encountered in ontology development is to extract useful information from domain experts. By using structured information from the GMA session, this difficulty was addressed.

Challenges

- Although the MGMA used in this research reduced the workshop contact time considerably as opposed to the classic GMA methodology, the information acquisition is still a time consuming exercise. In addition, the original GMA challenge that domain experts tend to have very busy schedules and find it difficult to attend the entire workshop still stands.
- The balance of the selected expert group is very important and it is necessary to include experts from all relevant domains or fields in order to restrict the possibility of bias. This may lead to a group that is too large, or a lack of contribution from all perspectives. Electing to do a series of MGMAs in order to reach a well-balanced group of experts will be an expensive and time consuming endeavour.
- The requirement that the development team needs to include a combination of subject experts, a GMA facilitator and ontology engineers may be difficult to satisfy.

Limitation

The interpretation of the MGMA resulting artefact, i.e. ontology, may be complex and thus it may prove difficult to explain why they look as they do at the end of the process and how they can be applied to solve problems. Such an explanation often requires a session where the audience is exposed to the underlying relationships in both the model and the ontology in order to form an understanding of the outcomes and possible interpretations. In other words, it is often not sufficient to present the artefacts without guiding the audience through their inner constructs.

CONCLUSION

This chapter proposed a new hybrid methodology to ensure a more efficient way of extracting meaningful information from domain experts, and presenting this information in an expressive model. The MODE methodology combines principles from three different research methods, i.e. design science, modified general morphological analysis, and ontology based development, to maximise the respective benefits and to address the limitations posed by each of the core individual methodologies.

This chapter demonstrated the proposed hybrid MODE methodology by applying it to the cybersecurity domain, resulting in a cybersecurity ontology which can be used in support of implementing a South African cybersecurity policy. Within the South African environment, cybersecurity can be regarded as a complex, multi-dimensional problem. By applying the GMA principles, the knowledge acquisition applied in Stage 1 of the MODE methodology presented a well-defined input to the ontology based developed in Stage 2 of the MODE methodology. This proposed MODE methodology appears to be an effective research method to acquire, represent and manage knowledge effectively within any complex, multi-dimensional problem space. The authors are of the opinion that the benefits largely outweigh the challenges this methodology pose. The research team plans to apply the methodology to resolve other Cyber Defence problems in the future. These exercises will allow a more in-depth evaluation of the methodology.

REFERENCES

Ashburner, M., Ball, C. A., Blake, J. A., Botstein, D., Butler, H., Cherry, J. M., & Sherlock, G. T. et al. (2000). Gene Ontology: Tool for the unification of biology. The Gene Ontology Consortium. *Nature Genetics*, *25*(1), 25–29. doi:10.1038/75556 PMID:10802651

Barab, S., & Squire, K. (2004). Design-based research: Putting a stake in the ground. *Journal of the Learning Sciences*, *13*(1), 1–14. doi:10.1207/s15327809jls1301_1

Dube, B. (2012). *Statement on the approval by Cabinet of the Cyber Security Policy Framework for South Africa*. South African Government Information: South African Government Information. Retrieved from http://www.info.gov.za/speech/DynamicAction?pageid=461&sid=25751&tid=59794

Ghidini, C., Kump, B., Lindstaedt, S., Mahbub, N., Pammer, V., Rospocher, M., & Serafini, L. (2009). Moki: The Enterprise Modelling Wiki. In *Proceedings of the Semantic Web: Research and Applications* (pp. 831-835). Springer.

SA Goverment Gazette. (2010). *South African National Cyber Security Policy*. Communication. Pretoria, Republic of South Africa. 536.

Grobler, M., Jansen van Vuuren, J., & Leenen, L. (2012). *Implementation of a Cyber Security Policy in South Africa: Reflection on Progress and the Way Forward. In ICT Critical Infrastructures and Society* (pp. 215–225). Springer.

Gruber, T. R. (1993). A translation approach to portable ontology specifications. *Knowledge Acquisition*, *5*(2), 199–220. doi:10.1006/knac.1993.1008

Guarino, N. (1998). *Formal ontology in information systems.* Paper presented at the Formal ontology in information systems, Trento, Italy.

Horrocks, I., Patel-Schneider, P. F., & Van Harmelen, F. (2003). From SHIQ and RDF to OWL: The making of a web ontology language. *Web Semantics: Science, Services, and Agents on the World Wide Web*, *1*(1), 7–26. doi:10.1016/j.websem.2003.07.001

Jansen van Vuuren, J., Leenen, L., Phahlamohlaka, J., & Zaaiman, J. (2013). *Development of a South African cybersecurity policy implementation framework*. Paper presented at the 8th International Conference on Information Warfare and Security, Denver, CO.

Jansen van Vuuren, J., Phahlamohlaka, J., & Brazzoli, M. (2010). The Impact of the Increase in Broadband Access on National Security and the Average citizen. *Journal of Information Warfare*, *5*, 171–181.

Jansen van Vuuren, J., Phahlamohlaka, J., & Leenen, L. (2012). *Governance of cybersecurity in South Africa*. Paper presented at the 11th European Conference on Information Warfare and Security, Laval, France.

Jansen van Vuuren, J. C., Leenen, L., Phahlamohlaka, J., & Zaaiman, J. J. (2012). An Approach to Governance of CyberSecurity in South Africa. *International Journal of Cyber Warfare & Terrorism*, *2*(4), 13–27. doi:10.4018/ijcwt.2012100102

Jansen van Vuuren, J. C., Leenen, L., & Zaaiman, J. J. (2014). *Using an Ontology as a Model for the Implementation of the National Cybersecurity Policy Framework for South Africa*. Paper presented at the 9th International Conference on Cyber Warfare & Security.

Kassel, G. (2005). *Integration of the DOLCE top-level ontology into the OntoSpec methodology*. Technical Report HAL: hal-00012203/arXiv: cs.AI/0510050, Laboratoire de Recherche en Informatique d'Amiens (LaRIA). Retrieved from http://hal.archives-ouvertes.fr/ccsd-00012203

Lewis, M. W., & Grimes, A. I. (1999). Metatriangulation: Building theory from multiple paradigms. *Academy of Management Review*, *24*(4), 672–690.

Martins, J., Santos, H., Nunes, P., & Silva, R. (2012). *Information Security Model to Military Organizations in Environment of Information Warfare*. Paper presented at the 11th European Conference on Information Warfare and Security, Laval, France.

Miller, J. L. (2012). *Design's Mechanism--General Morphological Analysis*. DTIC Document.

Mingers, J. (2001). Combining IS Research Methods: Towardsa Pluralist Methodology. *Information Systems Research, 12*(3), 240–259.

Peffers, K., Tuunanen, T., Rothenberger, M. A., & Chatterjee, S. (2007). A design science research methodology for information systems research. *Journal of Management Information Systems*, *24*(3), 45–77. doi:10.2753/MIS0742-1222240302

Phahlamohlaka, L. J., Jansen van Vuuren, J. C., & Coetzee, A. J. (2011). *Cyber security awareness toolkit for national security: an approach to South Africa's cyber security policy implementation*. Paper presented at the SACSAW, Gaberone, Botswana.

Ritchey, T. (1998). *Fritz Zwicky, morphologie and policy analysis*. Paper presented at the 16th EURO conference on operational analysis, Brussels, Belgium.

Ritchey, T. (2006). Problem structuring using computer-aided morphological analysis. *The Journal of the Operational Research Society*, *57*(7), 792–801. doi:10.1057/palgrave.jors.2602177

Ritchey, T. (2011). *Wicked problems–social messes: Decision support modelling with morphological analysis*. Springer. doi:10.1007/978-3-642-19653-9

Simperl, E. P. B., Mochol, M., & Bürger, T. (2010). Achieving Maturity: The State of Practice in Ontology Engineering in 2009. *IJCSA*, *7*(1), 45–65.

Stanford University. (2013). *Protégé ontology editor: Stanford University*. Retrieved from http://protege.stanford.edu/

Suarez-Figueroa, M. C., de Cea, G. A., Buil, C., Dellschaft, K., Fernandez-Lopez, M., Garcia, A., & Yufei, Z. et al. (2008). NeOn methodology for building contextualized ontology networks. *NeOn Deliverable D*, *5*, 4–1.

Tashakkori, A., & Teddlie, C. (1998). *Mixed methodology: Combining qualitative and quantitative approaches* (Vol. 46). Sage.

Teddlie, C., & Tashakkori, A. (2003). *Handbook of mixed methods in social & behavioral research*. Sage.

Section 7

Conducting Research from Start to Finish

The numerous steps to be considered when conducting a research study can be daunting for an inexperienced researcher. Similarly, teaching these skills to students requires considerable thought as to how to best provide an integrated approach to teaching both strands and then combining this orientation. This section provides a step by step linear approach to the research process. The mixed methods researcher needs to manage numerous competencies starting with the philosophical assumptions underlying the use of mixed methods. The researcher must gather considerable data from each component and make meaningful inferences. In addition, a mixed methods study needs to integrate, link and connect the two strands of research in order to provide a comprehensive understanding of the issue under investigation.

Chapter 15
Creating and Implementing a Mixed Method Research Study

Mette L. Baran
Cardinal Stritch University, USA

Janice E. Jones
Cardinal Stritch University, USA

ABSTRACT

This chapter serves as a guideline for outlining the core characteristics of mixed methods research (MMR) and the various steps researchers undertake in order to conduct a research study. The purpose is to create a worksheet assisting the researcher step by step from beginning to end following the seven steps to conducting research. While the focus is on MMR the steps are similar for any type of research methodology. It is important to note that MMR is not a limiting form of research. Researchers need a MMR question and a mixed methods purpose statement for the research project. This chapter will also help explain why mixed method research is one of the best approaches in answering a research question. Finally, the chapter includes a suggestion to the importance of adding a visual diagram of the MMR into the research project and into the final report.

INTRODUCTION

The following checklist is a list of decisions researchers need to consider at the start of any MMR project. The various prompts are displayed in a logical step by step fashion which is typically the outline followed when conducting empirical research. The authors envision practitioners to complete each section and then build on the previous decisions when completing the worksheet. The information can then be compiled to be used in a master's thesis, dissertation, or any research study.

STARTING YOUR RESEARCH JOURNEY

Use the guiding questions below to develop your research plan.

DOI: 10.4018/978-1-5225-0007-0.ch015

Reasons for Conducting Mixed Methods Research?

- The Compatibility of Quantitative and Qualitative Methods.
- Strengths and Weaknesses of Mixed Methods Research.
- Triangulation, Complementarity, Expansion.

Characteristics of Quantitative Research

- Deductive data analysis.
- Strength and Weaknesses of Quantitative Methods (Figure 1).

Characteristics of Qualitative Research

- Inductive data analysis.
- Strengths and Weaknesses of Qualitative Methods (Figure 2).

Order of Methodology: Which Comes First: Quantitative or Qualitative?

- **Explanatory:** Quantitative data are collected first with qualitative data collection following.
- **Exploratory:** Qualitative data are collected first with quantitative data collection following.
- **Descriptive:** Either Quantitative or Qualitative comes first.

Describe your methodology:

Figure 1. Steps in deductive research
Source: Trochim, 2006

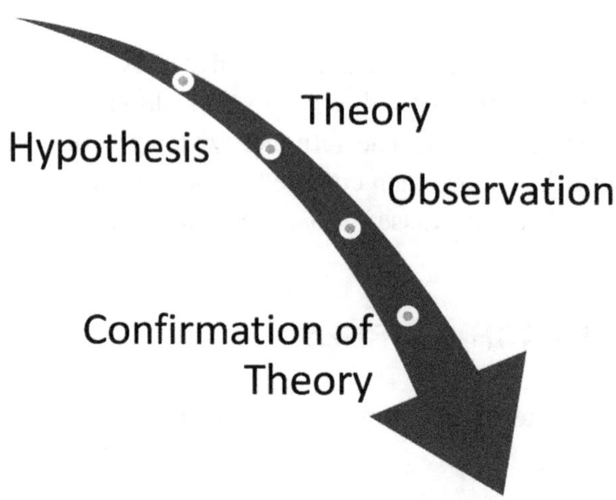

Figure 2. Steps in inductive research
Source: Trochim, 2006

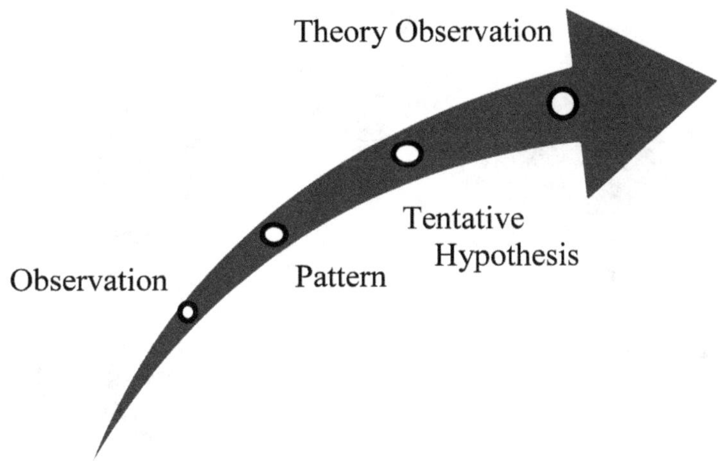

ASSUMPTIONS THAT DIFFERENTIATE QUALITATIVE AND QUANTITATIVE STUDIES

Epistemology

- Qualitative researchers believe there are multiple realities represented by the participants' perspectives.
- Quantitative researchers believe a single, objective reality exists.

Context

- Qualitative researchers believe context is critical to understanding the phenomena being studied.
- Quantitative researchers do not believe context is an important factor.

Describe your worldview and epistemology including researcher assumptions about the research:

Researcher Bias

- Qualitative researchers believe the researcher's biases and perspectives must be understood to interpret the results.
- Quantitative researchers believe researcher bias is controlled through the control of internal validity threats.

Describe your bias towards the research study:

Figure 3. Steps to conducting a Mixed Methods Research Study

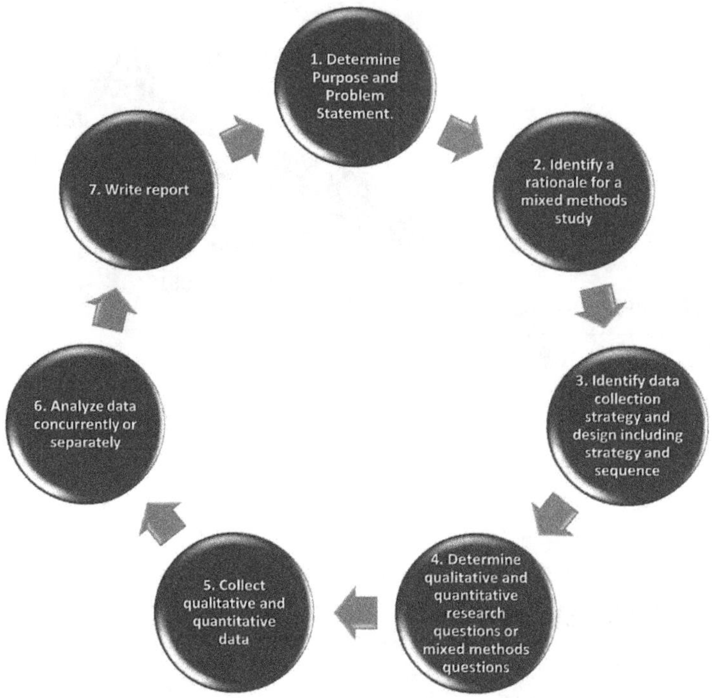

STEPS IN CONDUCTING A MIXED METHODS RESEARCH STUDY

The following 7 decision steps demonstrate the research path when conducting a research study as outlined in Figure 3.

1. Determine purpose and problem statement.
2. Identify a rationale for a mixed methods study.
3. Identify data collection strategy and design including strategy and sequence.
4. Determine qualitative and quantitative research question or mixed methods questions? Is mixed methods the best approach?
5. Collect qualitative and quantitative data.
6. Analyze data concurrently or separately.
7. Write report.

Researcher Work Sheets Using the 7 Decision Steps

1. Determine Purpose and Problem Statement

What is the purpose of the study?
What is the problem statement?

2. Identify a Rationale for a Mixed Methods Study

Describe your worldview and epistemological stance?
Describe your theoretical framework?

3. Identify Data Collection Strategy and Design Including Strategy and Sequence

Research design refers to the plan of action that links the philosophical assumptions to specific methods (Creswell, 2003; Crotty, 1998).

Quantitative Strategies

- Experimental
- Quasi-Experimental

Qualitative Strategies

1. **Narrative:** Narrative research is a form of inquiry in with the researcher studies the lives of individuals and asks one or more individuals to provide stories about their lives. This information is then retold or restoried by the researcher into narrative chronology. In the end, the narrative combines views from the participant's life with those of the researcher's observations in a collaborative narrative.
2. **Ethnography:** An in-depth description and interpretation of cultural patterns and meanings within a culture or social group with culture being a shared patterns of beliefs, normative expectations, behaviors, and meanings.
3. **Phenomenology:** Describes and interprets the experiences of participants to understand their perspectives based on the belief that there are multiple ways of interpreting the same experience and the meaning of that experience is what constitutes reality.
4. **Grounded Theory:** Grounded theory offers the researcher an opportunity to derive a general, abstract theory of a process, action, or interaction grounded in the views of participants in a study. This process involves using multiple stages of data collection and the refinement and interrelationship of categories of information. Two primary characteristics of this design are the constant comparison of data with emerging categories and theoretical sampling of different groups to maximize the similarities and the differences of information.
5. **Case Study:** An in-depth analysis of one or more events, settings, programs, groups, or other "bounded systems." Types of case studies:
   a. **Historical Organizational:** Focus on the development of an organization over time.
   b. **Observational:** Study of a single entity using participant observation.
   c. **Life History (i.e., Oral History):** A first-person narrative completed with one person.
   d. **Situation Analysis:** A study of a specific event from multiple perspectives.
   e. **Multi-Case:** A study of several different independent entities.
   f. **Multi-Site:** A study of many sites and participants the main purpose of which is to develop theory.

Mixed Methods Strategies

- Parallel Convergent
- Sequential
- Embedded

Describe your research design?

4. Determine Qualitative and Quantitative Research Questions or Mixed Methods Questions: Is Mixed Methods the Best Approach?

Explain clearly and explain why MM is necessary. Provide a compelling reason why MM is needed and how a holistic understanding of the research problem will be gained through MM. Explain how both approaches make it possible to draw inferences that would have been different if only one approach or strand had been used.

Fill in:

- Quantitative Research Question(s):
  - Null Hypothesis(es:)
  - Alternative Hypothesis(es):
- Qualitative Research Question(s):
- Mixed Methods Question(s):

5. Collect Qualitative and Quantitative Data

What data to use?

- Existing data:
- Archival data:
- Longitudinal data:

Quantitative Data Collection

Who are your participants and how will they be selected and contacted?:

Which research question will this data help answer?:

Qualitative Data Collection

Who are your participants and how will they be selected and contacted?:

Which research question will this data help answer?:

- Observation.
  - Unstructured in nature.
  - Comprehensive: Continuous over an extended period of time.
  - Participant: Observer role of the researcher.
  - Continuum between complete participant and complete observer:
    - Passive participant.
    - Moderate participant.
    - Active participant.
    - Complete participant.
  - Use of field notes to record observations.
  - Two types of information:
    - Descriptions of what occurred.

        - Reflections of what the descriptions mean (i.e., speculations, emerging themes, patterns, problems).
    - Accuracy.
    - Extensive nature of notes.
- Interviews.
    - Unstructured in nature.
    - Begins with a general idea of what needs to be asked and moves to specific questions based on what the respondent says.
    - Types of interviews:
        - Key informant.
        - Life history.
    - Tape recording and transcribing interviews afford the opportunity to study the data carefully.
- Focus Groups.
- Document Analysis.
    - Written records:
        - Print (e.g., minutes from meetings, reports, yearbooks, articles, diaries).
        - Non-print (e.g., recordings, videotapes, pictures).
    - Types of sources:
        - Primary: Original work.
        - Secondary: Secondhand interpretations of original work.
    - Commonly used to verify other observations or interview data.

Mixed Methods Data Collection

Who are your participants and how will they be selected and contacted?:

Which research question will this data help answer?:

Identify the Research Site(s).

Describe access to physical locations, participants, documents, physical locations, etc.:

Determine Sampling Techniques

- Probability Sampling.
- Non-Probability sampling:
    - Use of purposeful sampling strategies to select "information rich" participants.
    - Purposeful sampling strategies:
        - Maximum Variation: Selecting individuals or cases to represent extremes (e.g., very positive or very negative attitudes, highest and lowest achieving students).
        - Snowball (i.e., Network): Initially selected participants recommend others for involvement.
        - Key Informant: Selecting an individual(s) particularly knowledgeable about the setting and or topic.
        - Comprehensive: Selecting all relevant individuals or cases.

Describe your sampling technique:

Reliability and Validity in Mixed Methods Research

This section covers reliability and validity issues that are apparent in MMR.

Define Validity:

Define Reliability:

- **Quantitative Validity and Reliability:**
    - Internal Validity.
    - External Validity.
    - Construct Validity.
    - Reliability.
    - Objectivity.
- **Qualitative Validity and Reliability:**
    - Credibility.
    - Transferability.
    - Dependability.
    - Confirmability.
- **Techniques to Enhance Validity and Credibility:**
    - Triangulation.
    - Prolonged and persistent field work.
    - Copious field notes.
    - Low inference descriptors.
    - Mechanically recorded data.
    - Member checking / Peer Review.
    - Verbatim accounts.
    - Abundant use of detail.
    - Data collection in natural settings.
    - Researcher's role as participant observer.

Describe how you will address quantitative and qualitative validity and reliability in your research?

6. Analyze Data Concurrently or Separately

Analyzing Quantitative Data

Describe statistical procedures to be used:

1. Levels of Measurement.
2. Hypothesis Testing.

3. T-tests inferences concerning two means.
4. ANOVA: Inferences using a one-way ANOVA.
5. Correlational research.

Analyzing Qualitative Data

Observations, interviews, and document analyses result in large quantities of narrative data.

Analysis includes critically examining, summarizing, and synthesizing the data.

Three stages of analysis:

1. Transcription:
    a. Transcribe all your documents carefully and store in an organized fashion.
2. Coding:
    a. Organizing the data into reasonable, meaningful units that are coded with words or very short phrases that signify a category.
    b. Use of major codes and sub-codes is common.
    c. Summarizing the coded data.
    d. Examining all similarly coded data and summarizing it with a sentence or two that reflects its essence.
    e. Computerized sorting of data is common and effective.
    f. Pattern seeking and synthesizing.
    g. Synthesizing identifies the relationships among the categories and patterns that suggest generalization.
    h. The researcher interprets findings inductively, synthesizes the information, and draws inferences.
3. Pattern Seeking and Themes:
    a. Begins with the researcher's informed hunches and ideas.
    b. Tentative patterns are identified and additional data collected to determine if they are consistent with those patterns.
    c. Characterized by enlarging, combining, subsuming, and creating new categories that make sense.

Additional Decisions pertaining to Data Analysis:

- Timing (Sequencing).
- Weight (Priority) of Data.
- Where and how to mix the strands:
    - **Concurrent Design:** How will the quantitative and qualitative data be merged? What strategies will be used if findings conflict or are contradictory from the two data? A strategy of

resolving differences needs to be considered, such as gathering more data or revisiting the databases.

- **Sequential Design:** What results from the first phase will be used in the follow-up phase? Key issues surround the "point of interface" in which the investigator needs to decide what results from the first phase will be the focus of attention for the follow-up data collection.

7. Write Report

How will you use visual modeling in your study?

- Graphic representation of literature review (Chapter 2).
- Graphic representation of Methodology (Chapter 3).
- Graphic representation of Findings (Chapter 4).
- Graphic representation of findings connecting with existing literature and/or new theory (Chapter 5).

REFERENCES

Creswell, J. W. (2003). *Research design: Qualitative, quantitative, and mixed methods approaches* (2nd ed.). Thousand Oaks, CA: Sage.

Crotty, M. (1998). *The foundations of social research: Meaning and perspective in the research process.* Thousand Oaks, CA: Sage.

Troschim, W. M. K. (2006). *Deduction and Induction. Research Methods Knowledge Base.* Retrieved from: http://www.socialresearchmethods.net/kb/dedind.php

Compilation of References

AECT. (n.d.). *The Handbook of Research for Educational Communications and Technology*. Retrieved from http://www.aect.org/edtech/ed1/41/41-01.html

Allen, M. T., Angie, A. D., Davis, J. L., Byrne, C. L., O'Hair, H. D., Connelly, S., & Mumford, M. D. (2008). Virtual risk: The role of new media in violent and nonviolent ideological groups. In R. L. Heath & H. D. O'Hair (Eds.), *The handbook of risk & crisis communication* (pp. 446–447). Mahwah, NJ: Erlbaum.

American Association of Colleges for Teacher Education & Stanford University. (2013). *edTPA General Information*. Retrieved from http://edtpa.aacte.org

American Association of Colleges for Teacher Education & Stanford University. (2013). *Using edTPA Handout*. Retrieved from http://edtpa.aacte.org

Andrew, S., & Halcomb, E. J. (2011). Mixed Methods Research. *Navigating the Maze of Research: Enhancing Nursing and Midwifery Practice*, 147.

Andrew, S., & Halcomb, E. (Eds.). (2009). *Mixed methods research for nursing and the health sciences*. Wiley-Blackwell Pub. doi:10.1002/9781444316490

Angie, A. D., Davis, J. L., Allen, M. T., Byrne, C. L., Ruark, G., Cunningham, C. B., & Mumford, M. D. et al. (2011). Studying ideological groups online: Identification and assessment of risk factors for violence. *Journal of Applied Social Psychology*, *41*(3), 627–657. doi:10.1111/j.1559-1816.2011.00730.x

Arbnor, I., & Bjerke, B. (2008). *Methodology for creating business knowledge*. London: Sage Publications.

Arora, R. (2011). Bundling or unbundling frequently purchased products: A mixed method approach. *Journal of Consumer Marketing*, *28*(1), 67–75. doi:10.1108/07363761111101967

Ashburner, M., Ball, C. A., Blake, J. A., Botstein, D., Butler, H., Cherry, J. M., & Sherlock, G. T. et al. (2000). Gene Ontology: Tool for the unification of biology. The Gene Ontology Consortium. *Nature Genetics*, *25*(1), 25–29. doi:10.1038/75556 PMID:10802651

Babbie, E. (1990). *Survey research methods*. Belmont, CA: Academic Press.

Baker, B., Sciarra, D., & Farrie, D. (2010). *Is school funding fair? A national report card*. Newark, NJ: Education Law Center.

Barab, S., & Squire, K. (2004). Design-based research: Putting a stake in the ground. *Journal of the Learning Sciences*, *13*(1), 1–14. doi:10.1207/s15327809jls1301_1

Baron, R. M., & Kenny, D. A. (1986). Moderator-mediator Variables Distinction in Social psychological Research: Conceptual, Strategic, and Strategic Considerations. *Journal of Personality and Social Psychology*, *51*(6), 1173–1182. doi:10.1037/0022-3514.51.6.1173 PMID:3806354

Baskerville, N. B., Hogg, W., & Lemelin, J. (2001). Process evaluation of a tailored multifaceted approach to changing family physician practice patterns improving preventive care. *The Journal of Family Practice*, *50*(3), W242–9. PMID:11252222

Bazeley, P. (2004). Issues in mixing qualitative and quantitative approaches to research. In R. Buber, J. Gadner, & L. Richards (Eds.), *Applying qualitative methods to marketing management research* (pp. 141–156). Palgrave Macmillan.

Berelson, B. (1952). *Content analysis in communication research.* New York: Free Press.

Bergman, M. M. (2011). The politics, fashions, and conventions of research methods. *Journal of Mixed Methods Research*, *5*(2), 99–102. doi:10.1177/1558689811408318

Berinstein, P. (2003). *Business statistics on the web: Find them fast – at little or no cost.* Cyber Age Books.

Bernardin, H. J., & Buckley, M. R. (1981). Strategies in rater training. *Academy of Management Review*, *6*, 205–212.

Berry, A. J., & Otley, D. T. (2004). Case-based research in accounting. In C. Humprey & B. Lee (Eds.), *The real life guide to accounting research: A behind-the-scenes view of using qualitative research methods* (pp. 231–256). Oxford, UK: Elsevier. doi:10.1016/B978-008043972-3/50016-5

Bhaskar, R. (1975). *A realist theory of science.* Leeds, UK: Leeds Books.

Biernacki, P., & Waldorf, D. (1981). Snowball sampling: Problems and techniques of chain referral sampling. *Sociological Methods & Research*, *10*(2), 141–163.

Biesta, G. (2010). Pragmatism and the philosophical foundations of mixed methods research. In A. Tashakkori & C. Teddlie (Eds.), *SAGE Handbook of Mixed Methods in Social & Behavioral Research* (2nd ed.; pp. 95–118). Thousand Oaks, CA: Sage. doi:10.4135/9781506335193.n4

Boeije, H. (2010). *Analysis in qualitative research.* London: Sage Publications.

Bouchard, T. J. Jr. (1976). Unobtrusive measures: An inventory of uses. *Sociological Methods & Research*, *4*(3), 267–300. doi:10.1177/004912417600400301

Boucher, E. M., Hancock, J. T., & Dunham, P. J. (2008). Interpersonal sensitivity in computer-mediated and face-to-face conversations. *Media Psychology*, *11*(2), 235–258. doi:10.1080/15213260801906471

Brannen, J. (2005). Mixing methods: The entry of qualitative and quantitative approaches into the research process. *International Journal of Social Research Methodology*, *8*(3), 173–184. doi:10.1080/13645570500154642

Brewer, J., & Hunter, A. (1989). *Multi-method research: A synthesis of styles.* London: Sage Publications, Inc.

Brodsky, A. E. (2008). Negative case analysis. In L. M. Given (Ed.), *The Sage encyclopedia of qualitative research methods.* Thousand Oaks, CA: Sage Publications. doi:10.4135/9781412963909.n283

Brown, J. D. (2004). Yates correction factor. SHIKEN. *The JALT Testing & Evaluation SIG Newsletter*, *8*(1), 22–27.

Bryman, A. (1992). Quantitative and qualitative research: further reflections on their integration. *Mixing methods: Qualitative and quantitative research*, 57-78.

Bryman, A. (1988). *Quantity and Quality in Social Research.* London: Unwin Hyman. doi:10.4324/9780203410028

Bryman, A. (2003). *Quantity and quality in social research.* London: Routledge.

Bryman, A. (2006). Paradigm peace and the implications for quality. *International Journal of Social Research Methodology Theory and Practice*, *9*(2), 111–126. doi:10.1080/13645570600595280

Bryman, A. (2012). *Social research methods* (4th ed.). New York: Oxford University Press.

Bryman, A., & Cramer, D. (1990). *Quantitative Data Analysis for Social Scientists*. London: Routledge.

Bryson, J. (n.d.). *Current and emerging trends in qualitative market research*. Retrieved on March 4, 2015. Available at http://qualblog.com/current-and-emerging-trends-in-qualitative-market-research/

Burns, N., & Grove, S. K. (2005). *The practice of nursing research: conduct, critique, and utilization* (5th ed.). St. Louis, MO: Elsevier Saunders.

Byrne, C. L., Nei, D. S., Barrett, J. D., Hughes, M. G., Davis, J. L., Griffith, J. A., & Mumford, M. D. et al. (2013). Online ideology: A comparison of website communication and media use. *Journal of Computer-Mediated Communication*, *18*, 25–39. doi:10.1111/jcc4.12003

Cameron, R., & Miller, P. (2007). Mixed methods research: Phoenix of the paradigm wars. In R. Chapman (Ed.), *Managing our intellectual and social capital: Proceedings of the 21st ANZAM Conference*. Sydney, Australia: Promaco.

Cameron, R. (2009). A sequential mixed model research design: Design, analytical and display issues. *International Journal of Multiple Research Approaches*, *3*(2), 140–152. doi:10.5172/mra.3.2.140

Cameron, R. (2011). Mixed methods research: The Five Ps framework. *The Electronic Journal of Business Research Methods*, *9*(2), 96–108.

Campbell, D. T., & Fiske, D. W. (1959). Convergent and discriminant validation by the multitrait-multimethod matrix. *Psychological Bulletin*, *56*(2), 81–105. doi:10.1037/h0046016 PMID:13634291

Campbell, D. T., & Stanley, J. C. (1966). *Experimental and quasi-experimental designs for research*. Chicago, IL: Rand McNally & Co.

Caplan, S. E. (2003). Preference for online social interaction: A theory of problematic Internet use and psychosocial well-being. *Communication Research*, *30*(6), 625–648. doi:10.1177/0093650203257842

Caracelli, V. J., & Greene, J. C. (1993). Data analysis strategies for mixed-method evaluation designs. *Educational Evaluation and Policy Analysis*, *15*(2), 195–207. doi:10.3102/01623737015002195

Carmen, J. M. (1990). Consumer perceptions of service quality: An assessment of the SERVQUAL dimensions. *Journal of Retailing*, *66*(1), 33–55.

Charmaz, K. (2006). *Constructing grounded theory: A practical guide through qualitative analysis*. Thousand Oaks, CA: SAGE Publications.

Charmaz, K. (2006). *Grounded theory: A practical guide through qualitative analysis*. Thousand Oaks, CA: Sage Publications.

Chicago Teacher's Union. (2014). *The schools Chicago's students deserve: Research-based proposals to strengthen elementary and secondary education in the Chicago public schools*. Retrieved from: http://www.ctunet.com/blog/text/SCSD_Report-02-16-2012-1.pdf

Christmann, G. B. (2008). The power of photographs of buildings in the Dresden urban discourse. Towards a visual discourse analysis. Forum Qualitative Sozialforschung / Forum: Qualitative. *Social Research*, *9*(3), 11. Retrieved from http://nbn-resolving.de/urn:nbn:de:0114-fqs0803115

City of Chicago (2014). *Census data: Selected socioeconomic indicators 2008 – 2012*. Author.

City of Chicago Data Portal. (2014). *Crimes 2001 – present*. Retrieved from:https://data.cityofchicago.org/Public-Safety/Crimes-2001-to-present/ijzp-q8t2

Clark, L. A., & Watson, D. (1995). Constructing validity: Basic issues in objective scale development. *Psychological Assessment*, *7*(3), 309–319. doi:10.1037/1040-3590.7.3.309

Cochran, W. G. (1953). *Sampling techniques*. New York: John Wiley & Sons.

Coghlan, D. (2011). Action research: Exploring perspectives on a philosophy of practical knowing. *The Academy of Management Annals*, *5*(1), 53–87. doi:10.1080/19416520.2011.571520

Coghlan, D., & Brannick, T. (2014). *Doing action research in your own organization*. London: Sage.

Cohen, L., & Manion, L. (1989). *Research methods in education* (3rd ed.). London, England: Routledge.

Cohen, L., Manion, L., & Morrison, K. (2007). *Research Methods in Education*. London, New York: Routledge.

Collins, K. M. T., Onwuegbuzie, A. J., & Sutton, I. L. (2006). A model incorporating the rationale and purpose for conducting mixed methods research in special education and beyond. *Learning Disabilities (Weston, Mass.)*, *4*(1), 67–100.

Collins, K., Onwuegbuzie, A., & Jiao, Q. (2007). A mixed methods investigation of mixed methods sampling designs in social and health science research. *Journal of Mixed Methods Research*, *1*(3), 267–294. doi:10.1177/1558689807299526

Cook, T. D. (1985). Postpositivist critical multiplism. In L. Shotland & M. M. Mark (Eds.), *Social science and social policy* (pp. 21–62). Beverly Hills, CA: Sage Publications.

Corbin, J., & Strauss, A. (2008). *Basics of qualitative research: Techniques and procedures for developing grounded theory*. Thousand Oaks, CA: Sage Publications.

Cornwell, B., & Lundgren, D. C. (2001). Love on the Internet: Involvement and misrepresentation in romantic relationships in cyberspace vs. realspace. *Computers in Human Behavior*, *17*(2), 197–211. doi:10.1016/S0747-5632(00)00040-6

Cousineau, T., Rancourt, D., & Green, T. C. (2006). Web chatter before and after the women's health initiative results: A content analysis of in-line menopause message boards. *Journal of Health Communication*, *11*(2), 133–147. doi:10.1080/10810730500526661 PMID:16537284

Crawford, S. D., Couper, M. P., & Lamias, M. J. (2001). Web surveys: Perception of burden. *Social Science Computer Review*, *19*(2), 146–162. doi:10.1177/089443930101900202

Creswell, J. W., & Plano Clark, V. L. (2007). Choosing a mixed methods design. *Designing and Conducting Mixed Methods Research*, 58–88.

Creswell, J. (2003). *Research design: Qualitative, quantitative and mixed methods approaches* (2nd ed.). Thousand Oaks, CA: SAGE Publications.

Creswell, J. W. (1994). *Research Design: Qualitative and Quantitative Approaches*. London: Sage.

Creswell, J. W. (1994). *Research design: Qualitative and quantitative approaches*. Thousand Oaks, CA: Sage Publications.

Creswell, J. W. (1998). *Qualitative inquiry and research design: Choosing among five traditions*. London: Sage.

Creswell, J. W. (1998). *Qualitative inquiry and research design: Choosing among five*. New York: Sage.

Creswell, J. W. (2003). *Research design: Qualitative, quantitative, and mixed methods approaches* (2nd ed.). Thousand Oaks, CA: Sage Publications.

Creswell, J. W. (2009). Editorial: Mapping the field of mixed methods research. *Journal of Mixed Methods Research, 3*(2), 95–108. doi:10.1177/1558689808330883

Creswell, J. W. (2009). *Research design, qualitative, quantitative, and mixed methods / approaches* (3rd ed.). London: Sage Publications, Inc.

Creswell, J. W. (2009). *Research Design: Qualitative, Quantitative, and Mixed Methods Approaches* (3rd ed.). London: Sage Publications.

Creswell, J. W. (2013). Research design: Qualitative, quantitative, and mixed methods approaches. *Sage (Atlanta, Ga.).*

Creswell, J. W. (2014). *A concise introduction to mixed methods research.* SAGE Publications.

Creswell, J. W., Klassen, A. C., Plano Clark, V. L., & Smith, K. C. (2011). *Best practices for mixed methods research in the health sciences.* Bethesda, MD: National Institutes of Health. doi:10.1037/e566732013-001

Creswell, J. W., & Plano Clark, V. L. (2007). *Designing and Conducting Mixed Methods Research. United State of America.* SAGE.

Creswell, J. W., & Plano Clark, V. L. (2010). *Understanding research: A consumer's guide.* Thousand Oaks, CA: Sage.

Creswell, J. W., Plano Clark, V. L., Gutmann, M., & Hanson, W. (2003). Advanced mixed methods research designs. In A. Tashakkori & C. Teddlie (Eds.), *Handbook of mixed methods in social & behavioral research* (pp. 209–240). Thousand Oaks, CA: Sage.

Creswell, J. W., & Plano-Clark, V. L. (2007). *Designing and conducting mixed methods research.* Thousand Oaks, CA: Sage Publications.

Creswell, J. W., Plano-Clark, V. L., Guttmann, M. L., & Hanson, E. E. (2003). Advanced mixed methods research design. In A. Tashakkori & C. Teddlie (Eds.), *Handbook of mixed methods in social and behavioral research* (pp. 209–240). Thousand Oaks, CA: Sage Publications.

Crotty, M. (1998). *The foundations of social research.* Thousand Oaks, CA: Sage Publications.

Crotty, M. (1998). *The foundations of social research: Meaning and perspective in the research process.* London: Sage Publications.

Curwin, J., & Slater, R. (2007). *Quantitative Methods for Business Decisions* (5th ed.). London: Thomson Learning.

Datta, L.-E. (1994). Paradigm wars: A basis for peaceful coexistence and beyond. *New Directions for Program Evaluation, 61*, 53-70.

Davis, D. F., Golicic, S. L., & Boerstler, C. N. (2011). Benefits and challenges of conducting multiple methods research in marketing. *Journal of the Academy of Marketing Science, 39*(3), 467–479. doi:10.1007/s11747-010-0204-7

Denscombe, M. (2008). Communities of practice: A research paradigm for the mixed methods approach. *Journal of Mixed Methods Research, 2*(3), 270–283. doi:10.1177/1558689808316807

Denzin, N. K. (1978). *The research act: A theoretical introduction to sociological methods.* New York: Praeger.

Denzin, N. K. (2010). Moments, mixed methods, and paradigm dialogs. *Qualitative Inquiry, 16*(6), 419–427. doi:10.1177/1077800410364608

Denzin, N. K. (2012). Triangulation 2.0. *Journal of Mixed Methods Research, 6*(2), 80–88. doi:10.1177/1558689812437186

Denzin, N., & Lincoln, Y. (2005). *The sage handbook of qualitative research* (3rd ed.). London: Sage Publications.

Department of Public Instruction. (2014). *Teacher Evaluation Process Manual. Version 4.* Retrieved from http://dpi.gov

Deutschmann, M., & Panichi, L. (2009). Talking into empty space? Signaling involvement in a virtual language classroom in Second Life. *Language Awareness, 18*(3-4), 310–328. doi:10.1080/09658410903197306

Dewey, J. (1934). *Art as experience.* New York: Minton, Malach, and Co.

Dewey, J. (1938). *Logic: The theory of inquiry.* New York: Holt, Rinehart and Winston.

Dewey, J. (1966). *Logic: They theory of inquiry.* New York: Henry Holt.

Dillman, D. A., Tortora, R. D., & Bowker, D. (2001). *Principles for constructing web surveys.* Retrieved December 12, 2013, from Social and Economic Science Research Center.http://survey.sesrc.wsu.edu/dillman/papers.htm

Donovan, J., Mills, N., Smith, M., Brindle, L., Jacoby, A., & Peters, T. et al.. (2002). Improving design and conduct of randomised trials by embedding them in qualitative research: Protetc (prostate testing for cancer and treatment) study. *British Medical Journal, 325*, 766–769. doi:10.1136/bmj.325.7367.766 PMID:12364308

Driscoll, D. L., Appiah-Yeboah, A., Salib, P., & Rupert, D. J. (2007). Merging qualitative and quantitative data in mixed methods research: How to and why not. *Ecological and Environmental Anthropology, 3*(1), 19–28.

Dube, B. (2012). *Statement on the approval by Cabinet of the Cyber Security Policy Framework for South Africa.* South African Government Information: South African Government Information. Retrieved from http://www.info.gov.za/speech/DynamicAction?pageid=461&sid=25751&tid=59794

Dunbar, N. E., Connelly, S., Jensen, M. L., Adame, B. J., Rozzell, B., Griffith, J. A., & O'Hair, H. D. (2014). Fear appeals, message processing cues, and credibility in the websites of violent, ideological, and nonideological groups. *Journal of Computer-Mediated Communication, 19*(4), 871–899. doi:10.1111/jcc4.12083

Dzurec, L. C., & Abraham, J. L. (1993). The nature of inquiry: Linking quantitative and qualitative research. *ANS. Advances in Nursing Science, 16*(1), 73–79. doi:10.1097/00012272-199309000-00009 PMID:8311427

Echambadi, R., & Hess, J. D. (2007). Mean-Centering Does Not Alleviate Collinearity Problems in Multiple Regression Models. *Marketing Science, 26*(3), 438–445. doi:10.1287/mksc.1060.0263

Eisner, E. W. (1991). *The enlightened eye: Qualitative inquiry and the enhancement of educational practice.* New York: Macmillan.

Elliott, J. (2005). *Using narrative in social research: Qualitative and quantitative approaches.* Thousand Oaks, CA: Sage Publications.

Elliott, M. S., & Williams, D. I. (2002). A qualitative evaluation of an employee counselling service from the perspective of client, counsellor and organization. *Counselling Psychology Quarterly, 15*(2), 201–208. doi:10.1080/09515070210128991

Ellison, N., Heino, R., & Gibbs, J. (2006). Managing impressions online: Self-presentation processes in the online dating environment. *Journal of Computer-Mediated Communication, 11,* article 2.

Elms, A. C. (1994). *Uncovering lives: The uneasy alliance of biography and psychology.* New York, NY: Oxford University Press.

Elo, S., Kaarlainen, M., Kanste, O., Polkki, T., Utriainen, K. & Kyngas, H. (2014). Qualitative content analysis: A focus on trustworthiness. *Sage Open,* 1-10. DOI: 10.1177/2158244014522633

Faugier, J., & Sargeant, M. (1997). Sampling hard to reach populations. *Journal of Advanced Nursing*, *26*(4), 790–797. doi:10.1046/j.1365-2648.1997.00371.x PMID:9354993

Feilzer, M. Y. (2010). Doing mixed methods research pragmatically: Implications for the rediscovery of pragmatism as a research paradigm. *Journal of Mixed Methods Research*, *4*(1), 6–16. doi:10.1177/1558689809349691

Feldon, D. F., & Kafai, Y. B. (2008). Mixed methods for mixed reality: Understanding users' avatar activities in virtual worlds. *Educational Technology Research and Development*, *56*(5-6), 575–593. doi:10.1007/s11423-007-9081-2

Fetterman, D. M. (1994). Empowerment evaluation. *Evaluation Practice*, *15*(1), 1–15. doi:10.1016/0886-1633(94)90055-8

Feuer, M. J., Towne, L., & Shavelson, R. J. (2002). Scientific culture and educational research. *Educational Researcher*, *31*(8), 4–14. doi:10.3102/0013189X031008004

Field, A. (2009). *Discovering Statistics Using SPSS* (3rd ed.). Thousand Oaks, CA: Sage Publications.

Field, A. (2013). *Discovering statistics using SPSS* (4th ed.). London: Sage Publication.

Fielding, N. G., & Fielding, J. L. (1986). *Linking data: Sage university paper series on qualitative research methods*. Beverly Hills, CA: Sage Publications. doi:10.4135/9781412984775

Fink, A. (1995). *How to sample in surveys* (Vol. 6). London: Sage Publications.

Finn, J., & Lavitt, M. (1994). Computer-based self help groups for sexual abuse survivors. *Social Work with Groups*, *17*(1-2), 21–46. doi:10.1300/J009v17n01_03

Fitzsimons, G. J. (2008). Death to Dichotomizing. *The Journal of Consumer Research*, *35*(1), 5–8. doi:10.1086/589561

Frey, L. R., Botan, C. H., & Kreps, G. L. (2000). *Investigating communication: An Introduction to research methods* (2nd ed.). Boston: Allyn and Bacon.

Friedman, L., & Wall, M. (2005). Graphical Views of Suppression and Multicollinearity in Multiple Linear Regression. *The American Statistician*, *59*(2), 127–136. doi:10.1198/000313005X41337

Galton, M., & Wilcocks, J. (1983). *Moving from the primary school*. London: Routledge and Kegan Paul.

Gans, H. J. (1963). *Urban villagers: Group life and class in the life of Italian-Americans*. New York: Free Press.

Garfinkel, H. (1967). *Studies in ethnomethodology*. Englewood Cliffs, NJ: Prentice-Hall.

Garfinkel, H. (2002). *Ethnomethodology's program: Working out Durkheim's aphorism*. Lanham, MD: Rowman & Littlefield.

Gay, L. R. (1976). *Educational research: Competencies for analysis and application*. Ohio: Bell & Howell Company.

George, D., & Mallery, P. (2014). *IBM SPSS Statistics 21 Step by Step: A Simple Guide and Reference* (13th ed.). Boston, MA: Allyn and Bacon.

Gerstenfeld, P. B., Grant, D. R., & Chang, C. (2003). Hate online: A content analysis of extremist Internet sites. *Analyses of Social Issues and Public Policy (ASAP)*, *3*(1), 29–44. doi:10.1111/j.1530-2415.2003.00013.x

Ghidini, C., Kump, B., Lindstaedt, S., Mahbub, N., Pammer, V., Rospocher, M., & Serafini, L. (2009). Moki: The Enterprise Modelling Wiki. In *Proceedings of the Semantic Web: Research and Applications* (pp. 831-835). Springer.

Giddings, L. S. (2006). Mixed-methods research: Positivism dressed in drag? *Journal of Research in Nursing*, *11*(3), 195–203. doi:10.1177/1744987106064635

Giorgi, A. (1970). *Psychology as a human science*. New York, NY: Harper & Row.

Glaser, B., & Strauss, A. (1967). *The discovery of grounded theory: Strategies for qualitative research*. New Brunswick, NJ: Aldine Transaction.

Goldenberg, C., Gallimore, R., & Reese, L. (2005). Using mixed methods to explore Latino children's literacy development. *Discovering Successful Pathways in Children's Development: Mixed Methods in the Study of Childhood and Family Life*, 21–46.

Goldkuhl, G. (2012). Pragmatism vs interpretivism in qualitative information systems research. *European Journal of Information Systems*, *21*(2), 135–146. doi:10.1057/ejis.2011.54

Graff, J. C. (2013). Mixed Methods Research. In Evidence-Based Practice: An Integrative Approach to Research, Administration and Practice (pp. 44-64). Academic Press.

Gravetter, F. J., & Forzano, L. B. (2006). *Research methods for the behavioural sciences* (2nd ed.). Mason, OH: Thompson.

Gray, D. E. (2009). *Doing research in the real world.* London: Sage.

Greene, J. C. (2007). *Mixed methods in social inquiry*. San Francisco, CA: John Wiley & Sons.

Greene, J. C. (2007). *Mixing methods in social inquiry*. San Francisco: Jossey-Bass.

Greene, J. C., & Caracelli, V. J. (1997). *Advances in mixed-method evaluation: The challenges and benefits of integrating diverse paradigms*. Jossey-Bass Publishers.

Greene, J. C., & Caracelli, V. J. (1997). Defining and describing the paradigm issue in mixed- method evaluation. In J. C. Green & V. J. Caracelli (Eds.), *Advances in Mixed-Method Evaluation: The Challenges and Benefits of Integrating Diverse Paradigms* (pp. 5–18). San Francisco: Jossey-Bass. doi:10.1002/ev.1068

Greene, J. C., & Caracelli, V. J. (2003). Making paradigmatic sense of mixed methods practice. In A. Tashakkori & C. Teddlie (Eds.), *Handbook of mixed methods in social and behavioral research* (pp. 91–110). Thousand Oaks, CA: Sage Publications.

Green, J. C., Caracelli, V. J., & Graham, W. F. (1989). Toward a conceptual framework for mixed-method evaluation designs. *Educational Evaluation and Policy Analysis*, *11*(3), 255–274. doi:10.3102/01623737011003255

Grewal, R., Cote, J. A., & Baumgartner, H. (2004). Multicollinearity and Measurement Error in Structural Equation Models: Implications for Theory Testing. *Marketing Science*, *23*(4), 519–529. doi:10.1287/mksc.1040.0070

Grobler, M., Jansen van Vuuren, J., & Leenen, L. (2012). *Implementation of a Cyber Security Policy in South Africa: Reflection on Progress and the Way Forward. In ICT Critical Infrastructures and Society* (pp. 215–225). Springer.

Groenewald, T. (2004). A phenomenological research design illustrated. *International Journal of Qualitative Methods*, *3*(1). Retrieved, from http://www.ualberta.ca/~iiqm/backissues/3_1/pdf/groenewald.pdf

Gruber, T. R. (1993). A translation approach to portable ontology specifications. *Knowledge Acquisition*, *5*(2), 199–220. doi:10.1006/knac.1993.1008

Guarino, N. (1998). *Formal ontology in information systems.* Paper presented at the Formal ontology in information systems, Trento, Italy.

Guba, E. C. (1990). The alternative paradigm dialogue. In E. C. Guba (Ed.), *The Paradigm Dialogue* (pp. 17–27). Newbury Park, CA: Sage.

Guba, E. G. (1990). *The paradigm dialog*. Newbury Park, CA: Sage Publications.

Guba, E. G., & Lincoln, Y. S. (1989). *Fourth generation evaluation*. Newbury Park, CA: Sage Publications.

Guba, E. G., & Lincoln, Y. S. (1994). Competing paradigms in qualitative research. In N. K. Dezin & Y. S. Lincoln (Eds.), *Handbook of qualitative research* (pp. 105–117). Thousand Oaks, CA: Sage Publications.

Guba, E. G., & Lincoln, Y. S. (2005). Paradigmatic controversies, contradictions, and emerging confluences. In N. K. Denzin & Y. S. Lincoln (Eds.), *The Sage handbook of qualitative research* (3rd ed., pp. 191–215). Thousand Oaks, CA: Sage Publications.

Guest, G., MacQueen, K., & Namey, E. (2012). *Applied thematic analysis*. Thousand Oaks, CA: Sage Publications. doi:10.4135/9781483384436

Guest, G., Namey, E. E., & Mitchell, M. L. (2012). Collecting qualitative data: A field manual for applied research. *Sage (Atlanta, Ga.)*.

Gummesson, E. (2000). *Qualitative methods in management research*. Thousand Oaks, CA: Sage.

Hall, R. F. (2013). Mixed methods: In search of a paradigm. In T. Le & Q. Le (Eds.), *Conducting research in a changing and challenging world* (pp. 71–78). Nova Science Publishers, Inc.

Halton, E. (2004). In G. Ritzer (Ed.), *Pragmatism. Encyclopedia of Social Theory*. Thousand Oaks, CA: Sage.

Hammersley, M. (2002). *Educational research, policymaking and practice*. London: Sage.

Harrigan, P., Ramsey, E., & Ibbotson, P. (2012). Exploring and explaining SME marketing: Investigating e-CRM using a mixed methods approach. *Journal of Strategic Marketing*, *20*(2), 127–163. doi:10.1080/0965254X.2011.606911

Harrison, R. L., & Reilly, T. M. (2011). Mixed methods designs in marketing research. *Qualitative Market Research: An International Journal*, *14*(1), 7–26. doi:10.1108/13522751111099300

Harrits, G. S. (2011). More than method?: A discussion of paradigm differences within mixed methods research. *Journal of Mixed Methods Research*, *5*(2), 150–166. doi:10.1177/1558689811402506

Hartley & Muhit. (2003). Using qualitative research methods for disability research in majority world countries. *Asia Pacific Disability Rehabilitation Journal*, *103*(14).

Hawkins, D. (2004). Explaining costly international institutions: Persuasion and enforceable human rights norms. *International Studies Quarterly*, *48*(4), 779–804. doi:10.1111/j.0020-8833.2004.00325.x

Hayes, B., Bonner, A., & Douglas, C. (2013). An introduction to mixed methods research for nephrology nurses. *Ren Soc Australasia J*, *9*(1), 8–14.

Henry, G. T. (1990). *Practical sampling* (Vol. 21). London: Sage Publications. doi:10.4135/9781412985451

Henry, G. T., Julnes, G., & Mark, M. M. (1998). *Realist evaluation: An emerging theory in support of practice: New directions for evaluation, No. 78*. San Francisco, CA: Jossey-Bass Publishers.

Hewege, C. R. (2011). *Resolving structure-egency dichotomy in management research: Case for adaptive theory research methodology*. Paper presented at the 24th Annual Australian and New Zealand Academy of Management Conference, North Terrace, Australia.

Hilton, B., Budgen, C., Molzahn, A. E., & Attridge, C. B. (2001). Developing and testing instruments to measure client outcomes at the Comox Valley Nursing Center. *Public Health Nursing (Boston, Mass.)*, *18*(5), 327–339. doi:10.1046/j.1525-1446.2001.00327.x PMID:11559416

Hollongshead, A. B. (1949). *Elmtown's youth*. New York: John Wiley.

Horkheimer, M. (1972). Critical Theory: Selected Essays. New York: Herder and Herder.

Horrocks, I., Patel-Schneider, P. F., & Van Harmelen, F. (2003). From SHIQ and RDF to OWL: The making of a web ontology language. *Web Semantics: Science, Services, and Agents on the World Wide Web*, *1*(1), 7–26. doi:10.1016/j.websem.2003.07.001

Houts, A. C., Cook, T. D., & Shadish, W. R. (1986). The person-situation debate: A critical multiplist perspective. *Journal of Personality*, *54*(1), 52–105. doi:10.1111/j.1467-6494.1986.tb00390.x

Howe, K. R. (1988). Against the quantitative-qualitative incompatibility thesis or dogmas die hard. *Educational Researcher, 17*(8), 10-16.

Howe, K. R. (1988). Against the quantitative-qualitative incompatibility thesis or, Dogmas die hard. *Educational Researcher*, *17*, 10–16.

Hsieh, H. F., & Shannon, S. E. (2005). Three approaches to qualitative content analysis. *Qualitative Health Research*, *15*(9), 1277–1288. doi:10.1177/1049732305276687 PMID:16204405

Hudson, J. M., & Bruckman, A. (2004). "Go away": Participant objections to being studied and the ethics of chatroom research. *The Information Society*, *20*(2), 127–139. doi:10.1080/01972240490423030

Hunt, S. D. (1992). For reason and realism in marketing. *Journal of Marketing*, *56*(2), 89–102. doi:10.2307/1252045

Iacobucci, D. (2008). *Mediation Analysis*. Thousand Oaks, CA: Sage.

Iacobucci, D., Saldhana, N., & Deng, X. (2007). A Meditation on Mediation: Evidence that Structural Equations Models Perform Better Than Regressions. *Journal of Consumer Psychology*, *17*(2), 139–153. doi:10.1016/S1057-7408(07)70020-7

Igo, L. B., Bruning, R. A., & Riccomini, P. J. (2009). Should Middle School Students with Learning Problems Copy and Paste Notes from the Internet? Mixed-Methods Evidence of Study Barriers. *RMLE Online: Research in Middle Level Education*, *33*(2), n2. doi:10.1080/19404476.2009.11462065

Internet World Stats. (2011). Internet Usage Statistics: The Internet big picture -World Internet users and population stats. Retrieved from http://www.Internetworldstats.com/stats.htm

Irwin, J. R., & McClelland, G. H. (2001). Misleading Heuristics and Moderated Multiple Regression Models. *JMR, Journal of Marketing Research*, *38*(1), 100–109. doi:10.1509/jmkr.38.1.100.18835

Isaac, S., & Michael, W. B. (1995). *Handbook in research and evaluation: A collection of principles, methods, and strategies useful in the planning, design, and evaluation of studies in education and the behavioral sciences* (3rd ed.). San Diego, CA: Educational and Industrial Testing Services.

Ivankova, N. V. (2013). Implementing quality criteria in designing and conducting a sequential QUAN→ QUAL mixed methods study of student engagement with learning applied research methods online. *Journal of Mixed Methods Research*, 1–27. doi:10.1177/1558689813487945

Ivankova, N. V., Creswell, J. W., & Stick, S. L. (2006). Using mixed-methods sequential explanatory design: From theory to practice. *Field Methods*, *18*(1), 3–20. doi:10.1177/1525822X05282260

Jaeger, R. M. (1997). *Complementary research methods for research in education* (2nd ed.). Washington, DC: American Educational Research Association.

Jahoda, M., Lazarsfeld, P. F., & Zeisel, H. (2003). *Marienthal: The sociography of an unemployed community*. New Brunswick, NJ: Transaction Publishers. (Original work published 1931)

Janesick, V. J. (2011). *Stretching exercises for qualitative researchers* (3rd ed.). Thousand Oaks: Sage.

Jansen van Vuuren, J. C., Leenen, L., & Zaaiman, J. J. (2014). *Using an Ontology as a Model for the Implementation of the National Cybersecurity Policy Framework for South Africa.* Paper presented at the 9th International Conference on Cyber Warfare & Security.

Jansen van Vuuren, J., Leenen, L., Phahlamohlaka, J., & Zaaiman, J. (2013). *Development of a South African cybersecurity policy implementation framework.* Paper presented at the 8th International Conference on Information Warfare and Security, Denver, CO.

Jansen van Vuuren, J., Phahlamohlaka, J., & Leenen, L. (2012). *Governance of cybersecurity in South Africa.* Paper presented at the 11th European Conference on Information Warfare and Security, Laval, France.

Jansen van Vuuren, J. C., Leenen, L., Phahlamohlaka, J., & Zaaiman, J. J. (2012). An Approach to Governance of Cyber-Security in South Africa. *International Journal of Cyber Warfare & Terrorism*, *2*(4), 13–27. doi:10.4018/ijcwt.2012100102

Jansen van Vuuren, J., Phahlamohlaka, J., & Brazzoli, M. (2010). The Impact of the Increase in Broadband Access on National Security and the Average citizen. *Journal of Information Warfare*, *5*, 171–181.

Jenkins, J. E. (2001). Rural adolescent perceptions of alcohol and other drug resistance. *Child Study Journal*, *31*(4), 211–224.

Jick, T. D. (1979). Mixing qualitative and quantitative methods: Triangulation in action. *Administrative Science Quarterly*, *24*(4), 602–611. doi:10.2307/2392366

Johnson, B., & Gray, R. (2010). A history of philosophical and theoretical issues for mixed methods research. In A. Tashakkori & C. Teddlie (Eds.), *SAGE Handbook of Mixed Methods in Social & Behavioral Research* (2nd ed.; pp. 69–94). Thousand Oaks, CA: Sage. doi:10.4135/9781506335193.n3

Johnson, R. B., & Onwuegbuzie, A. J. (2004). Mixed methods research: A research paradigm whose time has come. *Educational Researcher*, *33*(7), 14–26. doi:10.3102/0013189X033007014

Johnson, R. B., Onwuegbuzie, A. J., & Turner, L. A. (2007). Toward a definition of mixed methods research. *Journal of Mixed Methods Research*, *1*(2), 112–133. doi:10.1177/1558689806298224

Johnson, R. B., & Turner, L. A. (2003). Data collection strategies in mixed methods research. In A. Tashakkori & C. Teddlie (Eds.), *Handbook of mixed methods in social and behavioral research* (pp. 297–319). Thousand Oaks, CA: Sage.

Joinson, A. N. (2001). Knowing me, knowing you: Reciprocal self-disclosure in Internet-based surveys. *Cyberpsychology & Behavior*, *4*(5), 587–591. doi:10.1089/109493101753235179 PMID:11725651

Jones, H. L. (1955). The Application of Sampling Procedures to Business Operations. *Journal of the American Statistical Association*, *50*(271), 763–774.

Jonker, J., & Pennink, B. (2010). *The essence of research methodology: A consice guide for master and PhD students in management science.* Heidelberg, Germany: Springers.

Judd, C. M., & Kenny, D. A. (2010). In D. Gilbert, S. T. Fiske, & G. Lindzey (Eds.), *Data Analysis in Social Psychology: Recent and Recurring Issues. In the Handbook of Social Psychology* (5th ed.; pp. 233–265). New York: McGraw-Hill.

Kaid, L. L. & Wadsworth. (1989). Content analysis. In P. Emmert & L. L. Barker (Eds.), *Measurement of communication behavior* (pp. 197-217). New York, NY: Longman.

Kalaian, S. A. (2008). Frequency Distribution. In P. J. Lavarkas (Ed.), *Encyclopedia of Survey Research Methods* (pp. 292–294). Thousand Oaks, CA: Sage Publications. doi:10.4135/9781412963947.n195

Kalaian, S. A., & Kasim, R. M. (2008). Research Hypothesis. In P. J. Lavarkas (Ed.), *Encyclopedia of Survey Research Methods* (pp. 334–335). Thousand Oaks, CA: Sage Publications. doi:10.4135/9781412963947.n472

Kalaian, S. A., & Kasim, R. M. (2015). Predictive Analytics. In M. Tavana, S. B. Zhou, & S. K. Puranam (Eds.), *Organizational Transformations through Big Data Analytic* (pp. 12–29). Hershey, PA: IGI Global.

Kalman, Y. M., Ravid, G., Raban, D. R., & Rafaeli, S. (2006). Pauses and response latencies: A chronemic analysis of asynchronous CMC. *Journal of Computer-Mediated Communication*, *12*(1), 1–23. doi:10.1111/j.1083-6101.2006.00312.x

Kalof, L., Dan, A., & Dietz, T. (2008). *Essential of social research*. New York: McGraw-Hill.

Kassel, G. (2005). *Integration of the DOLCE top-level ontology into the OntoSpec methodology*. Technical Report HAL: hal-00012203/arXiv: cs.AI/0510050, Laboratoire de Recherche en Informatique d'Amiens (LaRIA). Retrieved from http://hal.archives-ouvertes.fr/ccsd-00012203

Kaye, B. K., & Johnson, T. J. (1999). Research methodology: Taming the cyber frontier. *Social Science Computer Review*, *17*(3), 323–337. doi:10.1177/089443939901700307

Kiesler, S., Siegel, J., & Mcguire, T. W. (1984). Social psychological aspects of computer-mediated communication. *The American Psychologist*, *39*(10), 1123–1134. doi:10.1037/0003-066X.39.10.1123

King, S. A. (1996). Researching Internet communities: Proposed ethical guidelines for the reporting of results. *The Information Society*, *12*(2), 119–128. doi:10.1080/713856145

Kirk, R. E. (1999). *Statistics: An Introduction*. London: Harcourt Brace.

Kish, L. (1965). *Survey Sampling*. New York: John Wiley & Sons.

Klassen, A. C., Creswell, J., Clark, V. L. P., Smith, K. C., & Meissner, H. I. (2012). Best practices in mixed methods for quality of life research. *Quality of Life Research: An International Journal of Quality of Life Aspects of Treatment, Care and Rehabilitation*, *21*(3), 377–380. doi:10.1007/s11136-012-0122-x PMID:22311251

Knoblauch, H., Baer, A., Laurier, E., Petschke, S., & Schnettler, B. (2008). Visual analysis. *Qualitative Social Research, 9*(3), Art. 14. Retrieved from http://nbn-resolving.de/urn:nbn:de:0114-fqs0803148

Koller, M. (2008). A future research agenda for mixed-method-design in business research. *International Journal of Business Research*, *8*(4), 50–60.

Kothari, C. R. (2004). *Research methodology, Methods and Techniques* (2nd ed.). New Delhi: New Age International.

Kothari, C. R. (2007). *Quantitative techniques*. New Delhi: UBS Publishers Limited.

Krauss, S. E. (2005). Research paradigms and meaning making: A primer. *Qualitative Report*, *10*(4), 758–770.

Krippendorff, K. (2004). Reliability in content analysis: Some common misconceptions and recommendations. *Human Communication Research*, *30*, 411–433.

Krippendorf, K. (1980). *Content analysis: An introduction to its methodology* (2nd ed.). London, GB: Sage.

Kuhn, T. S. (1962). *The structure of scientific revolutions*. Chicago: University of Chicago Press.

Kuhn, T. S. (1970). *The structure of scientific revolution* (2nd ed.). Chicago: University of Chicago Press.

Kutner, J. S., Steiner, J. F., Corbett, K. K., Jahnigen, D. W., & Barton, P. L. (1999). Information needs in terminal illness. *Social Science & Medicine*, *48*(10), 1341–1352. doi:10.1016/S0277-9536(98)00453-5 PMID:10369435

Landis, J. R., & Koch, G. G. (1977). The measurement of observer agreement for categorical data. *Biometrics*, *33*(1), 159–174. doi:10.2307/2529310 PMID:843571

Lapin, L. L. (1987). *Statistics for modern business decisions*. Harcourt, Jovanovich, Inc.

Lather, P. (2004). This *IS* your father's paradigm: Government intrusion and the case of qualitative research in education. *Qualitative Inquiry*, *10*(1), 15–34. doi:10.1177/1077800403256154

Laughlin, R. (1995). Empirical research in accounting: Alternative approaches and a case for "middle-range" thinking. *Accounting, Auditing & Accountability Journal*, *8*(1), 63–87. doi:10.1108/09513579510146707

Lave, J., & Wenger, E. (1991). *Situated learning: Legitimate peripheral participation*. Cambridge, UK: Cambridge University Press. doi:10.1017/CBO9780511815355

Lavidge, R. J., & Steiner, G. A. (1961, October). A model for predictive measurements of advertising effectiveness. *Journal of Marketing*, *25*(6), 59–62. doi:10.2307/1248516

Lee, A., & Nickerson, J. (2010). Theory as a case of design: Lessons for design from the philosophy of science. In *Proc. of the 43rd Hawaii International Conference on System Sciences*. doi:10.1109/HICSS.2010.484

Leech, N. L., & Onwuegbuzie, A. J. (2009). A typology of mixed methods research designs. *Quality & Quantity*, *43*(2), 265–275. doi:10.1007/s11135-007-9105-3

Leedy, P. D., & Ormrod, J. E. (2013). *Practical research: Planning and design* (10th ed.). Boston: Pearson.

Lewin, K. (1947). Group decision and social change. In T. M. Newcombe & E. L. Hartley (Eds.), *Readings in Social Psychology*. New York: Holt.

Lewis, M. W., & Grimes, A. I. (1999). Metatriangulation: Building theory from multiple paradigms. *Academy of Management Review*, *24*(4), 672–690.

Lieber, E., (2009). Mixing qualitative and quantitative methods: Insights into design and analysis issues. *Journal of Ethnographic and Qualitative Research*, (3), 218-227.

Lincoln, Y. S., & Guba, E. G. (1985). *Naturalistic inquiry*. Beverly Hills, CA: Sage Publications.

Lincoln, Y. S., & Guba, E. G. (1985). *Naturalistic inquiry.* Técnicas.

Lin, X., & Lehman, J. D. (1999). Supporting learning of variable control in a computer-based biology environment: Effects of prompting college students to reflect on their own thinking. *Journal of Research in Science Teaching*, *36*(7), 837–858. doi:10.1002/(SICI)1098-2736(199909)36:7<837::AID-TEA6>3.0.CO;2-U

Lohr, S. L. (1999). *Sampling: Design and Analysis*. Albany, NY: Duxbury Press.

Lombard, M., Snyder-Duch, J., & Bracken, C. C. (2002). Content analysis in mass communication: Assessment and reporting of intercoder reliability. *Human Communication Research*, *28*(4), 587–604. doi:10.1111/j.1468-2958.2002.tb00826.x

Lyman, H. R. (1983). *An introduction to statistical methods and data analysis* (4th ed.). Dioxburg Press.

Lynd, R. S., & Lynd, H. M. (1929/1959). *Middletown: A study in modern American culture*. Orlando, FL: Harcourt Brace.

Mackenzie, N., & Knipe, S. (2006). Research dilemmas: Paradigms, methods and methodology. *Issues in Educational Research*, *16*(2), 193–205.

MacNealy, M. S. (1999). *Strategies for empirical research in writing*. New York: Longman.

Mak, L., & Marshall, S. K. (2004). Perceived mattering in young adults' romantic relationships. *Journal of Social and Personal Relationships*, *21*(4), 469–486. doi:10.1177/0265407504044842

Malhotra, N. K. (2004). *Marketing research: an applied orientation* (4th ed.). London: Prentice-Hall International.

Martins, J., Santos, H., Nunes, P., & Silva, R. (2012). *Information Security Model to Military Organizations in Environment of Information Warfare*. Paper presented at the 11th European Conference on Information Warfare and Security, Laval, France.

Maxcy, S. J. (2003). Pragmatic threads in mixed methods research in the social sciences: The search for multiple modes of inquiry and the end of the philosophy of formalism. In A. Tashakkori & C. Teddlie (Eds.), *Handbook of Mixed Methods in Social & Behavioral Research* (pp. 51–90). Thousand Oaks, CA: Sage.

Maxcy, S. J. (2003). Pragmatic threads in mixed methods research in the social sciences: The search for multiple modes of inquiry and the end of the philosophy of formalism. In A. Tashakkori & C. Teddlie (Eds.), *Handbook of mixed methods in social and behavioral research* (pp. 51–90). Thousand Oaks, CA: Sage Publications.

Maxwell, J. A. (2005). *Qualitative research design: An interactive approach* (2nd ed.). Thousand Oaks, CA: Sage.

Maxwell, J. A., & Loomis, D. M. (2003). Mixed methods design: An alternative approach. In A. Tashakkori & C. Teddlie (Eds.), *Handbook of Mixed Methods in Social & Behavioral Research* (pp. 241–271). Thousand Oaks, CA: Sage.

Maxwell, J. A., & Mittapalli, K. (2010). Realism as a stance for mixed methods research. In A. Tashakkori & C. Teddlie (Eds.), *SAGE Handbook of Mixed Methods in Social & Behavioral Research* (2nd ed., pp. 145–162). Thousand Oaks, CA: Sage. doi:10.4135/9781506335193.n6

May, D. B., & Etkina, E. (2002). College physics students' epistemological self-reflection and its relationship to conceptual learning. *American Journal of Physics*, *70*(12), 1249–1258. doi:10.1119/1.1503377

Mayring, P. (2007). Introduction: Arguments for mixed methodology. In P. Mayring, G. L. Huber, L. Gurtler, & M. Kiegelmann (Eds.), *Mixed methodology in psychological research* (pp. 1–4). Rotterdam, Taipei: Sense Publishers.

McGawley, J. (1982). *Thirty million theories of grammar*. Chicago: University of Chicago Press.

McKenna, K. Y. A., Green, A. S., & Gleason, M. E. J. (2002). Relationship formation on the Internet: What's the big attraction? *The Journal of Social Issues*, *58*(1), 9–31. doi:10.1111/1540-4560.00246

McNiff, J. (1988). *Action Research: Principles and Practice*. Basingstoke, UK: Macmillan. doi:10.4324/9780203305676

McNiff, J., & Whitehead, J. (2006). *All you need to know about action research*. London: Sage.

McNiff, J., & Whitehead, J. (2011). *All You Need to Know About Action Research*. London: Sage.

McVea, K., Crabtree, B. F., Medder, J. D., Susman, J. L., Lukas, L., McIlvain, H. E., & Hawver, M. et al. (1996). An ounce of prevention? Evaluation of the'Put Prevention into Practice'program. *The Journal of Family Practice*, *43*(4), 361–369. PMID:8874371

Mertens, D. M. (2003). Mixed methods and the politics of human research: The transformative-emancipatory perspective. In A. Tashakkori & C. Teddlie (Eds.), *Handbook of mixed methods in social and behavioral research* (pp. 135–166). Thousand Oaks, CA: Sage Publications.

Mertens, D. M. (2010). *Research and evaluation in education and psychology: Integrating diversity with quantitative, qualitative, and mixed methods* (3rd ed.). Thousand Oaks, CA: Sage.

Mertler, C. A., & Vannatta, R. A. (2013). *Advanced and Multivariate Statistical Methods* (3rd ed.). Glendale, CA: Pyrczak Publishing.

Miller, J. L. (2012). *Design's Mechanism--General Morphological Analysis.* DTIC Document.

Mingers, J. (2001). Combining IS Research Methods: Towardsa Pluralist Methodology. *Information Systems Research, 12*(3), 240–259.

Morahan-Martin, J., & Schumacher, P. (2000). Incidence and correlates of pathological Internet use among college students. *Computers in Human Behavior, 16*(1), 13–29. doi:10.1016/S0747-5632(99)00049-7

Morese, J. M. (2003). Principles of mixed methods and multimethod research design. In A. Tashakkori & C. Teddlie (Eds.), *Handbook of mixed methods in social and behavioral research* (pp. 189–208). Thousand Oaks, CA: Sage Publications.

Morgan, D. L. (1998). Practical strategies for combining qualitative and quantitative methods: Applications to health research. *Qualitative Health Research, 8*(3), 362–376. doi:10.1177/104973239800800307 PMID:10558337

Morgan, D. L. (2007). Paradigms lost and pragmatism regained. Methodological implications of combining qualitative and quantitative methods. *Journal of Mixed Methods Research, 1*(1), 48–76. doi:10.1177/2345678906292462

Morio, H., & Buchholz, C. (2009). How anonymous are you online? Examining online social behaviors from a cross-cultural perspective. *AI & Society, 23*(2), 297–307. doi:10.1007/s00146-007-0143-0

Morse, J. M. (1991a). Strategies for sampling. In J. M. Morse (Ed.), *Qualitative nursing research: A contemporary dialogue* (pp. 127–145). Newbury Park, CA: Sage Publications. doi:10.4135/9781483349015.n16

Morse, J. M. (1991b). Approaches to qualitative-quantitative methodological triangulation. *Nursing Research, 40*(2), 120–123. doi:10.1097/00006199-199103000-00014 PMID:2003072

Morse, J. M., & Niehaus, L. (2009). *Mixed methods design: Principles and procedures.* Walnut Creek, CA: Left Coast Press.

Mosteller, F., & Boruch, R. F. (2002). *Evidence matters: Randomized trials in education research.* Washington, DC: Brookings Institution Press.

Moustakes, C. (1994). *Phenomenological research methods.* Thousand Oaks, CA: Sage Publications.

Mujis, D. (2004). *Doing Quantitative Research in Education with SPSS.* London: Sage Publications.

Mumford, A. (2006). Management Development - Strategies for Action. Development and Learning in Organizations: An International Journal, 20(2).

Mumford, M. D., Bedell-Avers, K. E., Hunter, S. T., Espejo, J., Eubanks, D., & Connelly, M. S. (2008). Violence in ideological and non-ideological groups: A quantitative analysis of qualitative data. *Journal of Applied Social Psychology, 38*(6), 1521–1561. doi:10.1111/j.1559-1816.2008.00358.x

Murphy, R. (2008, September 14). *Post-Positivist Perspectives on Theory Development.* Retrieved from https://davidlobato.wordpress.com/2008/09/14/chapter-3-post-positivist-perspectives-on-theory-development/

Mutai, B. K. (2000). *How to write quality research proposal: A complete and simplified recipe.* New Delhi: Thelley Publications.

Myers, K. K., & Oetzel, J. G. (2003). Exploring the dimensions of organizational assimilation: Creating and validating a measure. *Communication Quarterly, 51*(4), 438–457. doi:10.1080/01463370309370166

National Center for Educational Statistics. (2014). *College navigator.* Retrieved from: http://nces.ed.gov/collegenavigator/?q=East+West+University&s=all&id=144883

Neuman, S. (2006). Speak up. *Scholastic Early Childhood Today, 20*(4), 12–13.

Neuman. (2000). *Social research methods: qualitative and quantitative approaches* (4th ed.). Boston: Allyn & Bacon.

Neuman, W. L. (2011). *Social research methods: Qualitative and quantitative approaches* (7th ed.). Boston, MA: Pearson/Allyn and Bacon.

New York University. (n.d.). *What is research design? The context of design*. Retrieved from http://www.nyu.edu/classes/bkg/methods/005847ch1.pdf

Newman, I., Ridenour, C. S., & Newman, C. D., Jr. (2003). A typology of research purposes and its relationship to mixed methods. In Handbook of Mixed Methods in Social and Behavioural Research, (pp. 167-188). Academic Press.

Nonnecke, B., Andrews, D., & Preece, J. (2006). Non-public and public online community participation: Needs, attitudes, and behavior. *Electronic Commerce Research, 6*(1), 7–20. doi:10.1007/s10660-006-5985-x

Onwuegbuzie, A. J., & Leech, N. L. (2006). Linking research questions to mixed methods data analysis procedures. *Qualitative Report, 11*(3), 474–498.

Onwuegbuzie, A. J., & Teddlie, C. (2003). A framework for analyzing data in mixed methods research. In A. Tashakkori & C. Teddlie (Eds.), *Handbook of mixed methods in social and behavioral research* (pp. 351–383). Thousand Oaks, CA: Sage Publications.

Owens, J. M., & Rogers, P. J. (2003). *Program evaluation forms and approaches* (3rd ed.). London: Sage Publication.

Pagano, M. E., Hirsch, B. J., Deutsch, N. L., & McAdams, D. P. (2003). The transmission of values to school-age and young adult offspring: Race and gender differences in parenting. *Journal of Feminist Family Therapy, 14*(3-4), 13–36. doi:10.1300/J086v14n03_02 PMID:22323854

Parker, L. D. (2012). Qualitative management accounting research: Assessing deliverables and relevance. *Critical Perspectives on Accounting, 23*(1), 54–70. doi:10.1016/j.cpa.2011.06.002

Parry, S., Kupiec-Teahan, B., & Rowley, J. (2012). Exploring marketing and relationships in software SMEsA mixed methods approach. *Management Research Review, 35*(1), 52–68. doi:10.1108/01409171211190805

Pasick, R. J., Burke, N. J., Barker, J. C., Galen, J., Bird, J. A., & Otero-Sabogal, R. (2009). Behavioral theory in a diverse society: Like a compass on Mars. *Health Education & Behavior, 36*(5), 11S–35S. doi:10.1177/1090198109338917 PMID:19805789

Patton, M. Q. (2005). *Qualitative research.* John Wiley & Sons, Ltd.

Patton, M. Q. (1990). *Qualitative evaluation and research methods* (2nd ed.). Newbury Park, CA: Sage Publications.

Patton, M. Q. (1996). A world larger than formative and summative. *Evaluation Practice, 17*(2), 131–144. doi:10.1016/S0886-1633(96)90018-5

Patton, M. Q. (2002). *Qualitative research & evaluation methods* (3rd ed.). Thousand Oaks, CA: Sage.

Pawson, R., & Tilly, N. (1997). *Realistic evaluation.* London: Sage Publications.

Peffers, K., Tuunanen, T., Rothenberger, M. A., & Chatterjee, S. (2007). A design science research methodology for information systems research. *Journal of Management Information Systems, 24*(3), 45–77. doi:10.2753/MIS0742-1222240302

Perry, C., Riege, A., & Brown, L. (1998). *Realism rules OK: Scientific paradigms in marketing research about networks.* Paper presented at the 1998 Annual Australian and New Zealand Academy of Management Conference, Dunedin, New Zealand.

PEW Internet and American Life Project. (2010). *Teenage trend data.* Retrieved from http://pewInternet.org/Reports/2010/Social-Media-and-Young-Adults.aspx

PEW Internet and American Life Project. (2012). *Adult trend data.* Retrieved from http://pewInternet.org/Static-Pages/Trend-Data-%28Adults%29/Online-Activites-Total.aspx

Phahlamohlaka, L. J., Jansen van Vuuren, J. C., & Coetzee, A. J. (2011). *Cyber security awareness toolkit for national security: an approach to South Africa's cyber security policy implementation.* Paper presented at the SACSAW, Gaberone, Botswana.

Phillips, D. C., & Burbules, N. C. (2000). *Postpositivism and educational research.* Lanham, MA: Rowman & Littlefield Publishers, Inc.

Pitman, M. A., & Maxwell, J. A. (1990). Qualitative approaches to evaluation. In M. D. LeCompte, W. L. Milroy, & J. Preissle (Eds.), *The Handbook of Qualitative Research in Education* (pp. 729–770). San Diego, CA: Academic Press.

Plano-Clark, V. I., & Badiee, M. (2010). Research questions in mixed methods research. In A. Tashakkori & C. Teddlie (Eds.), *SAGE Handbook of Mixed Methods in Social & Behavioral Research* (2nd ed., pp. 275–300). Thousand Oaks, CA: Sage. doi:10.4135/9781506335193.n12

Polit, D. F., & Hungler, B. P. (1999). *Nursing research: Principles and Methods* (6th ed.). Philadelphia: Lippincott.

Ponterotto, J. G. (2005). Qualitative research in counseling psychology: A primer on research paradigms and philosophy of science. *Journal of Counseling Psychology, 52*(2), 126–136. doi:10.1037/0022-0167.52.2.126

Popper, K. R. (1959). *The logic of scientific discovery.* London: Hutchinson.

Posavac, E. J., & Carey, R. G. (2003). *Program evaluation methods and case studies* (6th ed.). Upper Saddle River, NJ: Prentice Hall.

Postmes, T., Spears, R., & Lea, M. (1998). Breaching or building social boundaries? SIDE-Effects of computer-mediated communication. *Communication Research, 25*(6), 689–715. doi:10.1177/009365098025006006

Raban, D. R., Rafaeli, S., Ravid, G., & Kalman, Y. M. (2006). Speak *now* or forever hold your peace: Power law chronemics of turn-taking and response in asynchronous CMC.*Conference Papers – International Communication Association, 2006 Annual Meeting*, 1-24.

Ragin, C. C., Nagel, J., & White, P. (2004). *Workshop on scientific foundations of qualitative research.* National Science Foundation.

Ragin, C., Nagel, J., & White, P. (2004). *Sociology program methodology, measurement & statistics program directorate for social, behavioral & economic sciences.* Presented at the Workshop on Scientific Foundations of Qualitative Research, Arlington, VA.

Reason, P., & Bradbury, H. (Eds.). (2001). *Handbook of action research: Participative inquiry and practice.* London: Sage.

Redmond, M. R., Mumford, M. D., & Teach, R. (1993). Putting creativity to work: Effects of leader behavior on subordinate creativity. *Organizational Behavior and Human Decision Processes, 55*(1), 120–151. doi:10.1006/obhd.1993.1027

Reichhardt, C. S., & Rallis, S. F. (1994). Qualitative and quantitative inquiries are not incompatible: A call for a new partnership. In C. S. Reichhardt & S. F. Rallies (Eds.), *The qualitative-quantitative debate: New perspectives.* San Francisco, CA: Jossey Bass. doi:10.1002/ev.1670

Ritchey, T. (1998). *Fritz Zwicky, morphologie and policy analysis.* Paper presented at the 16th EURO conference on operational analysis, Brussels, Belgium.

Ritchey, T. (2006). Problem structuring using computer-aided morphological analysis. *The Journal of the Operational Research Society*, *57*(7), 792–801. doi:10.1057/palgrave.jors.2602177

Ritchey, T. (2011). *Wicked problems–social messes: Decision support modelling with morphological analysis*. Springer. doi:10.1007/978-3-642-19653-9

Rogers, A., Day, J., Randall, F., & Bentall, R. P. (2003). Patients' understanding and participation in a trial designed to improve the management of anti-psychotic medication. *Social Psychiatry and Psychiatric Epidemiology*, *38*(12), 720–727. doi:10.1007/s00127-003-0693-5 PMID:14689177

Rossman, G. B., & Wilson, B. L. (1985). Numbers and words: Combining quantitative and qualitative methods in a single large-scale evaluation study. *Evaluation Review*, *9*(5), 627–643. doi:10.1177/0193841X8500900505

Rourke, L., & Anderson, T. (2004). Validity in quantitative content analysis. *Educational Technology Research and Development*, *52*(1), 5–18. doi:10.1007/BF02504769

Rourke, L., Anderson, T., Garrison, D. R., & Archer, W. (1999). Assessing social presence in asynchronous text-based computer conferencing. *Journal of Distance Education*, *14*, 50–71.

Russell, B. (1910). *Philosophical essays*. London.

Russell, B. (1945). *A history of western philosophy*. Forage Village, MA: Simon & Schuster.

SA Goverment Gazette. (2010). *South African National Cyber Security Policy*. Communication. Pretoria, Republic of South Africa. 536.

Salant, P., & Dillman, D. A. (1994). *How to conduct your own survey*. London: John Wiley & Sons, Inc.

Salkind, N. J. (2006). *Statistics for people who think they hate statistics*. Thousand Oaks, CA: Sage Publications.

Saunders, M., Lewis, P., & Thornhill, A. (2009). *Research methods for business students*. London: Pearson Education.

Sayer, A. (1992). *Method in social science: A realist approach*. London: Routledge.

Sayer, A. (2000). *Realism and social science*. London: Sage Publications. doi:10.4135/9781446218730

Schafer, J. A. (2002). Spinning the web of hate: Web-based hate propagation by extremist organizations. *The Journal of Criminal Justice and Popular Culture*, *9*, 69–88.

Schütz, A. (1967). *The Phenomenology of the social world*. Evanston, IL: Northwestern University Press.

Schwartz, R., Slate, J., & Onwuegbuzie, A. J. (1999). Empowering teachers: Actingupon action research. *GATEways to Teacher Education*, *11*(2), 44–59.

Sechrest, L., & Sidana, S. (1995). Quantitative and qualitative methods: Is there an alternative? *Evaluation and Program Planning*, *18*(1), 77–87. doi:10.1016/0149-7189(94)00051-X

Shadish, W. R., Cook, T. D., & Campbell, D. T. (2002). *Experimental and quasi-experimental designs for generalized causal inference*. Wadsworth Cengage learning.

Shadish, W. R. (1995). The quantitative-qualitative debates: "Dekuhnifying" the conceptual context. *Evaluation and Program Planning*, *18*(1), 47–49. doi:10.1016/0149-7189(94)00048-3

Shadish, W. R. (1998). Evaluation theory is who we are. *The American Journal of Evaluation*, *19*(1), 1–19. doi:10.1177/109821409801900102

Shrout, P. E., & Fleiss, J. L. (1979). Intraclass correlations: Uses in assessing rater reliability. *Psychological Bulletin*, *86*(2), 420–428. doi:10.1037/0033-2909.86.2.420 PMID:18839484

Shuttleworth, M. (n.d.). True Experimental Design. Retrieved from: https://explorable.com/true-experimental-design

Sieber, S. D. (1973). The integration of fieldwork and survey methods. *American Journal of Sociology*, *78*(6), 1335–1359. doi:10.1086/225467

Silva, C. N. (2008). Review: Catherine Marshall & Gretchen B. Rossman (2006). Designing qualitative research. Forum Qualitative Sozialforschung / Forum: Qualitative. *Social Research*, *9*(3), 13. http://nbn-resolving.de/urn:nbn:de:0114-fqs0803137

Simonton, D. K. (1999). Significant samples: The psychological study of eminent individuals. *Psychological Methods*, *4*(4), 425–451. doi:10.1037/1082-989X.4.4.425

Simonton, D. K. (2003). Qualitative and quantitative analyses of historical data. *Annual Review of Psychology*, *54*(1), 617–640. doi:10.1146/annurev.psych.54.101601.145034 PMID:12171999

Simperl, E. P. B., Mochol, M., & Bürger, T. (2010). Achieving Maturity: The State of Practice in Ontology Engineering in 2009. *IJCSA*, *7*(1), 45–65.

Skinner, D., Matthews, S., & Burton, L. (2005). Combining ethnography and GIS technology to examine constructions of developmental opportunities in contexts of poverty and disability. *Discovering Successful Pathways in Children's Development: Mixed Methods in the Study of Childhood and Family Life*, 223–239.

Smith, J. K. (1983). Quantitative versus qualitative research: An attempt to clarify the issue. *Educational Researcher*, *12*(3), 6–13. doi:10.3102/0013189X012003006

Smith, J. K., & Heshusius, L. (1986). Closing down the conversation: The end of the quantitative-qualitative debate among educational inquirers. *Educational Researcher*, *15*(1), 4–12. doi:10.3102/0013189X015001004

Smith, J., Flowers, P., & Larkin, M. (2009). *Interpretive phenomenological analysis: Theory, method and research.* Thousand Oaks, CA: Sage Publications.

Smith, S. (2013). *Determining sample size: How to ensure you get the correct sample size.* LLC, Qualtrics.

Smith, T. M., & Ingersoll, R. M. (2004). What are the effects of induction and mentoring on beginning teacher turnover? *American Educational Research Journal*, *41*(3), 681–714. doi:10.3102/00028312041003681

Sobh, R., & Perry, C. (2006). Research design and data analysis in realism research. *European Journal of Marketing*, *40*(11/12), 1194–1209. doi:10.1108/03090560610702777

Sokal, R. R., & Rohlf, F. J. (1981). *Biometry: The Principles and Practice of Statistics in Biological Research* (3rd ed.). Oxford, UK: W. H. Freeman.

Solomon, D. J. (2001). Conducting web surveys. *Practical Assessment, Research & Evaluation, 7*(19).

Somekh, B., & Lewin, C. (2005). *Research methods in social science.* London: Sage.

Stake, R. E. (1978). The case study method in social inquiry. *Educational Researcher*, *7*(2), 5–8. doi:10.3102/0013189X007002005

Stake, R. E. (1995). *The Art of Case Study Research.* Thousand Oaks, CA: Sage.

Stanford University. (2013). *Protégé ontology editor: Stanford University.* Retrieved from http://protege.stanford.edu/

Steckler, A., McLeroy, K. R., Goodman, R. M., Bird, S. T., & McCormick, L. (1992). Toward integrating qualitative and quantitative methods: An introduction. *Health Education Quarterly*, *19*(1), 1–8. doi:10.1177/109019819201900101 PMID:1568869

Strauss, A., & Corbin, J. (1990). *Basics of qualitative research: Grounded theory procedures and techniques*. London: Sage.

Stringer, E. T. (1999). *Action research* (2nd ed.). London: Sage.

Strong, E. K. (1925). *The Psychology of selling*. New York: McGraw-Hill.

Suarez-Figueroa, M. C., de Cea, G. A., Buil, C., Dellschaft, K., Fernandez-Lopez, M., Garcia, A., & Yufei, Z. et al. (2008). NeOn methodology for building contextualized ontology networks. *NeOn Deliverable D*, *5*, 4–1.

Sweetman, D., Badiee, M., & Creswell, J. W. (2010). Use of the transformative framework in mixed methods studies. *Qualitative Inquiry*, *16*(6), 441–454. doi:10.1177/1077800410364610

Tabachnick, B. G., & Fidell, L. S. (2012). *Using Multivariate Statistics* (6th ed.). Pearson Publications.

Tashakkori, A., & Teddlie, C. (1998). *Mixed Methodology: Combining Qualitative and Quantitative Approaches in Applied Social Research methods Series* (Vol. 46). Thousand Oaks, CA: SAGE.

Tashakkori, A., & Teddlie, C. (1998). *Mixed methodology: Combining qualitative and quantitative approaches.* Sage.

Tashakkori, A., & Teddlie, C. (1998). *Mixed methodology: Combining qualitative and quantitative approaches*. Thousand Oaks, CA: Sage Publications.

Tashakkori, A., & Teddlie, C. (1998). *Mixed Methodology: Combining Qualitative and Quantitative Approaches*. Thousand Oaks, CA: Sage.

Tashakkori, A., & Teddlie, C. (2002). *Handbook of mixed methods in social & behavioral research*. Sage Publications, Inc.

Tashakkori, A., & Teddlie, C. (2003a). *Handbook of mixed methods in social and behavioral research*. Thousand Oaks, CA: Sage Publications.

Tashakkori, A., & Teddlie, C. (2003b). The past and future of mixed methods research: From data triangulation to mixed model designs. In A. Tashakkori & C. Teddlie (Eds.), *Handbook of mixed methods in social and behavioral research* (pp. 671–702). Thousand Oaks, CA: Sage Publications.

Tashakorri, A., & Teddlie, C. (1998). *Mixed methods methodology: Combining the qualitative and quantitative approaches*. Thousand Oaks, CA: Sage.

Tashakorri, A., & Teddlie, C. (2010). Epilogue: Current developments and emerging trends in integrated research methodology. In A. Tashakkori & C. Teddlie (Eds.), *SAGE Handbook of Mixed Methods in Social & Behavioral Research* (2nd ed., pp. 803–825). Thousand Oaks, CA: Sage. doi:10.4135/9781506335193.n31

Teddlie, C., & Tashakkori, A. (2011). Mixed methods research. The Sage Handbook of Qualitative Research, 285.

Teddlie, C., & Johnson, R. B. (2009a). Methodological thought before the 20th century. In C. Teddlie & A. Tashakkori (Eds.), *Foundations of mixed methods research: Integrating quantitative and quaitative approaches in the social and behavioral sciences* (pp. 40–61). Thousand Oaks, CA: Sage.

Teddlie, C., & Johnson, R. B. (2009b). Methodological thought since the 20th century. In C. Teddlie & A. Tashakkori (Eds.), *Foundations of mixed methods research: Integrating quantitative and qualitative approaches in the social and behavioral sciences* (pp. 62–82). Thousand Oaks, CA: Sage.

Teddlie, C., & Tashakkori, A. (2000). *Foundations of mixed methods research*. Thousand Oaks, CA: Sage Publications.

Teddlie, C., & Tashakkori, A. (2003). Major issues and controversies in the use of mixed methods in the social and behavioral sciences. In A. Tashakkori & C. Teddlie (Eds.), *Handbook of mixed methods in social & behavioral research* (pp. 3–50). Thousand Oaks, CA: Sage.

Teddlie, C., & Tashakkori, A. (2003). Major issues and controversies in the use of mixed methods in the social and behavioral sciences. In A. Tashakkori & C. Teddlie (Eds.), *Handbook of mixed methods in social and behavioral research* (pp. 3–50). Thousand Oaks, CA: Sage Publications.

Teddlie, C., & Tashakkori, A. (2009). *Foundations of mixed methods research: Integrating quantitative and qualitative approaches in the social and behavioral sciences*. Thousand Oaks, CA: Sage.

Teddlie, C., & Tashakkori, A. (2010). Overview of contemporary issues in mixed methods research. In A. Tashakkori & C. Teddlie (Eds.), *SAGE Handbook of Mixed Methods in Social & Behavioral Research* (2nd ed.; pp. 1–41). Thousand Oaks, CA: Sage. doi:10.4135/9781506335193.n1

Teddlie, C., & Yu, F. (2006). Mixed methods sampling: A typology with examples. *Journal of Mixed Methods Research*, *1*(1), 77–100. doi:10.1177/2345678906292430

Tran, B. (2014b). Rhetoric of play: Utilizing the gamer factor in selecting and training employees. In T. M. Connolly, L. Boyle, T. Hainey, G. Baxter, & P. Moreno-Ger (Eds.), Psychology, pedagogy and assessment in serious games (pp. 175-203). Hershey, PA: Information Science Reference/IGI Global.

Tran, B. (2014a). Triangulation in organizational research: Validating knowledge in human competence at work. In A. Takhar & A. Ghorbani (Eds.), *Market research methodologies: Multi-method and qualitative approaches* (pp. 93–117). Hersey, PA: Premier Reference Source/IGI Global.

Trochim, W. M. K. (2006). *Research methods knowledge base. Deduction and induction.* Retrieved from: http://www.socialresearchmethods.net/kb/dedind.php

Troschim, W. M. K. (2006). *Deduction and Induction. Research Methods Knowledge Base*. Retrieved from: http://www.socialresearchmethods.net/kb/dedind.php

U.S. Department of Health & Human Services. (2015). *The nature and design of mixed methods research.* Retrieved from ttps://obssr.od.nih.gov/scientific_areas/methodology/mixed_methods_research/section2.aspx

Van Dalen, D. B. (1979). *Understanding educational research* (4th ed.). New York: McGraw-Hill.

Van Dijk, T. A. (2006). Ideology and discourse analysis. *Journal of Political Ideologies*, *11*(2), 115–140. doi:10.1080/13569310600687908

van Manen, M. (1990). *Research lived experience: Human science for an action sensitive pedagogy*. London, Canada: State University of New York Press.

Vel de, V. E., Vermeir, W., & Corten, F. (2005). Corporate Social Responsibility and financial performance. *Journal of Finance and Accounting*, *5*, 129–138.

Venkatesh, V., Brown, S. A., & Bala, H. (2013). Bridging the qualitative-quantitative divide: Guidelines for conducting mixed methods research in information systems. *Management Information Systems Quarterly*, *37*(1), 21–54.

Victor, C. R., Ross, F., & Axford, J. (2004). Capturing lay perspectives in a randomized control trial of a health promotion intervention for people with osteoarthritis of the knee. *Journal of Evaluation in Clinical Practice*, *10*(1), 63–70. doi:10.1111/j.1365-2753.2003.00395.x PMID:14731152

Wahyuni, D. (2012). The research design maze: Understanding paradigms, cases, methods and methodologies. *Journal of Applied Management Accounting Research, 10*(1), 69–80.

Walther, J. B. (1996). Computer-mediated communication: Impersonal, interpersonal, and hyperpersonal interaction. *Communication Research, 23*(1), 3–43. doi:10.1177/009365096023001001

Webb, D. A., Sweet, D., & Pretty, A. (2002). The emotional and psychological impact of mass casualty incidents on forensic odontologists. *Journal of Forensic Sciences, 47*(3), 539–541. PMID:12051333

Webb, E. J., Campbell, D. T., Schwartz, R. D., & Sechrest, L. (1966). *Unobtrusive measures.* Chicago: Rand McNally.

Weber, R. P. (1990). *Basic content analysis* (2nd ed.). London, GB: Sage.

Weitzman, P. F., & Levkoff, S. E. (2000). Combining qualitative and quantitative methods in health research with minority elders: Lessons from a study of dementia caregiving. *Field Methods, 12*(3), 195–208. doi:10.1177/1525822X0001200302

Welsh, W. (2007). *Transformation changes everything: Exploring organizational leadership challenges in an "aspiring - to" transformative environment.* (Doctoral dissertation, Regent University). Available from Dissertations & Theses: Full Text. ProQuest

Wicks, A. C., & Freeman, R. E. (1998). Organization studies and the new pragmatism: Positivism, anti-positivism, and the search for ethics. *Organization Science, 9*(2), 123–140. doi:10.1287/orsc.9.2.123

Willig, C. (2008). *Introducing Qualitative Research in Psychology: Adventures in Theory and Method.* London: Open University Press.

Witte, R. S., & Witte, J. S. (2013). *Statistics* (10th ed.). John Wiley & Sons, Inc.

Woodrum, E. (1984). "Mainstreaming" content analysis in social sciences: Methodological advantages, obstacles, and solution. *Social Science Research, 13*(1), 1–19. doi:10.1016/0049-089X(84)90001-2

Wright, D. B. (2003). Making friends with your data: Improving how statistics are conducted and reported. *The British Journal of Educational Psychology, 73*(1), 123–136. doi:10.1348/000709903762869950 PMID:12639281

Yaping, Y. (2011). A review on the research of microblog. *Journal of Intelligence.* Retrieved from: http://en.cnki.com.cn/Article_en/CJFDTOTAL-QBZZ201109013.htm

Yates, F. (1934). Contingency table involving small numbers and the $\chi 2$ test: Supplement. *Journal of the Royal Statistical Society, 1*(2), 217–235.

Ye, K., & Zhang, R. (2011). Do Lenders Value CSR? Evidence from China. *Journal of Business Ethics, 104*, 197–206. doi:10.1007/s10551-011-0898-6

Yendol-Hoppey, D., Jacobs, J., & Dana, N. F. (2009). Critical concepts of mentoring in the urban context. *New Educator, 5*(1), 25–44. doi:10.1080/1547688X.2009.10399562

Yin, R. K. (1981). The case study crisis: Some answers. *Administrative Science Quarterly, 26*(1), 58–65. doi:10.2307/2392599

Yin, R. K. (2003). *Case study research, design and methods* (3rd ed.; Vol. 5). Thousand Oaks, CA: Sage.

Yin, R. K. (2003). *Case study research: Design and methods* (3rd ed.). Thousand Oaks, CA: Sage.

Yin, R. K. (2012). *Applications for case study research* (3rd ed.). Thousand Oaks, CA: Sage Publications.

Yin, R. K. (2016). *Qualitative research from start to finish.* New York, NY: Guilford Press.

Zhao, X., Lynch, J. G. Jr, & Chen, Q. (2010). Reconsidering Baron and Kenny: Myths and Truths about Mediation Analysis. *The Journal of Consumer Research*, *37*(2), 197–206. doi:10.1086/651257

Zhu, E. (2006). Interaction and cognitive engagement: An analysis of four asynchronous online discussions. *Instructional Science*, *34*(6), 451–480. doi:10.1007/s11251-006-0004-0

Zikmund, W. G., Babin, B. J., Carr, J. C., & Griffin, M. (2006). Business Research Methods (8th ed.). Cengage Learning.

About the Contributors

Mette L. Baran completed her Ed.D. in Administrative Leadership and Supervision from DePaul University. She obtained an M.B.A. in International Business and a baccalaureate degree in Marketing from DePaul University. Mette is a tenured associate professor in the School of Leadership-Doctoral Studies Department within the College of Education and Leadership at Cardinal Stritch University and teaches leadership, learning, higher education, and research courses. Her background includes being a faculty member and senior executive at Robert Morris University in Chicago including the positions as campus director, director of education, and director of development. She is an international consultant preparing U.S. professionals for their overseas assignments. Dr. Baran's research interests and expertise include looping, student attitudes and achievement, charter schools, middle school education, higher education administration and access, international family policy, and peace study. She has authored the book, The Impact of Looping in Middle School. She is a member of the Board of Trustees to Robert Morris University. In addition, she is a Board member of several not-for-profit organizations.

Janice E. Jones received her Ph.D. in Counseling Psychology from the University of Wisconsin-Milwaukee, her Master's in Educational Psychology from the University of Wisconsin-Milwaukee and her bachelor's in Business Administration from Mt. Senario College. Janice is a tenured associate professor in the School of Leadership-Doctoral Studies Department within the College of Education and Leadership at Cardinal Stritch University. Janice teaches in the Doctoral Leadership Studies Department in both leadership and higher education courses in addition to the Master's In Educational leadership programs at Cardinal Stritch University. Janice has worked as a consultant for the Department of Public Instruction for the state of Wisconsin and has provided workshops, trainings and professional staff development for a wide variety of K-12 schools throughout the state. Janice has also worked on a number of grants for both the private and public sector. Janice's areas of expertise and research interests overlap and include issues around early childcare, students with disabilities, vocational development across the lifespan and work/family conflict. Janice has presented locally, regionally, nationally and internationally at research conferences. Having a strong commitment to improve the lives of children, Janice is a member of the Executive Board of Kids Matter, Inc. which is working to improve the lives of foster children in Milwaukee County.

* * *

Roger J. Baran received his Ph.D. and MBA degrees from the University of Chicago Graduate School of Business and his BBA (cum laude) from the University of Notre Dame. His dissertation was awarded first prize by the American Marketing Association in the North America competition. He is a fellow of the National Opinion Research Center, has served on the U.S. Department of Commerce Census Advisory Committee of the American Marketing Association, and was Chair of the Bank Marketing Association National Research and Planning Council. Dr. Baran serves as a consultant in the area of marketing strategy, marketing research and customer relationship management for many well-known companies in the U.S., Europe, Asia, and Middle-East. He is currently Executive Vice-President of the Asian Forum on Business Education based in Bangkok, Thailand. Dr. Baran joined DePaul University after serving as Director of Marketing Research at Continental Bank of Chicago. His teaching and publishing specialties are marketing research, global marketing management, marketing of services, marketing management and customer relationship management. He has served as visiting associate professor of marketing at the University of Chicago Graduate School of Business, Helsinki School of Economics and Business Administration; University of Hamburg; University of the Thai Chamber of Commerce, Siam University, and Mahidol University in Bangkok, Thailand; KIMEP University in Kazakhstan and Prague School of Economics. For seven years he served as DePaul's Director of Asian and Middle East Graduate Programs, managing its MBA programs in Hong Kong and Bahrain and establishing partner relationships with schools in Thailand and China Dr. Baran wrote the first textbook on Customer Relationship Management with colleagues R. Galka and D. Strunk for Thomson Southwestern which was published in 2008. Along with R. Galka he published CRM: The Foundation of Contemporary Marketing Strategy in 2013 and they are currently writing a second edition (Routledge: Taylor & Francis Group). Cengage Publishing commissioned his book Principles of Marketing: MBA Primer for their MBA Business Series and published it in 2011. His book, Practical Bank Marketing Research was the first book published by the Bank Marketing Association dealing exclusively with the topic. His book International Joint Ventures in East Asia with Yigang Pan and Erdener Kaynak was published in 1995. His monograph, The CEO's Guide to Maximizing Managers' Daily Use of MBA Concepts, was awarded 1st prize by the American Association of Collegiate Schools of Business Mid-Continent East Association. He has published in numerous marketing, banking and international journals.

Lauren Harkrider Beechly is a Consultant for IBM's Smarter Workforce team where she develops and validates assessments, helping clients to identify the right candidates for their jobs. Dr. Beechly leads many projects including test development, selection process consulting, job analysis, competency modeling, predictive studies, and return on investment studies. Previously, Dr. Beechly worked as an internal human capital management consultant where she led company-wide job evaluations, developed career progressions, and analyzed human capital data to recommend strategic HR initiatives and inform talent decisions. She earned Bachelors' degrees in Human Resource Management and Psychology, a Master's and Ph.D. in Industrial-Organizational Psychology, and a minor in Quantitative Psychology from the University of Oklahoma.

Christopher Boachie is a lecturer of Central Business School, Central University College in Accra Ghana, an academic and a practising Chartered Accountant (ACCA-UK) with specialization in corporate finance, international economics and trade and financial risk management. He was educated in Kwame Nkrumah University of Science and Technology in Kumasi Ghana, Technical University of Freiberg in Germany and the London School of Accountancy (UK). He is currently reading his PhD at Open University of Malaysia. He has considerable, teaching, consulting and practise experience in the application of accountancy and finance theory and financing of international trade and risk management. He is a Chartered Accountant with the Association of Certified Chartered Accountants of UK. He was an

Investment Manager and in charge of oil and gas unit of International Energy Insurance. He has worked and consulted for Stephens and Co in London, LD and sons both in Ghana and Italy. He is a founding Director of Premia consulting firm in Ghana. His professional focus is on the corporate financial analysis, financial accounting and reporting and has a strong passion for financial risk management.

Clavon Byrd is an Assistant Professor and Department Chair of Teacher Education at Cardinal Stritch University in Milwaukee, WI. Clavon worked in K – 12 education for 17 years, including 10 years as a principal. Clavon earned a doctorate in educational leadership. His areas of interest are teacher preparation, edTPA support, and experiences of minority teachers and students.

Cristina Byrne graduated from the University of Oklahoma with a PhD in Industrial/Organizational Psychology where she studied leadership, creativity, innovation, communications, and ideological groups.

Ka Fai Peter Chan is a researcher at the Council of Scientific and Industrial Research (CSIR) in the department of Defence Peace Safety and Security with the focus on cyber defence. His research interest lies in formal methods, cybersecurity awareness and network security.

Preston B. Cosgrove, Ph.D., is an Assistant Professor at Cardinal Stritch University in the Doctoral Leadership program where he teaches Research Methods, Survey Research, and courses in Higher Education. His research interests include the philosophy of science, validity in qualitative inquiry, survey development, Christian higher education, and adoption.

Jennifer A. Griffith, PhD, is an Assistant Professor of Management at Alfred University. Her research interests include leadership, emotions, gender, and communication, specifically computer-mediated communication.

Marthie Grobler has been working as a Cyber Security Researcher at the Council for Scientific and Industrial Research (CSIR) since January 2008. She has a PhD Computer Science (Live Digital Forensics), and a MSc Computer Science (Information Security Governance), both from the University of Johannesburg. Her research focus is on cyber security awareness, strategic data management and incident management and response. She is co-editor of the now published ISO/IEC 27037, Guidelines for identification, collection and acquisition and preservation of digital evidence; and ISO/IEC 27035, Incident management. Marthie is an ISACA Certified Information Security Manager and is appointed as a visiting Professor at the University of Johannesburg, Academy for Computer Science and Software Engineering. She is currently supervising a number of post graduate students, and Managing Editor of the Journal of Contemporary Management.

Michael Hughes is an I/O psychologist whose primary areas of expertise include test development and validation, and analyses of high-stakes testing data. His research interests also include complex skill acquisition, training, and ideological groups.

Peter M. Jonas is a tenured Professor in the Doctoral Leadership Department at Cardinal Stritch University in Milwaukee WI. Dr. Jonas has a doctorate in History from Marquette University but has been working in administration and as a faculty member in higher education for more than 35 years. He has been at Cardinal Stritch University for 30 years teaching, researching, and serving in various leadership capacities (e.g., Director of Institutional Research, Dean in the College of Business, Director of Strategic Planning and Assessment). For the past 18 years Dr. Jonas has been teaching research and statistics in the doctoral program at Stritch and has served as the department chairperson for 16 years. Over the years Dr. Jonas has written three books in support of his research: Outcomes Assessment in Higher Education Linked with Strategic Planning and Budgeting [2nd ed.] (2013), Laughing and Learning: An Alternative to Shut-up and Listen (2009), Secrets of Connecting Leadership and Learning to Humor (2004). He has made more than 100 presentations across the country talking about humor, research, leadership, and assessment. In addition, he has authored more than 40 books, manuals, and articles in professional periodicals, in addition to serving as a consultant (typically in the area of professional development and program evaluation) for more than 25 different organizations and projects.

Sema Kalaian is a Professor of Statistics and Research Methods in the College of Technology at Eastern Michigan University. Professor Kalaian was a recipient of the (1) "Best Paper" award from the American Educational Research Association (AERA), and (2) "Distinguished Paper Award" from the Society for the Advancement of Information Systems (SAIS). Over the years, Dr. Kalaian taught introductory and advanced statistical courses such as Statistical and Research Methods, Multivariate Statistics, Survey Research, Multilevel Modeling, Structural Equation Modeling, Meta-Analysis, and Program Evaluation. Professor Kalaian's research interests focus on the development of new statistical methods and its applications. Much of her methodological developments and applications have focused on the (a) development of the multivariate meta-analytic techniques for combining evidence from multiple primary studies; (b) applications of the meta-analysis methods to multi-site studies; (c) developments of statistical methods for analyzing Delphi survey data; and (d) applications of multilevel modeling methods for meta-analysis to Science, Technology, Engineering, and Mathematics (STEM) teaching and learning research. Recently, Professor Kalaian completed a major grant project funded by the National Science Foundation (NSF) to investigate the effectiveness of various forms of active small-group learning methods in STEM disciplines. For more information about the project visit https://arc.uchicago.edu/reese/users/skalaian and http://people.emich.edu/skalaian/stem/index.htm.

Nitty Hirawaty Kamarulzaman is a senior lecturer in the Department of Agribusiness and Information Systems, Universiti Putra Malaysia. Her research interest include supply chain management, sustainable logistics, reverse logistics, agribusiness marketing, and consumer purchasing behavior.

Rafa Kasim is a professor of statistics and research methods at Indiana Tech University. Prior to that, Dr. Kasim served as a professor of statistics and research design in the College of Education at Kent State University. He was also a senior statistician at the Evaluation, Management & Training Associates Inc. (EMT). His research focused on the application of multilevel analysis to study the effects of educational and social contexts on educational outcomes and human development in large-scale longitudinal data sets. Some of Dr. Kasim work has also addressed the issues of selection and attrition bias in multi-site large studies. He has collaborated on numerous studies in fields such as small groups versus lecture-based traditional learning in STEM, adult literacy, education, and substance abuse treatments. Some of his work appears in book chapters in Application of Multilevel Models, Multilevel Meta-analysis: Effectiveness of Small-group Learning Methods Compared to Lecture-based Instruction in Science, Technology, Engineering, and Mathematics College classrooms, Small-group versus Competitive Learning in Computer Science Classrooms: A Meta-Analytic Review, Predictive Analytics, Journal of Educational and Behavioral Statistics, Harvard Educational Review and Advances in Health Sciences Education.

Louise Leenen is a Principal Scientist in the Cyber Defence Research Group at the CSIR in South Africa. She holds a PhD Computer Science from the University of Wollongong in Australia. Her research focus is on AI applications in cyber defence.

Dale MacKrell has a long track record of professional contributions as a researcher and educator in the Information Systems discipline. Her scholarly interests include business intelligence and gender relations, and a cooperative learning project for a not-for-profit organisation operating in the homelessness sector. Dale has published in Information Systems Journal, Decision Support Systems journal and the Australasian Journal of Information Systems as well as numerous Australian and international conferences. Dale is an Assistant Professor in Information Systems at the University of Canberra in Australia.

Amir Manzoor holds a bachelor's degree in engineering from NED University, Karachi, an MBA from Lahore University of Management Sciences (LUMS), and an MBA from Bangor University, United Kingdom. He has many years of diverse professional and teaching experience working at many renowned national and internal organizations and higher education institutions. His research interests include electronic commerce and technology applications in business. He is a member of Chartered Banker Institute of UK and Project Management Institute, USA.

A. J. Metz is an Assistant Professor in the Department of Educational Psychology at the University of Utah. She earned a M.Ed. in Vocational Rehabilitation Counseling and a Ph.D. in Urban Education (specialization in Counseling Psychology) from the University of Wisconsin-Milwaukee. Her research examining factors related to academic and career success in underrepresented and underserved student populations has led to numerous journal articles, book chapters, conference presentations, workshops, and most recently a student success textbook. Dr. Metz has extensive teaching, counseling, and career advising experience in high schools, community colleges, and four-year public and private institutions of higher education. She is passionate about mentoring students and received an Early Career Teaching Award in 2015. She has served on the board of directors of the Utah Psychological Association for six years most notably as president.

Never Mujere is a lecturer and Doctor of Philosophy (DPhil) candidate specializing in Water Resources in the Department of Geography and Environmental Science at the University of Zimbabwe (UZ). Since joining the University in 2006 as a staff member, he has been involved the teaching and research. He teaches undergraduate and post-graduate courses in water resources, waste management, research methods, disaster management and climate change and environmental issues. Never Mujere has published 10 journals articles, 3 book chapters, 3 books and presented more than 10 papers at local and international conferences. Never is a member of the International Association of Hydrological Sciences (IAHS). He is an early career scientist and researcher who is a team player, highly versatile, creative and results-oriented.

Aroop Mukherjee is a PhD Scholar in the Department of Agribusiness and Information Systems, Universiti Putra Malaysia. His research interest include agility, agribusiness, supply chain management, supply chain strategies, sustainability, innovation, knowledge management, and agri-informatics.

Carolyn Stevenson is a veteran educator currently working as a faculty member for OC@KU (Open College at Kaplan University). Carolyn has over 17 years teaching and administrative experience in higher education. She holds a Master of Arts degree in Communication, Master of Business Administration, and Doctor of Education with an emphasis in Higher Education. Prior to pursuing a career in higher education, she worked in the publishing field and served as a technical writing consultant. She currently serves as Associate Editor for the International Journal of Technologies and Educational Marketing (IJTEM), published by IGI-Global; Editorial Board Member and Reviewer for the Journal of Education and Learning published by the Canadian Center of Science and Education; and Membership Committee Member for the Qualitative Research Special Interest Group (AERA). Recent publications include a chapter entitled: "Leading across Generations: Issues for Higher Education Administrators" published in the Handbook of Research on Transnational Higher Education Management, by IGI Global; Technical Writing: A Comprehensive Resource for Technical Writers at all Levels, (Martinez, Hannigan, Wells, Peterson and Stevenson) Revised and Updated Edition, Kaplan Publishing; and Building Online Communities in Higher Education Institutions: Creating Collaborative Experience (with co-editor Joanna Bauer).

Randa Suleiman currently works as an assistant professor at Cardinal Stritch University Teacher Preparation Program. She worked in K-12 education for fifteen years in private, public school districts in USA and Internationally. She graduated from Cardinal Stritch University in May 2010 with a PhD degree in Leadership for the Advancement of Learning and Service in Higher Education. Randa earned National Board certification in early adolescent science in 2008. Currently working on developing a teacher assessment mentoring and support program. Areas of interest are teacher preparation, assessment, educator effectiveness, edTPA, and science education.

Ben Tran received his Doctor of Psychology (Psy.D) in Organizational Consulting/Organizational Psychology from California School of Professional Psychology at Alliant International University in San Francisco, California, United States of America. Dr. Tran's research interests include domestic and expatriate recruitment, selection, retention, evaluation, & training, CSR, business and organizational ethics, organizational/international organizational behavior, knowledge management, and minorities in multinational corporations. Dr. Tran has presented articles on topics of business and management ethics, expatriate, and gender and minorities in multinational corporations at the Academy of Management, Society for the Advancement of Management, and International Standing Conference on Organizational Symbolism. Dr. Tran has also published articles and book chapters with the Social Responsibility Journal, Journal of International Trade Law and Policy, Journal of Economics, Finance and Administrative Science, Financial Management Institute of Canada, and IGI Global.

Joey Jansen van Vuuren is the Research Group Leader for Cyber Defence for Scientific Research at the CSIR South Africa. She gives the strategic research direction for the research group that is mainly involved in research on network forensics, social media, and national security for the SANDF and Government sectors on Cyber Defence. As Cyber threats became extremely important for South Africa with the recent broadband changes, she focused her research around cyber security and government policies required to ensure national security. In particular her group is involved in contract research for all the main players responsible for the implementation of the National Cybersecurity Policy Framework for South Africa. She already presented this research on several national and international conferences and published journal articles on cyber security in South Africa. She was also a keynote speaker on international conferences and was invited for several radio interviews on this topic and published an article scientific magazine on the implementation of Cyber policies in South Africa.

Thomas A. Zeni is an Assistant Professor of Management, and the Chickasaw Nation Professor of Business Administration at East Central University in Ada, OK. He holds a Ph.D. in Industrial & Organizational Psychology, as well as an MS and MBA degree from the University of Oklahoma. His undergraduate degree in Psychology is from Mercy College in Dobbs Ferry, NY. In addition, he maintains a Senior Professional in Human Resources (SPHR) certification and is a SHRM Senior Certified Professional. Dr. Zeni teaches undergraduate and graduate coursework in Business Communication, Human Resource Management, Personnel Selection & Assessment, Compensation & Benefits, Training & Development, Principles of Management, Strategic Management, Organizational Behavior, Leadership, and Applied Statistics. His research interests include leadership, business ethics, emotions in organizations, and quantitative design and methodology.

Index

A

action research 39, 45, 50-57, 60, 64, 146, 166, 183
analytical methods 149, 163
ANOVA 8, 149, 162-163, 182
artifacts 9, 128-129, 140, 240, 243, 258, 260
axiology 14, 45

B

Between-Methods Triangulation 3, 27
bias 4, 58, 90, 111, 203-204, 295

C

case study 5, 11, 71-72, 83, 124-125, 127-129, 140, 146, 262, 264, 268, 271-272, 274, 286
Categories and Themes 200-201
causality 68-69, 171
coding 10, 146, 199-201, 204, 216, 219, 221-222, 227-228, 246
commentaries 237, 241, 243, 258, 260
computer-mediated communication 213-214
Confidentiality 146
consent 129, 146, 203, 228-229
constructivism 13-14, 30-32, 43-46
content analysis 146, 203, 213-216, 219, 221-227, 229, 234
contingency table 174, 186
Continuous Data 186
critical realism 16, 33
Cronbach's alpha 260
cross tabulation 173-174

D

data analysis 4, 10, 12, 42, 67, 74-75, 81-84, 91, 129, 146, 150, 165-167, 169, 172, 183, 197-204, 216, 236, 246, 253, 256, 301
dependent variable 6, 8, 156-162, 171, 176, 179-182, 184, 186
descriptive statistics 7-8, 150, 153, 170, 247-250
design science 262-265, 270-271, 289
Discrete Data 186
Document Analysis 146

E

edTPA 236-250, 252-259, 261
Educator Teacher Performance Assessment 236
Embedded Mixed Methods Design 81, 83, 98, 105
epistemology 14, 16, 35, 44-45, 204, 295
Ethnography 5, 9-10, 71, 146
explanatory mixed method design 85-86, 89, 92, 98, 105
Exploratory Mixed Methods Design 81, 86-87, 98, 105

F

Field Observation 146
Framework Analysis 146
frequencies 150-151, 170, 172, 183, 186, 227

G

graphs 150, 152, 165, 169, 174-175
Grounded Theory 10, 72, 146, 199

H

higher education 125, 141, 238-239
histograms 149, 151, 174
history 1-3, 6, 15, 91, 239, 255
holistic 9, 34, 51, 71-72, 128, 146, 239-240, 263-264, 272, 298
Hybridized Content Analysis 234
hypothesis testing 165-166, 170, 175, 177

I

inferential statistics 8, 108, 156, 166, 169-170, 229

L

Likert scale 221, 223, 245, 247, 254, 260

M

mean 1, 14, 117-118, 149, 151-157, 159-162, 176-177, 183, 196, 203, 247-250, 252-256, 260-261
memoing 199, 201
message board forums 214-216, 220, 222
mixed action research 54, 56-57, 64
mixed method 12, 39-40, 42, 45, 47, 54-57, 59-60, 64, 73, 79-92, 98-99, 105, 165, 167, 178, 183, 188, 236, 264, 293
mixed method design 47, 59, 64, 81-92, 98-99, 105, 188
mixed method research 12, 39-40, 59, 64, 73, 79-81, 83, 86-87, 98, 165, 167, 178, 183, 293
mixed methods 1-6, 12-15, 18, 27-29, 32-35, 39-42, 45, 47-48, 50, 54, 56-60, 66-67, 72-74, 77, 79, 81, 83, 85-88, 90-92, 98-99, 105, 117, 149, 165-166, 187-188, 192, 194-195, 199, 205, 236, 243-245, 262-264, 293-294, 296-300
mixed-methods 14, 28-29, 32-35, 188, 192, 195-196
mixed methods research 1-3, 5-6, 12, 14-15, 18, 27-28, 32-35, 39-42, 45, 47, 54, 56, 59-60, 66-67, 72-73, 77, 79, 81, 88, 90, 92, 98, 105, 117, 165-166, 187-188, 194-195, 199, 205, 243-244, 262-264, 293-294, 296, 300
Mixing Data 27
morphological analysis 262-266, 268, 289
multiple regression 8, 149, 158-159, 170-171, 179-180

N

non-probability 105, 110, 114-115

O

online communication 213, 215-217, 219-222, 225-229
online communities 226, 228
online community 217, 228-229, 234
ontology 14, 16, 45, 262-264, 268-272, 274, 277-286, 288-289

P

paradigm 1-5, 13-16, 27-35, 39, 42-43, 45-47, 51, 64, 264
Paradigm (Research) 27
Pearson 155, 178, 255-256, 259-260
photography 123-124, 127, 129-130, 141
population 7-8, 11, 17, 68-69, 75, 89, 92, 107-121, 129, 156-162, 167, 169, 175-178, 220, 229
post-positivism 31-32, 43-44
pragmatism 1-2, 14-15, 32-33, 42-46, 64
probability 110-112, 114, 118, 169, 175, 178, 186, 244, 252-253, 266-267
program evaluation 236, 245
purposive sampling 105, 114, 117-119, 220

Q

qualitative analysis 85-88, 165, 183, 197, 219, 234
qualitative data 2, 4, 9-12, 17-18, 27, 40, 42, 53, 67, 71, 73-76, 79-81, 83-92, 98, 105, 123-124, 127, 149, 167, 183, 197-205, 213-214, 219, 226, 229, 245-246, 254, 256, 259, 298, 300
qualitative research 1-5, 8-9, 13, 16-18, 27, 29, 31, 35, 39, 42, 48, 71-73, 79-80, 82, 86, 88, 90-91, 105, 117-119, 123-124, 127-128, 187-189, 192-196, 201-204, 263, 294
quantitative analysis 8, 75, 86, 88, 163, 165, 167, 171, 183-184, 198, 202, 213, 229, 234
quantitative data 3-5, 7, 13, 27, 69, 73-75, 80, 82, 84-86, 88, 90, 92, 149-157, 159-160, 162-163, 165-167, 183, 199, 205, 214, 226, 229, 244, 256, 298, 300
quantitative methods 2, 4, 15, 29, 80, 149, 165, 213, 266
quantitative research 1-5, 7-8, 13, 15, 27-29, 40, 43, 45, 69-70, 73, 87, 90, 105, 114, 165, 177, 186-189, 192-193, 195-196, 264, 294, 298

R

regression 6, 8, 149, 156-160, 163, 165-166, 170-171, 175, 178-183
reliability 16-18, 57, 67, 90, 98, 171-172, 184, 186, 202, 222, 224, 226, 246-247, 260, 300
research approach 32, 39, 46-47, 57, 64
research methodologies 1-3, 18, 79, 262-264, 270, 274, 286
research paradigm 2-3, 5, 16, 28-31, 34, 39, 43, 45-47, 64
rubrics 258, 261

S

sample 4, 7-8, 10, 56-57, 68, 74-75, 83, 86, 88-90, 92, 107-121, 147, 156, 161, 167, 173-179, 187, 189, 191, 198, 200, 204-205, 209-210, 216, 218-226, 244, 247, 252-255

sample frame 121
sample size 7, 56-57, 107, 109-111, 116, 119, 121, 177, 189, 247, 255
sampling 7-8, 10, 67, 72-73, 85, 105, 107-119, 121, 140, 189, 220, 223, 247, 299
significance 5, 80, 159, 166, 177, 182-184, 252-253
social media 17
sources of qualitative data 198
standard deviation 109-110, 149, 154-155, 182-183, 247-250, 252-254, 261
statistically significant 110, 160-161, 178, 182-183, 186, 243, 252-254
statistical methods 7, 149-151, 153, 156, 163, 177
strata 112-113, 115, 121, 178
survey 5, 7, 17, 50, 68, 70, 74, 98, 110, 113, 115, 120, 165, 174, 186, 188-193, 196, 199, 202, 224-227, 236, 239, 244-247, 252-256, 258-259, 270, 274, 280

T

teacher education 239, 241, 244, 257
terminologies 1-2, 5, 47, 270
Thomas Kuhn 5, 30
transferability 17-18, 33, 57, 147
transformative 14-15, 33-34, 43, 50-51
triangulation 1-4, 18, 27, 40, 47, 58, 80-82, 84-85, 88-89, 91-92, 99, 105, 119, 147, 202, 244, 259, 264
triangulation mixed method design 81-82, 84, 88-89, 91-92, 105

V

validity 2, 6, 16-18, 42, 57-58, 67, 80, 90, 98, 147, 169, 176-177, 202, 222, 300
variance 3, 80, 149, 154-157, 159-160, 162-163, 170-171, 177, 180, 182, 247

W

Within-Methods Triangulation 3, 27

www.ingramcontent.com/pod-product-compliance
Lightning Source LLC
LaVergne TN
LVHW062137300926
849161LV00010B/2555